To my students

Contents

Book of Proof

Richard Hammack
Virginia Commonwealth University

Richard Hammack (publisher)
Department of Mathematics & Applied Mathematics
P.O. Box 842014
Virginia Commonwealth University
Richmond, Virginia, 23284

Book of Proof

Edition 2.2

Typeset in 11pt TEX Gyre Schola using PDFLATEX

II How to Prove Conditional Statements

III More on Proof

IV Relations, Functions and Cardinality

Preface

In writing this book I have been motivated by the desire to create a high-quality textbook that costs almost nothing.

The book is available on my web page for free, and the paperback version (produced through an on-demand press) costs considerably less than comparable traditional textbooks. Any revisions or new editions will be issued solely for the purpose of correcting mistakes and clarifying exposition. New exercises may be added, but the existing ones will not be unnecessarily changed or renumbered.

This text is an expansion and refinement of lecture notes I developed while teaching proofs courses over the past fourteen years at Virginia Commonwealth University (a large state university) and Randolph-Macon College (a small liberal arts college). I found the needs of these two audiences to be nearly identical, and I wrote this book for them. But I am mindful of a larger audience. I believe this book is suitable for almost any undergraduate mathematics program.

This second edition incorporates many minor corrections and additions that were suggested by readers around the world. In addition, several new examples and exercises have been added, and a section on the Cantor-Bernstein-Schröeder theorem has been added to Chapter 13.

RICHARD HAMMACK

Richmond, Virginia
May 25, 2013

Introduction

This is a book about how to prove theorems.

Until this point in your education, mathematics has probably been presented as a primarily computational discipline. You have learned to solve equations, compute derivatives and integrals, multiply matrices and find determinants; and you have seen how these things can answer practical questions about the real world. In this setting, your primary goal in using mathematics has been to compute answers.

But there is another side of mathematics that is more theoretical than computational. Here the primary goal is to understand mathematical structures, to prove mathematical statements, and even to invent or discover new mathematical theorems and theories. The mathematical techniques and procedures that you have learned and used up until now are founded on this theoretical side of mathematics. For example, in computing the area under a curve, you use the fundamental theorem of calculus. It is because this theorem is true that your answer is correct. However, in learning calculus you were probably far more concerned with how that theorem could be applied than in understanding why it is true. But how do we *know* it is true? How can we convince ourselves or others of its validity? Questions of this nature belong to the theoretical realm of mathematics. This book is an introduction to that realm.

This book will initiate you into an esoteric world. You will learn and apply the methods of thought that mathematicians use to verify theorems, explore mathematical truth and create new mathematical theories. This will prepare you for advanced mathematics courses, for you will be better able to understand proofs, write your own proofs and think critically and inquisitively about mathematics.

The book is organized into four parts, as outlined below.

PART I Fundamentals

- Chapter 1: Sets
- Chapter 2: Logic
- Chapter 3: Counting

Chapters 1 and 2 lay out the language and conventions used in all advanced mathematics. Sets are fundamental because every mathematical structure, object or entity can be described as a set. Logic is fundamental because it allows us to understand the meanings of statements, to deduce information about mathematical structures and to uncover further structures. All subsequent chapters will build on these first two chapters. Chapter 3 is included partly because its topics are central to many branches of mathematics, but also because it is a source of many examples and exercises that occur throughout the book. (However, the course instructor may choose to omit Chapter 3.)

PART II Proving Conditional Statements

- Chapter 4: Direct Proof
- Chapter 5: Contrapositive Proof
- Chapter 6: Proof by Contradiction

Chapters 4 through 6 are concerned with three main techniques used for proving theorems that have the "conditional" form *"If P, then Q."*

PART III More on Proof

- Chapter 7: Proving Non-Conditional Statements
- Chapter 8: Proofs Involving Sets
- Chapter 9: Disproof
- Chapter 10: Mathematical Induction

These chapters deal with useful variations, embellishments and consequences of the proof techniques introduced in Chapters 4 through 6.

PART IV Relations, Functions and Cardinality

- Chapter 11: Relations
- Chapter 12: Functions
- Chapter 13: Cardinality of Sets

These final chapters are mainly concerned with the idea of *functions*, which are central to all of mathematics. Upon mastering this material you will be ready for advanced mathematics courses such as combinatorics, abstract algebra, theory of computation, analysis and topology.

To the instructor. The book is designed for a three credit course. Here is a possible timetable for a fourteen-week semester.

Week	Monday	Wednesday	Friday
1	Section 1.1	Section 1.2	Sections 1.3, 1.4
2	Sections 1.5, 1.6, 1.7	Section 1.8	Sections 1.9*, 2.1
3	Section 2.2	Sections 2.3, 2.4	Sections 2.5, 2.6
4	Section 2.7	Sections 2.8*, 2.9	Sections 2.10, 2.11*, 2.12*
5	Sections 3.1, 3.2	Section 3.3	Sections 3.4, 3.5*
6	EXAM	Sections 4.1, 4.2, 4.3	Sections 4.3, 4.4, 4.5*
7	Sections 5.1, 5.2, 5.3*	Section 6.1	Sections 6.2 6.3*
8	Sections 7.1, 7.2*, 7.3	Sections 8.1, 8.2	Section 8.3
9	Section 8.4	Sections 9.1, 9.2, 9.3*	Section 10.0
10	Sections 10.0, 10.3*	Sections 10.1, 10.2	EXAM
11	Sections 11.0, 11.1	Sections 11.2, 11.3	Sections 11.4, 11.5
12	Section 12.1	Section 12.2	Section 12.2
13	Sections 12.3, 12.4*	Section 12.5	Sections 12.5, 12.6*
14	Section 13.1	Section 13.2	Sections 13.3, 13.4*

Sections marked with $*$ may require only the briefest mention in class, or may be best left for the students to digest on their own. Some instructors may prefer to omit Chapter 3.

Acknowledgments. I thank my students in VCU's MATH 300 courses for offering feedback as they read the first edition of this book. Thanks especially to Cory Colbert and Lauren Pace for rooting out typographical mistakes and inconsistencies. I am especially indebted to Cory for reading early drafts of each chapter and catching numerous mistakes before I posted the final draft on my web page. Cory also created the index, suggested some interesting exercises, and wrote some solutions. Thanks to Andy Lewis and Sean Cox for suggesting many improvements while teaching from the book. I am indebted to Lon Mitchell, whose expertise with typesetting and on-demand publishing made the print version of this book a reality.

And thanks to countless readers all over the world who contacted me concerning errors and omissions. Because of you, this is a better book.

Part I

Fundamentals

CHAPTER 1

Sets

All of mathematics can be described with sets. This becomes more and more apparent the deeper into mathematics you go. It will be apparent in most of your upper level courses, and certainly in this course. The theory of sets is a language that is perfectly suited to describing and explaining all types of mathematical structures.

1.1 Introduction to Sets

A **set** is a collection of things. The things in the collection are called **elements** of the set. We are mainly concerned with sets whose elements are mathematical entities, such as numbers, points, functions, etc.

A set is often expressed by listing its elements between commas, enclosed by braces. For example, the collection $\{2,4,6,8\}$ is a set which has four elements, the numbers $2,4,6$ and 8. Some sets have infinitely many elements. For example, consider the collection of all integers,

$$\{\ldots,-4,-3,-2,-1,0,1,2,3,4,\ldots\}.$$

Here the dots indicate a pattern of numbers that continues forever in both the positive and negative directions. A set is called an **infinite** set if it has infinitely many elements; otherwise it is called a **finite** set.

Two sets are **equal** if they contain exactly the same elements. Thus $\{2,4,6,8\} = \{4,2,8,6\}$ because even though they are listed in a different order, the elements are identical; but $\{2,4,6,8\} \neq \{2,4,6,7\}$. Also

$$\{\ldots-4,-3,-2,-1,0,1,2,3,4\ldots\} = \{0,-1,1,-2,2,-3,3,-4,4,\ldots\}.$$

We often let uppercase letters stand for sets. In discussing the set $\{2,4,6,8\}$ we might declare $A = \{2,4,6,8\}$ and then use A to stand for $\{2,4,6,8\}$. To express that 2 is an element of the set A, we write $2 \in A$, and read this as "2 *is an element of* A," or "2 *is in* A," or just "2 *in* A." We also have $4 \in A$, $6 \in A$ and $8 \in A$, but $5 \notin A$. We read this last expression as "5 *is not an element of* A," or "5 *not in* A." Expressions like $6,2 \in A$ or $2,4,8 \in A$ are used to indicate that several things are in a set.

Some sets are so significant and prevalent that we reserve special symbols for them. The set of **natural numbers** (i.e., the positive whole numbers) is denoted by $\mathbb{N}$, that is,

$$\mathbb{N} = \{1,2,3,4,5,6,7,\ldots\}.$$

The set of **integers**

$$\mathbb{Z} = \{\ldots,-3,-2,-1,0,1,2,3,4,\ldots\}$$

is another fundamental set. The symbol $\mathbb{R}$ stands for the set of all **real numbers**, a set that is undoubtedly familiar to you from calculus. Other special sets will be listed later in this section.

Sets need not have just numbers as elements. The set $B = \{T,F\}$ consists of two letters, perhaps representing the values "true" and "false." The set $C = \{a,e,i,o,u\}$ consists of the lowercase vowels in the English alphabet. The set $D = \{(0,0),(1,0),(0,1),(1,1)\}$ has as elements the four corner points of a square on the x-y coordinate plane. Thus $(0,0) \in D$, $(1,0) \in D$, etc., but $(1,2) \notin D$ (for instance). It is even possible for a set to have other sets as elements. Consider $E = \{1,\{2,3\},\{2,4\}\}$, which has three elements: the number 1, the set $\{2,3\}$ and the set $\{2,4\}$. Thus $1 \in E$ and $\{2,3\} \in E$ and $\{2,4\} \in E$. But note that $2 \notin E$, $3 \notin E$ and $4 \notin E$.

Consider the set $M = \left\{\left[\begin{smallmatrix}0&0\\0&0\end{smallmatrix}\right],\left[\begin{smallmatrix}1&0\\0&1\end{smallmatrix}\right],\left[\begin{smallmatrix}1&0\\1&1\end{smallmatrix}\right]\right\}$ of three two-by-two matrices. We have $\left[\begin{smallmatrix}0&0\\0&0\end{smallmatrix}\right] \in M$, but $\left[\begin{smallmatrix}1&1\\0&1\end{smallmatrix}\right] \notin M$. Letters can serve as symbols denoting a set's elements: If $a = \left[\begin{smallmatrix}0&0\\0&0\end{smallmatrix}\right]$, $b = \left[\begin{smallmatrix}1&0\\0&1\end{smallmatrix}\right]$ and $c = \left[\begin{smallmatrix}1&0\\1&1\end{smallmatrix}\right]$, then $M = \{a,b,c\}$.

If X is a finite set, its **cardinality** or **size** is the number of elements it has, and this number is denoted as $|X|$. Thus for the sets above, $|A| = 4$, $|B| = 2$, $|C| = 5$, $|D| = 4$, $|E| = 3$ and $|M| = 3$.

There is a special set that, although small, plays a big role. The **empty set** is the set $\{\}$ that has no elements. We denote it as $\emptyset$, so $\emptyset = \{\}$. Whenever you see the symbol $\emptyset$, it stands for $\{\}$. Observe that $|\emptyset| = 0$. The empty set is the only set whose cardinality is zero.

Be careful in writing the empty set. Don't write $\{\emptyset\}$ when you mean $\emptyset$. These sets can't be equal because $\emptyset$ contains nothing while $\{\emptyset\}$ contains one thing, namely the empty set. If this is confusing, think of a set as a box with things in it, so, for example, $\{2,4,6,8\}$ is a "box" containing four numbers. The empty set $\emptyset = \{\}$ is an empty box. By contrast, $\{\emptyset\}$ is a box with an empty box inside it. Obviously, there's a difference: An empty box is not the same as a box with an empty box inside it. Thus $\emptyset \neq \{\emptyset\}$. (You might also note $|\emptyset| = 0$ and $|\{\emptyset\}| = 1$ as additional evidence that $\emptyset \neq \{\emptyset\}$.)

This box analogy can help us think about sets. The set $F = \{\emptyset, \{\emptyset\}, \{\{\emptyset\}\}\}$ may look strange but it is really very simple. Think of it as a box containing three things: an empty box, a box containing an empty box, and a box containing a box containing an empty box. Thus $|F| = 3$. The set $G = \{\mathbb{N}, \mathbb{Z}\}$ is a box containing two boxes, the box of natural numbers and the box of integers. Thus $|G| = 2$.

A special notation called **set-builder notation** is used to describe sets that are too big or complex to list between braces. Consider the infinite set of even integers $E = \{\ldots, -6, -4, -2, 0, 2, 4, 6, \ldots\}$. In set-builder notation this set is written as

$$E = \{2n : n \in \mathbb{Z}\}.$$

We read the first brace as "*the set of all things of form*," and the colon as "*such that.*" So the expression $E = \{2n : n \in \mathbb{Z}\}$ is read as "*E equals the set of all things of form* $2n$, *such that n is an element of* $\mathbb{Z}$." The idea is that E consists of all possible values of $2n$, where n takes on all values in $\mathbb{Z}$.

In general, a set X written with set-builder notation has the syntax

$$X = \{\text{expression} : \text{rule}\},$$

where the elements of X are understood to be all values of "expression" that are specified by "rule." For example, the set E above is the set of all values the expression $2n$ that satisfy the rule $n \in \mathbb{Z}$. There can be many ways to express the same set. For example, $E = \{2n : n \in \mathbb{Z}\} = \{n : n \text{ is an even integer}\} = \{n : n = 2k, k \in \mathbb{Z}\}$. Another common way of writing it is

$$E = \{n \in \mathbb{Z} : n \text{ is even}\},$$

read "*E is the set of all n in* $\mathbb{Z}$ *such that n is even.*" Some writers use a bar instead of a colon; for example, $E = \{n \in \mathbb{Z} \mid n \text{ is even}\}$. We use the colon.

Example 1.1 Here are some further illustrations of set-builder notation.

1. $\{n : n \text{ is a prime number}\} = \{2, 3, 5, 7, 11, 13, 17, \ldots\}$
2. $\{n \in \mathbb{N} : n \text{ is prime}\} = \{2, 3, 5, 7, 11, 13, 17, \ldots\}$
3. $\{n^2 : n \in \mathbb{Z}\} = \{0, 1, 4, 9, 16, 25, \ldots\}$
4. $\{x \in \mathbb{R} : x^2 - 2 = 0\} = \{\sqrt{2}, -\sqrt{2}\}$
5. $\{x \in \mathbb{Z} : x^2 - 2 = 0\} = \emptyset$
6. $\{x \in \mathbb{Z} : |x| < 4\} = \{-3, -2, -1, 0, 1, 2, 3\}$
7. $\{2x : x \in \mathbb{Z}, |x| < 4\} = \{-6, -4, -2, 0, 2, 4, 6\}$
8. $\{x \in \mathbb{Z} : |2x| < 4\} = \{-1, 0, 1\}$

These last three examples highlight a conflict of notation that we must always be alert to. The expression $|X|$ means *absolute value* if X is a number and *cardinality* if X is a set. The distinction should always be clear from context. Consider $\{x \in \mathbb{Z} : |x| < 4\}$ in Example 1.1 (6) above. Here $x \in \mathbb{Z}$, so x is a number (not a set), and thus the bars in $|x|$ must mean absolute value, not cardinality. On the other hand, suppose $A = \{\{1,2\},\{3,4,5,6\},\{7\}\}$ and $B = \{X \in A : |X| < 3\}$. The elements of A are sets (not numbers), so the $|X|$ in the expression for B must mean cardinality. Therefore $B = \{\{1,2\},\{7\}\}$.

We close this section with a summary of special sets. These are sets or types of sets that come up so often that they are given special names and symbols.

- The empty set: $\emptyset = \{\}$
- The natural numbers: $\mathbb{N} = \{1,2,3,4,5,\ldots\}$
- The integers: $\mathbb{Z} = \{\ldots,-3,-2,-1,0,1,2,3,4,5,\ldots\}$
- The rational numbers: $\mathbb{Q} = \{x : x = \frac{m}{n}$, where $m,n \in \mathbb{Z}$ and $n \neq 0\}$
- The real numbers: $\mathbb{R}$ (the set of all real numbers on the number line)

Notice that $\mathbb{Q}$ is the set of all numbers that can be expressed as a fraction of two integers. You are surely aware that $\mathbb{Q} \neq \mathbb{R}$, as $\sqrt{2} \notin \mathbb{Q}$ but $\sqrt{2} \in \mathbb{R}$.

Following are some other special sets that you will recall from your study of calculus. Given two numbers $a,b \in \mathbb{R}$ with $a < b$, we can form various intervals on the number line.

- Closed interval: $[a,b] = \{x \in \mathbb{R} : a \leq x \leq b\}$
- Half open interval: $(a,b] = \{x \in \mathbb{R} : a < x \leq b\}$
- Half open interval: $[a,b) = \{x \in \mathbb{R} : a \leq x < b\}$
- Open interval: $(a,b) = \{x \in \mathbb{R} : a < x < b\}$
- Infinite interval: $(a,\infty) = \{x \in \mathbb{R} : a < x\}$
- Infinite interval: $[a,\infty) = \{x \in \mathbb{R} : a \leq x\}$
- Infinite interval: $(-\infty,b) = \{x \in \mathbb{R} : x < b\}$
- Infinite interval: $(-\infty,b] = \{x \in \mathbb{R} : x \leq b\}$

Remember that these are intervals on the number line, so they have infinitely many elements. The set $(0.1,0.2)$ contains infinitely many numbers, even though the end points may be close together. It is an unfortunate notational accident that (a,b) can denote both an interval on the line and a point on the plane. The difference is usually clear from context. In the next section we will see still another meaning of (a,b).

Exercises for Section 1.1

A. Write each of the following sets by listing their elements between braces.

1. $\{5x-1 : x \in \mathbb{Z}\}$
2. $\{3x+2 : x \in \mathbb{Z}\}$
3. $\{x \in \mathbb{Z} : -2 \le x < 7\}$
4. $\{x \in \mathbb{N} : -2 < x \le 7\}$
5. $\{x \in \mathbb{R} : x^2 = 3\}$
6. $\{x \in \mathbb{R} : x^2 = 9\}$
7. $\{x \in \mathbb{R} : x^2 + 5x = -6\}$
8. $\{x \in \mathbb{R} : x^3 + 5x^2 = -6x\}$
9. $\{x \in \mathbb{R} : \sin \pi x = 0\}$
10. $\{x \in \mathbb{R} : \cos x = 1\}$
11. $\{x \in \mathbb{Z} : |x| < 5\}$
12. $\{x \in \mathbb{Z} : |2x| < 5\}$
13. $\{x \in \mathbb{Z} : |6x| < 5\}$
14. $\{5x : x \in \mathbb{Z}, |2x| \le 8\}$
15. $\{5a + 2b : a, b \in \mathbb{Z}\}$
16. $\{6a + 2b : a, b \in \mathbb{Z}\}$

B. Write each of the following sets in set-builder notation.

17. $\{2, 4, 8, 16, 32, 64 \ldots\}$
18. $\{0, 4, 16, 36, 64, 100, \ldots\}$
19. $\{\ldots, -6, -3, 0, 3, 6, 9, 12, 15, \ldots\}$
20. $\{\ldots, -8, -3, 2, 7, 12, 17, \ldots\}$
21. $\{0, 1, 4, 9, 16, 25, 36, \ldots\}$
22. $\{3, 6, 11, 18, 27, 38, \ldots\}$
23. $\{3, 4, 5, 6, 7, 8\}$
24. $\{-4, -3, -2, -1, 0, 1, 2\}$
25. $\{\ldots, \frac{1}{8}, \frac{1}{4}, \frac{1}{2}, 1, 2, 4, 8, \ldots\}$
26. $\{\ldots, \frac{1}{27}, \frac{1}{9}, \frac{1}{3}, 1, 3, 9, 27, \ldots\}$
27. $\{\ldots, -\pi, -\frac{\pi}{2}, 0, \frac{\pi}{2}, \pi, \frac{3\pi}{2}, 2\pi, \frac{5\pi}{2}, \ldots\}$
28. $\{\ldots, -\frac{3}{2}, -\frac{3}{4}, 0, \frac{3}{4}, \frac{3}{2}, \frac{9}{4}, 3, \frac{15}{4}, \frac{9}{2}, \ldots\}$

C. Find the following cardinalities.

29. $|\{\{1\}, \{2, \{3, 4\}\}, \emptyset\}|$
30. $|\{\{1, 4\}, a, b, \{\{3, 4\}\}, \{\emptyset\}\}|$
31. $|\{\{\{1\}, \{2, \{3, 4\}\}, \emptyset\}\}|$
32. $|\{\{\{1, 4\}, a, b, \{\{3, 4\}\}, \{\emptyset\}\}\}|$
33. $|\{x \in \mathbb{Z} : |x| < 10\}|$
34. $|\{x \in \mathbb{N} : |x| < 10\}|$
35. $|\{x \in \mathbb{Z} : x^2 < 10\}|$
36. $|\{x \in \mathbb{N} : x^2 < 10\}|$
37. $|\{x \in \mathbb{N} : x^2 < 0\}|$
38. $|\{x \in \mathbb{N} : 5x \le 20\}|$

D. Sketch the following sets of points in the x-y plane.

39. $\{(x, y) : x \in [1, 2], y \in [1, 2]\}$
40. $\{(x, y) : x \in [0, 1], y \in [1, 2]\}$
41. $\{(x, y) : x \in [-1, 1], y = 1\}$
42. $\{(x, y) : x = 2, y \in [0, 1]\}$
43. $\{(x, y) : |x| = 2, y \in [0, 1]\}$
44. $\{(x, x^2) : x \in \mathbb{R}\}$
45. $\{(x, y) : x, y \in \mathbb{R}, x^2 + y^2 = 1\}$
46. $\{(x, y) : x, y \in \mathbb{R}, x^2 + y^2 \le 1\}$
47. $\{(x, y) : x, y \in \mathbb{R}, y \ge x^2 - 1\}$
48. $\{(x, y) : x, y \in \mathbb{R}, x > 1\}$
49. $\{(x, x + y) : x \in \mathbb{R}, y \in \mathbb{Z}\}$
50. $\{(x, \frac{x^2}{y}) : x \in \mathbb{R}, y \in \mathbb{N}\}$
51. $\{(x, y) \in \mathbb{R}^2 : (y - x)(y + x) = 0\}$
52. $\{(x, y) \in \mathbb{R}^2 : (y - x^2)(y + x^2) = 0\}$

1.2 The Cartesian Product

Given two sets A and B, it is possible to "multiply" them to produce a new set denoted as $A \times B$. This operation is called the *Cartesian product*. To understand it, we must first understand the idea of an ordered pair.

Definition 1.1 An **ordered pair** is a list (x, y) of two things x and y, enclosed in parentheses and separated by a comma.

For example, $(2,4)$ is an ordered pair, as is $(4,2)$. These ordered pairs are different because even though they have the same things in them, the order is different. We write $(2,4) \neq (4,2)$. Right away you can see that ordered pairs can be used to describe points on the plane, as was done in calculus, but they are not limited to just that. The things in an ordered pair don't have to be numbers. You can have ordered pairs of letters, such as (m, ℓ), ordered pairs of sets such as $(\{2,5\},\{3,2\})$, even ordered pairs of ordered pairs like $((2,4),(4,2))$. The following are also ordered pairs: $(2,\{1,2,3\})$, $(\mathbb{R},(0,0))$. Any list of two things enclosed by parentheses is an ordered pair. Now we are ready to define the Cartesian product.

Definition 1.2 The **Cartesian product** of two sets A and B is another set, denoted as $A \times B$ and defined as $A \times B = \{(a,b) : a \in A, b \in B\}$.

Thus $A \times B$ is a set of ordered pairs of elements from A and B. For example, if $A = \{k, \ell, m\}$ and $B = \{q, r\}$, then

$$A \times B = \{(k,q),(k,r),(\ell,q),(\ell,r),(m,q),(m,r)\}.$$

Figure 1.1 shows how to make a schematic diagram of $A \times B$. Line up the elements of A horizontally and line up the elements of B vertically, as if A and B form an x- and y-axis. Then fill in the ordered pairs so that each element (x, y) is in the column headed by x and the row headed by y.

B		$A \times B$	
r	(k,r)	(ℓ,r)	(m,r)
q	(k,q)	(ℓ,q)	(m,q)
	k	ℓ	m A

Figure 1.1. A diagram of a Cartesian product

For another example, $\{0,1\} \times \{2,1\} = \{(0,2),(0,1),(1,2),(1,1)\}$. If you are a visual thinker, you may wish to draw a diagram similar to Figure 1.1. The rectangular array of such diagrams give us the following general fact.

Fact 1.1 If A and B are finite sets, then $|A \times B| = |A| \cdot |B|$.

The set $\mathbb{R} \times \mathbb{R} = \{(x,y) : x, y \in \mathbb{R}\}$ should be very familiar. It can be viewed as the set of points on the Cartesian plane, and is drawn in Figure 1.2(a). The set $\mathbb{R} \times \mathbb{N} = \{(x,y) : x \in \mathbb{R}, y \in \mathbb{N}\}$ can be regarded as all of the points on the Cartesian plane whose second coordinate is a natural number. This is illustrated in Figure 1.2(b), which shows that $\mathbb{R} \times \mathbb{N}$ looks like infinitely many horizontal lines at integer heights above the x axis. The set $\mathbb{N} \times \mathbb{N}$ can be visualized as the set of all points on the Cartesian plane whose coordinates are both natural numbers. It looks like a grid of dots in the first quadrant, as illustrated in Figure 1.2(c).

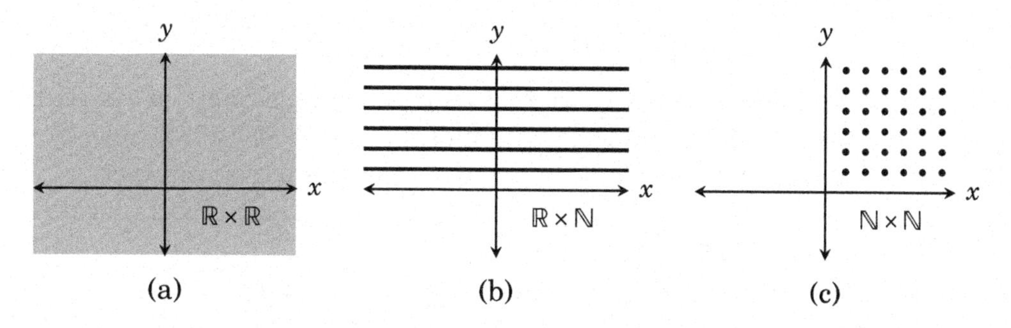

Figure 1.2. Drawings of some Cartesian products

It is even possible for one factor of a Cartesian product to be a Cartesian product itself, as in $\mathbb{R} \times (\mathbb{N} \times \mathbb{Z}) = \{(x,(y,z)) : x \in \mathbb{R}, (y,z) \in \mathbb{N} \times \mathbb{Z}\}$.

We can also define Cartesian products of three or more sets by moving beyond ordered pairs. An **ordered triple** is a list (x,y,z). The Cartesian product of the three sets $\mathbb{R}$, $\mathbb{N}$ and $\mathbb{Z}$ is $\mathbb{R} \times \mathbb{N} \times \mathbb{Z} = \{(x,y,z) : x \in \mathbb{R}, y \in \mathbb{N}, z \in \mathbb{Z}\}$. Of course there is no reason to stop with ordered triples. In general,

$$A_1 \times A_2 \times \cdots \times A_n = \{(x_1,x_2,\ldots,x_n) : x_i \in A_i \text{ for each } i = 1,2,\ldots,n\}.$$

Be mindful of parentheses. There is a slight difference between $\mathbb{R} \times (\mathbb{N} \times \mathbb{Z})$ and $\mathbb{R} \times \mathbb{N} \times \mathbb{Z}$. The first is a Cartesian product of two sets; its elements are ordered pairs $(x,(y,z))$. The second is a Cartesian product of three sets; its elements look like (x,y,z). To be sure, in many situations there is no harm in blurring the distinction between expressions like $(x,(y,z))$ and (x,y,z), but for now we consider them as different.

We can also take **Cartesian powers** of sets. For any set A and positive integer n, the power A^n is the Cartesian product of A with itself n times:

$$A^n = A \times A \times \cdots \times A = \{(x_1, x_2, \ldots, x_n) : x_1, x_2, \ldots, x_n \in A\}.$$

In this way, $\mathbb{R}^2$ is the familiar Cartesian plane and $\mathbb{R}^3$ is three-dimensional space. You can visualize how, if $\mathbb{R}^2$ is the plane, then $\mathbb{Z}^2 = \{(m,n) : m,n \in \mathbb{Z}\}$ is a grid of points on the plane. Likewise, as $\mathbb{R}^3$ is 3-dimensional space, $\mathbb{Z}^3 = \{(m,n,p) : m,n,p \in \mathbb{Z}\}$ is a grid of points in space.

In other courses you may encounter sets that are very similar to $\mathbb{R}^n$, but yet have slightly different shades of meaning. Consider, for example, the set of all two-by-three matrices with entries from $\mathbb{R}$:

$$M = \left\{ \left[\begin{smallmatrix} u & v & w \\ x & y & z \end{smallmatrix}\right] : u,v,w,x,y,z \in \mathbb{R} \right\}.$$

This is not really all that different from the set

$$\mathbb{R}^6 = \{(u,v,w,x,y,z) : u,v,w,x,y,z \in \mathbb{R}\}.$$

The elements of these sets are merely certain arrangements of six real numbers. Despite their similarity, we maintain that $M \neq \mathbb{R}^6$, for two-by-three matrices are not the same things as sequences of six numbers.

Exercises for Section 1.2

A. Write out the indicated sets by listing their elements between braces.

1. Suppose $A = \{1,2,3,4\}$ and $B = \{a,c\}$.

(a) $A \times B$ (b) $B \times A$ (c) $A \times A$ (d) $B \times B$ (e) $\emptyset \times B$ (f) $(A \times B) \times B$ (g) $A \times (B \times B)$ (h) B^3

2. Suppose $A = \{\pi, e, 0\}$ and $B = \{0,1\}$.

(a) $A \times B$ (b) $B \times A$ (c) $A \times A$ (d) $B \times B$ (e) $A \times \emptyset$ (f) $(A \times B) \times B$ (g) $A \times (B \times B)$ (h) $A \times B \times B$

3. $\{x \in \mathbb{R} : x^2 = 2\} \times \{a,c,e\}$

4. $\{n \in \mathbb{Z} : 2 < n < 5\} \times \{n \in \mathbb{Z} : |n| = 5\}$

5. $\{x \in \mathbb{R} : x^2 = 2\} \times \{x \in \mathbb{R} : |x| = 2\}$

6. $\{x \in \mathbb{R} : x^2 = x\} \times \{x \in \mathbb{N} : x^2 = x\}$

7. $\{\emptyset\} \times \{0, \emptyset\} \times \{0,1\}$

8. $\{0,1\}^4$

B. Sketch these Cartesian products on the x-y plane $\mathbb{R}^2$ (or $\mathbb{R}^3$ for the last two).

9. $\{1,2,3\} \times \{-1,0,1\}$

10. $\{-1,0,1\} \times \{1,2,3\}$

11. $[0,1] \times [0,1]$

12. $[-1,1] \times [1,2]$

13. $\{1,1.5,2\} \times [1,2]$

14. $[1,2] \times \{1,1.5,2\}$

15. $\{1\} \times [0,1]$

16. $[0,1] \times \{1\}$

17. $\mathbb{N} \times \mathbb{Z}$

18. $\mathbb{Z} \times \mathbb{Z}$

19. $[0,1] \times [0,1] \times [0,1]$

20. $\{(x,y) \in \mathbb{R}^2 : x^2 + y^2 \leq 1\} \times [0,1]$

1.3 Subsets

It can happen that every element of some set A is also an element of another set B. For example, each element of $A = \{0,2,4\}$ is also an element of $B = \{0,1,2,3,4\}$. When A and B are related this way we say that *A is a subset of B*.

Definition 1.3 Suppose A and B are sets. If every element of A is also an element of B, then we say A is a **subset** of B, and we denote this as $A \subseteq B$. We write $A \not\subseteq B$ if A is *not* a subset of B, that is, if it is *not* true that every element of A is also an element of B. Thus $A \not\subseteq B$ means that there is at least one element of A that is *not* an element of B.

Example 1.2 Be sure you understand why each of the following is true.

1. $\{2,3,7\} \subseteq \{2,3,4,5,6,7\}$
2. $\{2,3,7\} \not\subseteq \{2,4,5,6,7\}$
3. $\{2,3,7\} \subseteq \{2,3,7\}$
4. $\{2n : n \in \mathbb{Z}\} \subseteq \mathbb{Z}$
5. $\{(x, \sin(x)) : x \in \mathbb{R}\} \subseteq \mathbb{R}^2$
6. $\{2,3,5,7,11,13,17,\ldots\} \subseteq \mathbb{N}$
7. $\mathbb{N} \subseteq \mathbb{Z} \subseteq \mathbb{Q} \subseteq \mathbb{R}$
8. $\mathbb{R} \times \mathbb{N} \subseteq \mathbb{R} \times \mathbb{R}$

This brings us to a significant fact: If B is any set whatsoever, then $\emptyset \subseteq B$. To see why this is true, look at the last sentence of Definition 1.3. It says that $\emptyset \not\subseteq B$ would mean that there is at least one element of $\emptyset$ that is not an element of B. But this cannot be so because $\emptyset$ contains no elements! Thus it is not the case that $\emptyset \not\subseteq B$, so it must be that $\emptyset \subseteq B$.

Fact 1.2 The empty set is a subset of every set, that is, $\emptyset \subseteq B$ for any set B.

Here is another way to look at it. Imagine a subset of B as a thing you make by starting with braces $\{\}$, then filling them with selections from B. For example, to make one particular subset of $B = \{a,b,c\}$, start with $\{\}$, select b and c from B and insert them into $\{\}$ to form the subset $\{b,c\}$. Alternatively, you could have chosen just a to make $\{a\}$, and so on. But one option is to simply select nothing from B. This leaves you with the subset $\{\}$. Thus $\{\} \subseteq B$. More often we write it as $\emptyset \subseteq B$.

This idea of "making" a subset can help us list out all the subsets of a given set B. As an example, let $B = \{a,b,c\}$. Let's list all of its subsets. One way of approaching this is to make a tree-like structure. Begin with the subset $\{\}$, which is shown on the left of Figure 1.3. Considering the element a of B, we have a choice: insert it or not. The lines from $\{\}$ point to what we get depending whether or not we insert a, either $\{\}$ or $\{a\}$. Now move on to the element b of B. For each of the sets just formed we can either insert or not insert b, and the lines on the diagram point to the resulting sets $\{\}$, $\{b\}$,$\{a\}$, or $\{a,b\}$. Finally, to each of these sets, we can either insert c or not insert it, and this gives us, on the far right-hand column, the sets $\{\}$, $\{c\}$, $\{b\}$, $\{b,c\}$, $\{a\}$, $\{a,c\}$, $\{a,b\}$ and $\{a,b,c\}$. These are the eight subsets of $B = \{a,b,c\}$.

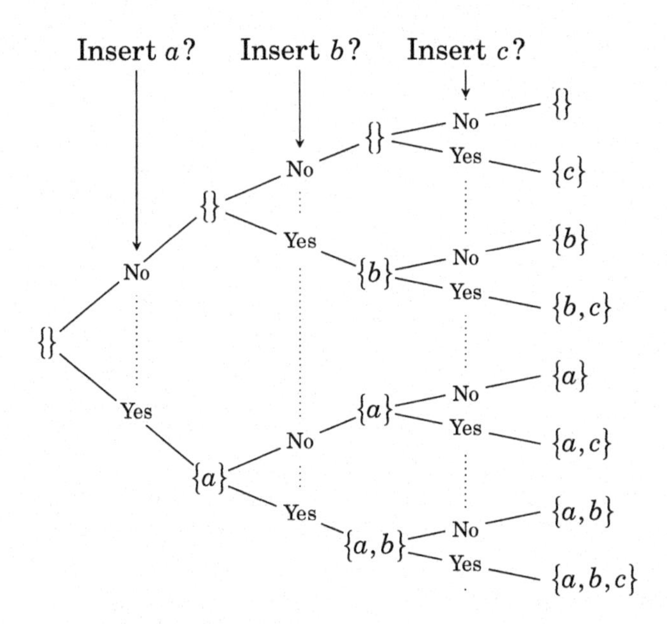

Figure 1.3. A "tree" for listing subsets

We can see from the way this tree branches out that if it happened that $B = \{a\}$, then B would have just two subsets, those in the second column of the diagram. If it happened that $B = \{a,b\}$, then B would have four subsets, those listed in the third column, and so on. At each branching of the tree, the number of subsets doubles. Thus in general, if $|B| = n$, then B must have 2^n subsets.

Fact 1.3 If a finite set has n elements, then it has 2^n subsets.

For a slightly more complex example, consider listing the subsets of $B = \{1,2,\{1,3\}\}$. This B has just three elements: 1, 2 and $\{1,3\}$. At this point you probably don't even have to draw a tree to list out B's subsets. You just make all the possible selections from B and put them between braces to get

$$\{\},\ \{1\},\ \{2\},\ \{\{1,3\}\},\ \{1,2\},\ \{1,\{1,3\}\},\ \{2,\{1,3\}\},\ \{1,2,\{1,3\}\}.$$

These are the eight subsets of B. Exercises like this help you identify what is and isn't a subset. You know immediately that a set such as $\{1,3\}$ is *not* a subset of B because it can't be made by selecting elements from B, as the 3 is not an element of B and thus is not a valid selection. Notice that although $\{1,3\} \not\subseteq B$, it *is* true that $\{1,3\} \in B$. Also, $\{\{1,3\}\} \subseteq B$.

Example 1.3 Be sure you understand why the following statements are true. Each illustrates an aspect of set theory that you've learned so far.

1. $1 \in \{1,\{1\}\}$ .... 1 is the first element listed in $\{1,\{1\}\}$
2. $1 \not\subseteq \{1,\{1\}\}$ .... because 1 is not a set
3. $\{1\} \in \{1,\{1\}\}$ .... $\{1\}$ is the second element listed in $\{1,\{1\}\}$
4. $\{1\} \subseteq \{1,\{1\}\}$ .... make subset $\{1\}$ by selecting 1 from $\{1,\{1\}\}$
5. $\{\{1\}\} \notin \{1,\{1\}\}$ .... because $\{1,\{1\}\}$ contains only 1 and $\{1\}$, and not $\{\{1\}\}$
6. $\{\{1\}\} \subseteq \{1,\{1\}\}$ .... make subset $\{\{1\}\}$ by selecting $\{1\}$ from $\{1,\{1\}\}$
7. $\mathbb{N} \notin \mathbb{N}$ .... because $\mathbb{N}$ is a set (not a number) and $\mathbb{N}$ contains only numbers
8. $\mathbb{N} \subseteq \mathbb{N}$ .... because $X \subseteq X$ for every set X
9. $\emptyset \notin \mathbb{N}$ .... because the set $\mathbb{N}$ contains only numbers and no sets
10. $\emptyset \subseteq \mathbb{N}$ .... because $\emptyset$ is a subset of every set
11. $\mathbb{N} \in \{\mathbb{N}\}$ .... because $\{\mathbb{N}\}$ has just one element, the set $\mathbb{N}$
12. $\mathbb{N} \not\subseteq \{\mathbb{N}\}$ .... because, for instance, $1 \in \mathbb{N}$ but $1 \notin \{\mathbb{N}\}$
13. $\emptyset \notin \{\mathbb{N}\}$ .... note that the only element of $\{\mathbb{N}\}$ is $\mathbb{N}$, and $\mathbb{N} \neq \emptyset$
14. $\emptyset \subseteq \{\mathbb{N}\}$ .... because $\emptyset$ is a subset of every set
15. $\emptyset \in \{\emptyset,\mathbb{N}\}$ .... $\emptyset$ is the first element listed in $\{\emptyset,\mathbb{N}\}$
16. $\emptyset \subseteq \{\emptyset,\mathbb{N}\}$ .... because $\emptyset$ is a subset of every set
17. $\{\mathbb{N}\} \subseteq \{\emptyset,\mathbb{N}\}$ .... make subset $\{\mathbb{N}\}$ by selecting $\mathbb{N}$ from $\{\emptyset,\mathbb{N}\}$
18. $\{\mathbb{N}\} \not\subseteq \{\emptyset,\{\mathbb{N}\}\}$ .... because $\mathbb{N} \notin \{\emptyset,\{\mathbb{N}\}\}$
19. $\{\mathbb{N}\} \in \{\emptyset,\{\mathbb{N}\}\}$ .... $\{\mathbb{N}\}$ is the second element listed in $\{\emptyset,\{\mathbb{N}\}\}$
20. $\{(1,2),(2,2),(7,1)\} \subseteq \mathbb{N} \times \mathbb{N}$ .... each of $(1,2)$, $(2,2)$, $(7,1)$ is in $\mathbb{N} \times \mathbb{N}$

Though they should help you understand the concept of subset, the above examples are somewhat artificial. But in general, subsets arise very naturally. For instance, consider the unit circle $C = \{(x,y) \in \mathbb{R}^2 : x^2 + y^2 = 1\}$. This is a subset $C \subseteq \mathbb{R}^2$. Likewise the graph of a function $y = f(x)$ is a set of points $G = \{(x, f(x)) : x \in \mathbb{R}\}$, and $G \subseteq \mathbb{R}^2$. Surely sets such as C and G are more easily understood or visualized when regarded as subsets of $\mathbb{R}^2$. Mathematics is filled with such instances where it is important to regard one set as a subset of another.

Exercises for Section 1.3

A. List all the subsets of the following sets.

1. $\{1,2,3,4\}$

2. $\{1,2,\emptyset\}$

3. $\{\{\mathbb{R}\}\}$

4. $\emptyset$

5. $\{\emptyset\}$

6. $\{\mathbb{R},\mathbb{Q},\mathbb{N}\}$

7. $\{\mathbb{R},\{\mathbb{Q},\mathbb{N}\}\}$

8. $\{\{0,1\},\{0,1,\{2\}\},\{0\}\}$

B. Write out the following sets by listing their elements between braces.

9. $\{X : X \subseteq \{3,2,a\} \text{ and } |X| = 2\}$

10. $\{X \subseteq \mathbb{N} : |X| \le 1\}$

11. $\{X : X \subseteq \{3,2,a\} \text{ and } |X| = 4\}$

12. $\{X : X \subseteq \{3,2,a\} \text{ and } |X| = 1\}$

C. Decide if the following statements are true or false. Explain.

13. $\mathbb{R}^3 \subseteq \mathbb{R}^3$

14. $\mathbb{R}^2 \subseteq \mathbb{R}^3$

15. $\{(x,y) : x - 1 = 0\} \subseteq \{(x,y) : x^2 - x = 0\}$

16. $\{(x,y) : x^2 - x = 0\} \subseteq \{(x,y) : x - 1 = 0\}$

1.4 Power Sets

Given a set, you can form a new set with the *power set* operation, defined as follows.

Definition 1.4 If A is a set, the **power set** of A is another set, denoted as $\mathscr{P}(A)$ and defined to be the set of all subsets of A. In symbols, $\mathscr{P}(A) = \{X : X \subseteq A\}$.

For example, suppose $A = \{1,2,3\}$. The power set of A is the set of all subsets of A. We learned how to find these subsets in the previous section, and they are $\{\}$, $\{1\}$, $\{2\}$, $\{3\}$, $\{1,2\}$, $\{1,3\}$, $\{2,3\}$ and $\{1,2,3\}$. Therefore the power set of A is

$$\mathscr{P}(A) = \{\ \emptyset, \{1\}, \{2\}, \{3\}, \{1,2\}, \{1,3\}, \{2,3\}, \{1,2,3\}\ \}.$$

As we saw in the previous section, if a finite set A has n elements, then it has 2^n subsets, and thus its power set has 2^n elements.

Fact 1.4 If A is a finite set, then $|\mathscr{P}(A)| = 2^{|A|}$.

Example 1.4 You should examine the following statements and make sure you understand how the answers were obtained. In particular, notice that in each instance the equation $|\mathscr{P}(A)| = 2^{|A|}$ is true.

1. $\mathscr{P}(\{0,1,3\}) = \{\emptyset, \{0\}, \{1\}, \{3\}, \{0,1\}, \{0,3\}, \{1,3\}, \{0,1,3\}\}$
2. $\mathscr{P}(\{1,2\}) = \{\emptyset, \{1\}, \{2\}, \{1,2\}\}$
3. $\mathscr{P}(\{1\}) = \{\emptyset, \{1\}\}$
4. $\mathscr{P}(\emptyset) = \{\emptyset\}$
5. $\mathscr{P}(\{a\}) = \{\emptyset, \{a\}\}$
6. $\mathscr{P}(\{\emptyset\}) = \{\emptyset, \{\emptyset\}\}$
7. $\mathscr{P}(\{a\}) \times \mathscr{P}(\{\emptyset\}) = \{(\emptyset,\emptyset), (\emptyset,\{\emptyset\}), (\{a\},\emptyset), (\{a\},\{\emptyset\})\}$
8. $\mathscr{P}(\mathscr{P}(\{\emptyset\})) = \{\emptyset, \{\emptyset\}, \{\{\emptyset\}\}, \{\emptyset,\{\emptyset\}\}\}$
9. $\mathscr{P}(\{1,\{1,2\}\}) = \{\emptyset, \{1\}, \{\{1,2\}\}, \{1,\{1,2\}\}\}$
10. $\mathscr{P}(\{\mathbb{Z},\mathbb{N}\}) = \{\emptyset, \{\mathbb{Z}\}, \{\mathbb{N}\}, \{\mathbb{Z},\mathbb{N}\}\}$

Next are some that are **wrong**. See if you can determine why they are wrong and make sure you understand the explanation on the right.

11. $\mathscr{P}(1) = \{\emptyset, \{1\}\}$ .... meaningless because 1 is not a set
12. $\mathscr{P}(\{1,\{1,2\}\}) = \{\emptyset,\{1\},\{1,2\},\{1,\{1,2\}\}\}$ .... wrong because $\{1,2\} \not\subseteq \{1,\{1,2\}\}$
13. $\mathscr{P}(\{1,\{1,2\}\}) = \{\emptyset,\{\{1\}\},\{\{1,2\}\},\{\emptyset,\{1,2\}\}\}$ ..... wrong because $\{\{1\}\} \not\subseteq \{1,\{1,2\}\}$

If A is finite, it is possible (though maybe not practical) to list out $\mathscr{P}(A)$ between braces as was done in the above example. That is not possible if A is infinite. For example, consider $\mathscr{P}(\mathbb{N})$. If you start listing its elements you quickly discover that $\mathbb{N}$ has infinitely many subsets, and it's not clear how (or if) they could be arranged as a list with a definite pattern:

$$\mathscr{P}(\mathbb{N}) = \{\emptyset,\{1\},\{2\},\ldots,\{1,2\},\{1,3\},\ldots,\{39,47\},\\ \ldots,\{3,87,131\},\ldots,\{2,4,6,8,\ldots\},\ldots\ ?\ \ldots\}.$$

The set $\mathscr{P}(\mathbb{R}^2)$ is mind boggling. Think of $\mathbb{R}^2 = \{(x,y) : x,y \in \mathbb{R}\}$ as the set of all points on the Cartesian plane. A subset of $\mathbb{R}^2$ (that is, an *element* of $\mathscr{P}(\mathbb{R}^2)$) is a set of points in the plane. Let's look at some of these sets. Since $\{(0,0),(1,1)\} \subseteq \mathbb{R}^2$, we know that $\{(0,0),(1,1)\} \in \mathscr{P}(\mathbb{R}^2)$. We can even draw a picture of this subset, as in Figure 1.4(a). For another example, the graph of the equation $y = x^2$ is the set of points $G = \{(x,x^2) : x \in \mathbb{R}\}$ and this is a subset of $\mathbb{R}^2$, so $G \in \mathscr{P}(\mathbb{R}^2)$. Figure 1.4(b) is a picture of G. Because this can be done for any function, the graph of any imaginable function $f : \mathbb{R} \to \mathbb{R}$ is an element of $\mathscr{P}(\mathbb{R}^2)$.

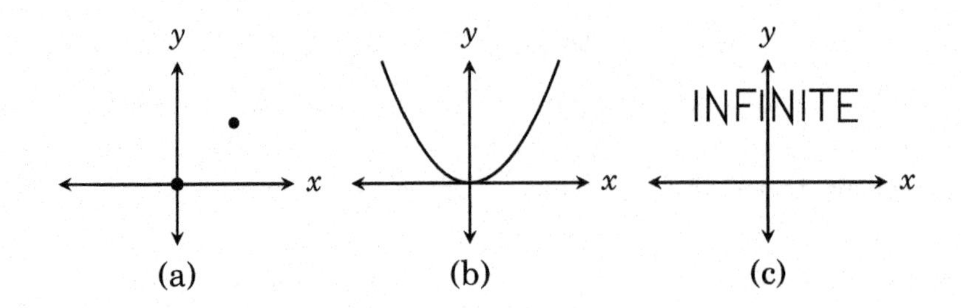

Figure 1.4. Three of the many, many sets in $\mathscr{P}(\mathbb{R}^2)$

In fact, any black-and-white image on the plane can be thought of as a subset of $\mathbb{R}^2$, where the black points belong to the subset and the white points do not. So the text "INFINITE" in Figure 1.4(c) is a subset of $\mathbb{R}^2$ and therefore an element of $\mathscr{P}(\mathbb{R}^2)$. By that token, $\mathscr{P}(\mathbb{R}^2)$ contains a copy of the page you are reading now.

Thus in addition to containing every imaginable function and every imaginable black-and-white image, $\mathscr{P}(\mathbb{R}^2)$ also contains the full text of every book that was ever written, those that are yet to be written and those that will never be written. Inside of $\mathscr{P}(\mathbb{R}^2)$ is a detailed biography of your life, from beginning to end, as well as the biographies of all of your unborn descendants. It is startling that the five symbols used to write $\mathscr{P}(\mathbb{R}^2)$ can express such an incomprehensibly large set.

Homework: Think about $\mathscr{P}(\mathscr{P}(\mathbb{R}^2))$.

Exercises for Section 1.4

A. Find the indicated sets.

1. $\mathscr{P}(\{\{a,b\},\{c\}\})$

2. $\mathscr{P}(\{1,2,3,4\})$

3. $\mathscr{P}(\{\{\emptyset\},5\})$

4. $\mathscr{P}(\{\mathbb{R},\mathbb{Q}\})$

5. $\mathscr{P}(\mathscr{P}(\{2\}))$

6. $\mathscr{P}(\{1,2\})\times\mathscr{P}(\{3\})$

7. $\mathscr{P}(\{a,b\})\times\mathscr{P}(\{0,1\})$

8. $\mathscr{P}(\{1,2\}\times\{3\})$

9. $\mathscr{P}(\{a,b\}\times\{0\})$

10. $\{X\in\mathscr{P}(\{1,2,3\}):|X|\le 1\}$

11. $\{X\subseteq\mathscr{P}(\{1,2,3\}):|X|\le 1\}$

12. $\{X\in\mathscr{P}(\{1,2,3\}):2\in X\}$

B. Suppose that $|A|=m$ and $|B|=n$. Find the following cardinalities.

13. $|\mathscr{P}(\mathscr{P}(\mathscr{P}(A)))|$

14. $|\mathscr{P}(\mathscr{P}(A))|$

15. $|\mathscr{P}(A\times B)|$

16. $|\mathscr{P}(A)\times\mathscr{P}(B)|$

17. $\left|\{X\in\mathscr{P}(A):|X|\le 1\}\right|$

18. $|\mathscr{P}(A\times\mathscr{P}(B))|$

19. $|\mathscr{P}(\mathscr{P}(\mathscr{P}(A\times\emptyset)))|$

20. $\left|\{X\subseteq\mathscr{P}(A):|X|\le 1\}\right|$

1.5 Union, Intersection, Difference

Just as numbers are combined with operations such as addition, subtraction and multiplication, there are various operations that can be applied to sets. The Cartesian product (defined in Section 1.2) is one such operation; given sets A and B, we can combine them with $\times$ to get a new set $A \times B$. Here are three new operations called union, intersection and difference.

Definition 1.5 Suppose A and B are sets.
The **union** of A and B is the set $A \cup B = \{x : x \in A \text{ or } x \in B\}$.
The **intersection** of A and B is the set $A \cap B = \{x : x \in A \text{ and } x \in B\}$.
The **difference** of A and B is the set $A - B = \{x : x \in A \text{ and } x \notin B\}$.

In words, the union $A \cup B$ is the set of all things that are in A or in B (or in both). The intersection $A \cap B$ is the set of all things in both A and B. The difference $A - B$ is the set of all things that are in A but not in B.

Example 1.5 Suppose $A = \{a,b,c,d,e\}$, $B = \{d,e,f\}$ and $C = \{1,2,3\}$.

1. $A \cup B = \{a,b,c,d,e,f\}$
2. $A \cap B = \{d,e\}$
3. $A - B = \{a,b,c\}$
4. $B - A = \{f\}$
5. $(A - B) \cup (B - A) = \{a,b,c,f\}$
6. $A \cup C = \{a,b,c,d,e,1,2,3\}$
7. $A \cap C = \emptyset$
8. $A - C = \{a,b,c,d,e\}$
9. $(A \cap C) \cup (A - C) = \{a,b,c,d,e\}$
10. $(A \cap B) \times B = \{(d,d),(d,e),(d,f),(e,d),(e,e),(e,f)\}$
11. $(A \times C) \cap (B \times C) = \{(d,1),(d,2),(d,3),(e,1),(e,2),(e,3)\}$

Observe that for any sets X and Y it is always true that $X \cup Y = Y \cup X$ and $X \cap Y = Y \cap X$, but in general $X - Y \neq Y - X$.

Continuing the example, parts 12–15 below use the interval notation discussed in Section 1.1, so $[2,5] = \{x \in \mathbb{R} : 2 \le x \le 5\}$, etc. Sketching these examples on the number line may help you understand them.

12. $[2,5] \cup [3,6] = [2,6]$
13. $[2,5] \cap [3,6] = [3,5]$
14. $[2,5] - [3,6] = [2,3)$
15. $[0,3] - [1,2] = [0,1) \cup (2,3]$

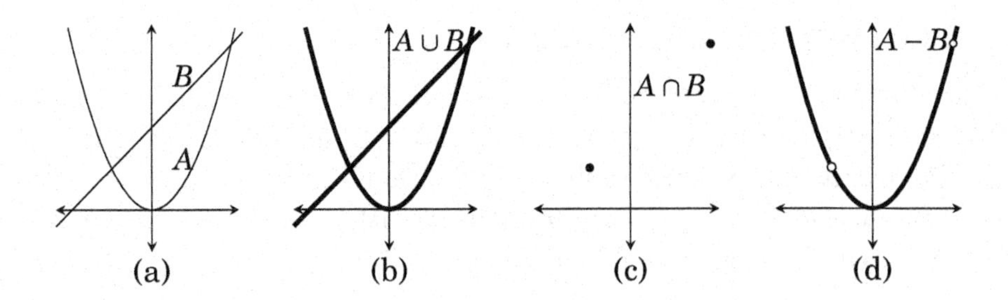

Figure 1.5. The union, intersection and difference of sets A and B

Example 1.6 Let $A = \{(x,x^2) : x \in \mathbb{R}\}$ be the graph of the equation $y = x^2$ and let $B = \{(x,x+2) : x \in \mathbb{R}\}$ be the graph of the equation $y = x+2$. These sets are subsets of $\mathbb{R}^2$. They are sketched together in Figure 1.5(a). Figure 1.5(b) shows $A \cup B$, the set of all points (x,y) that are on one (or both) of the two graphs. Observe that $A \cap B = \{(-1,1),(2,4)\}$ consists of just two elements, the two points where the graphs intersect, as illustrated in Figure 1.5(c). Figure 1.5(d) shows $A - B$, which is the set A with "holes" where B crossed it. In set builder notation, we could write $A \cup B = \{(x,y) : x \in \mathbb{R}, y = x^2 \text{ or } y = x+2\}$ and $A - B = \{(x,x^2) : x \in \mathbb{R} - \{-1,2\}\}$.

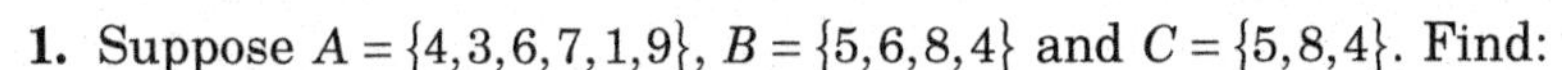

Exercises for Section 1.5

1. Suppose $A = \{4,3,6,7,1,9\}$, $B = \{5,6,8,4\}$ and $C = \{5,8,4\}$. Find:
   (a) $A \cup B$ (b) $A \cap B$ (c) $A - B$ (d) $A - C$ (e) $B - A$ (f) $A \cap C$ (g) $B \cap C$ (h) $B \cup C$ (i) $C - B$
2. Suppose $A = \{0,2,4,6,8\}$, $B = \{1,3,5,7\}$ and $C = \{2,8,4\}$. Find:
   (a) $A \cup B$ (b) $A \cap B$ (c) $A - B$ (d) $A - C$ (e) $B - A$ (f) $A \cap C$ (g) $B \cap C$ (h) $C - A$ (i) $C - B$
3. Suppose $A = \{0,1\}$ and $B = \{1,2\}$. Find:
   (a) $(A \times B) \cap (B \times B)$ (b) $(A \times B) \cup (B \times B)$ (c) $(A \times B) - (B \times B)$ (d) $(A \cap B) \times A$ (e) $(A \times B) \cap B$ (f) $\mathscr{P}(A) \cap \mathscr{P}(B)$ (g) $\mathscr{P}(A) - \mathscr{P}(B)$ (h) $\mathscr{P}(A \cap B)$ (i) $\mathscr{P}(A \times B)$
4. Suppose $A = \{b,c,d\}$ and $B = \{a,b\}$. Find:
   (a) $(A \times B) \cap (B \times B)$ (b) $(A \times B) \cup (B \times B)$ (c) $(A \times B) - (B \times B)$ (d) $(A \cap B) \times A$ (e) $(A \times B) \cap B$ (f) $\mathscr{P}(A) \cap \mathscr{P}(B)$ (g) $\mathscr{P}(A) - \mathscr{P}(B)$ (h) $\mathscr{P}(A \cap B)$ (i) $\mathscr{P}(A) \times \mathscr{P}(B)$

5. Sketch the sets $X = [1,3] \times [1,3]$ and $Y = [2,4] \times [2,4]$ on the plane $\mathbb{R}^2$. On separate drawings, shade in the sets $X \cup Y$, $X \cap Y$, $X - Y$ and $Y - X$. (Hint: X and Y are Cartesian products of intervals. You may wish to review how you drew sets like $[1,3] \times [1,3]$ in the exercises for Section 1.2.)
6. Sketch the sets $X = [-1,3] \times [0,2]$ and $Y = [0,3] \times [1,4]$ on the plane $\mathbb{R}^2$. On separate drawings, shade in the sets $X \cup Y$, $X \cap Y$, $X - Y$ and $Y - X$.
7. Sketch the sets $X = \{(x,y) \in \mathbb{R}^2 : x^2 + y^2 \le 1\}$ and $Y = \{(x,y) \in \mathbb{R}^2 : x \ge 0\}$ on $\mathbb{R}^2$. On separate drawings, shade in the sets $X \cup Y$, $X \cap Y$, $X - Y$ and $Y - X$.
8. Sketch the sets $X = \{(x,y) \in \mathbb{R}^2 : x^2 + y^2 \le 1\}$ and $Y = \{(x,y) \in \mathbb{R}^2 : -1 \le y \le 0\}$ on $\mathbb{R}^2$. On separate drawings, shade in the sets $X \cup Y$, $X \cap Y$, $X - Y$ and $Y - X$.
9. Is the statement $(\mathbb{R} \times \mathbb{Z}) \cap (\mathbb{Z} \times \mathbb{R}) = \mathbb{Z} \times \mathbb{Z}$ true or false? What about the statement $(\mathbb{R} \times \mathbb{Z}) \cup (\mathbb{Z} \times \mathbb{R}) = \mathbb{R} \times \mathbb{R}$?
10. Do you think the statement $(\mathbb{R} - \mathbb{Z}) \times \mathbb{N} = (\mathbb{R} \times \mathbb{N}) - (\mathbb{Z} \times \mathbb{N})$ is true, or false? Justify.

1.6 Complement

This section introduces yet another set operation, called the *set complement*. The definition requires the idea of a *universal set*, which we now discuss.

When dealing with a set, we almost always regard it as a subset of some larger set. For example, consider the set of prime numbers $P = \{2,3,5,7,11,13,\ldots\}$. If asked to name some things that are *not* in P, we might mention some composite numbers like 4 or 6 or 423. It probably would not occur to us to say that Vladimir Putin is not in P. True, Vladimir Putin is not in P, but he lies entirely outside of the discussion of what is a prime number and what is not. We have an unstated assumption that

$$P \subseteq \mathbb{N}$$

because $\mathbb{N}$ is the most natural setting in which to discuss prime numbers. In this context, anything not in P should still be in $\mathbb{N}$. This larger set $\mathbb{N}$ is called the **universal set** or **universe** for P.

Almost every useful set in mathematics can be regarded as having some natural universal set. For instance, the unit circle is the set $C = \{(x,y) \in \mathbb{R}^2 : x^2 + y^2 = 1\}$, and since all these points are in the plane $\mathbb{R}^2$ it is natural to regard $\mathbb{R}^2$ as the universal set for C. In the absence of specifics, if A is a set, then its universal set is often denoted as U. We are now ready to define the complement operation.

Definition 1.6 Let A be a set with a universal set U. The **complement** of A, denoted $\overline{A}$, is the set $\overline{A} = U - A$.

Example 1.7 If P is the set of prime numbers, then

$$\overline{P} = \mathbb{N} - P = \{1,4,6,8,9,10,12,\ldots\}.$$

Thus $\overline{P}$ is the set of composite numbers and 1.

Example 1.8 Let $A = \{(x,x^2) : x \in \mathbb{R}\}$ be the graph of the equation $y = x^2$. Figure 1.6(a) shows A in its universal set $\mathbb{R}^2$. The complement of A is $\overline{A} = \mathbb{R}^2 - A = \{(x,y) \in \mathbb{R}^2 : y \neq x^2\}$, illustrated by the shaded area in Figure 1.6(b).

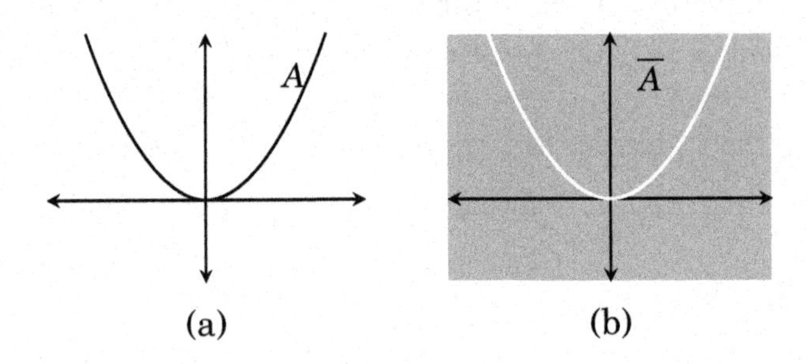

Figure 1.6. A set and its complement

Exercises for Section 1.6

1. Let $A = \{4,3,6,7,1,9\}$ and $B = \{5,6,8,4\}$ have universal set $U = \{0,1,2,\ldots,10\}$. Find:
   - **(a)** $\overline{A}$
   - **(b)** $\overline{B}$
   - **(c)** $A \cap \overline{A}$
   - **(d)** $A \cup \overline{A}$
   - **(e)** $A - \overline{A}$
   - **(f)** $A - \overline{B}$
   - **(g)** $\overline{A} - \overline{B}$
   - **(h)** $\overline{A} \cap B$
   - **(i)** $\overline{\overline{A} \cap B}$
2. Let $A = \{0,2,4,6,8\}$ and $B = \{1,3,5,7\}$ have universal set $U = \{0,1,2,\ldots,8\}$. Find:
   - **(a)** $\overline{A}$
   - **(b)** $\overline{B}$
   - **(c)** $A \cap \overline{A}$
   - **(d)** $A \cup \overline{A}$
   - **(e)** $A - \overline{A}$
   - **(f)** $\overline{A \cup B}$
   - **(g)** $\overline{A} \cap \overline{B}$
   - **(h)** $\overline{A \cap B}$
   - **(i)** $\overline{A} \times B$
3. Sketch the set $X = [1,3] \times [1,2]$ on the plane $\mathbb{R}^2$. On separate drawings, shade in the sets $\overline{X}$ and $\overline{X} \cap ([0,2] \times [0,3])$.
4. Sketch the set $X = [-1,3] \times [0,2]$ on the plane $\mathbb{R}^2$. On separate drawings, shade in the sets $\overline{X}$ and $\overline{X} \cap ([-2,4] \times [-1,3])$.
5. Sketch the set $X = \{(x,y) \in \mathbb{R}^2 : 1 \leq x^2 + y^2 \leq 4\}$ on the plane $\mathbb{R}^2$. On a separate drawing, shade in the set $\overline{X}$.
6. Sketch the set $X = \{(x,y) \in \mathbb{R}^2 : y < x^2\}$ on $\mathbb{R}^2$. Shade in the set $\overline{X}$.

1.7 Venn Diagrams

In thinking about sets, it is sometimes helpful to draw informal, schematic diagrams of them. In doing this we often represent a set with a circle (or oval), which we regard as enclosing all the elements of the set. Such diagrams can illustrate how sets combine using various operations. For example, Figures 1.7(a–c) show two sets A and B that overlap in a middle region. The sets $A \cup B$, $A \cap B$ and $A - B$ are shaded. Such graphical representations of sets are called **Venn diagrams**, after their inventor, British logician John Venn, 1834–1923.

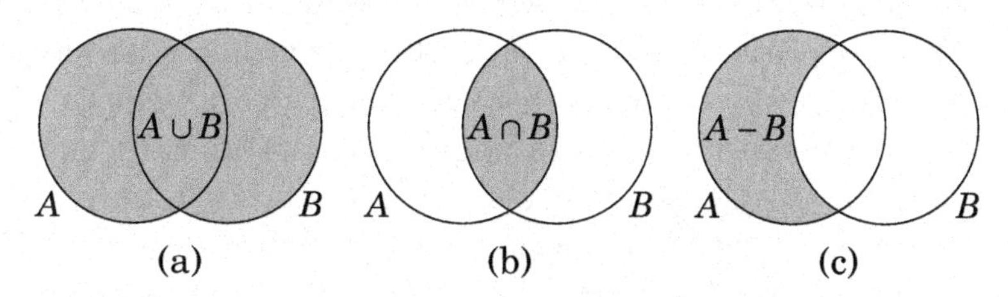

Figure 1.7. Venn diagrams for two sets

Though you are unlikely to draw Venn diagrams as a part of a proof of any theorem, you will probably find them to be useful "scratch work" devices that help you to understand how sets combine, and to develop strategies for proving certain theorems or solving certain problems. The remainder of this section uses Venn diagrams to explore how three sets can be combined using $\cup$ and $\cap$.

Let's begin with the set $A \cup B \cup C$. Our definitions suggest this should consist of all elements which are in one or more of the sets A, B and C. Figure 1.8(a) shows a Venn diagram for this. Similarly, we think of $A \cap B \cap C$ as all elements common to each of A, B and C, so in Figure 1.8(b) the region belonging to all three sets is shaded.

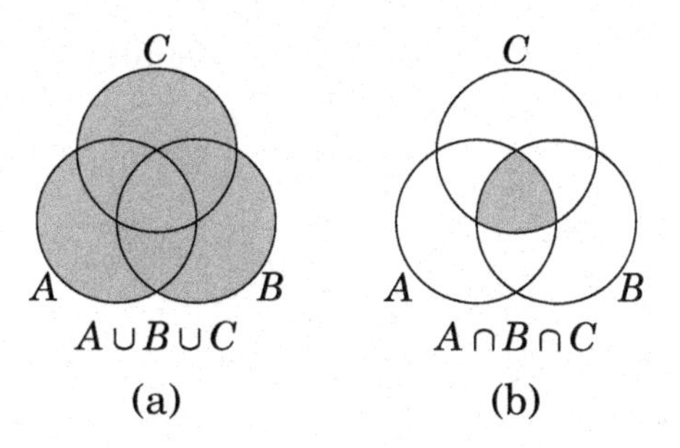

Figure 1.8. Venn diagrams for three sets

We can also think of $A \cap B \cap C$ as the two-step operation $(A \cap B) \cap C$. In this expression the set $A \cap B$ is represented by the region common to both A and B, and when we intersect *this* with C we get Figure 1.8(b). This is a visual representation of the fact that $A \cap B \cap C = (A \cap B) \cap C$. Similarly, we have $A \cap B \cap C = A \cap (B \cap C)$. Likewise, $A \cup B \cup C = (A \cup B) \cup C = A \cup (B \cup C)$.

Notice that in these examples, where the expression either contains only the symbol $\cup$ or only the symbol $\cap$, the placement of the parentheses is irrelevant, so we are free to drop them. It is analogous to the situations in algebra involving expressions $(a+b)+c = a+(b+c)$ or $(a \cdot b) \cdot c = a \cdot (b \cdot c)$. We tend to drop the parentheses and write simply $a+b+c$ or $a \cdot b \cdot c$. By contrast, in an expression like $(a+b) \cdot c$ the parentheses are absolutely essential because $(a+b) \cdot c$ and $a+(b \cdot c)$ are generally not equal.

Now let's use Venn diagrams to help us understand the expressions $(A \cup B) \cap C$ and $A \cup (B \cap C)$, which use a mix of $\cup$ and $\cap$. Figure 1.9 shows how to draw a Venn diagram for $(A \cup B) \cap C$. In the drawing on the left, the set $A \cup B$ is shaded with horizontal lines, while C is shaded with vertical lines. Thus the set $(A \cup B) \cap C$ is represented by the cross-hatched region where $A \cup B$ and C overlap. The superfluous shadings are omitted in the drawing on the right showing the set $(A \cup B) \cap C$.

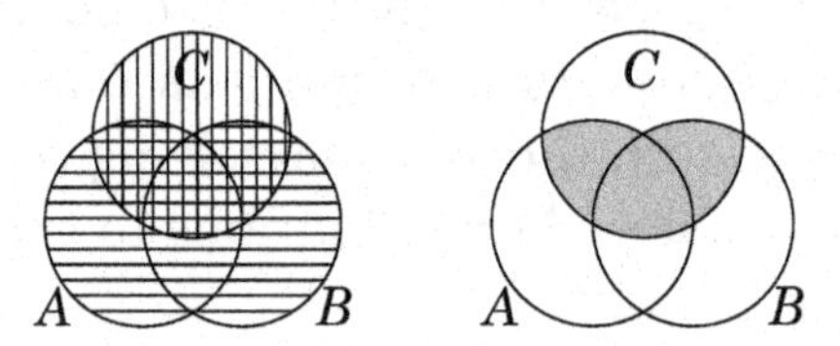

Figure 1.9. How to make a Venn diagram for $(A \cup B) \cap C$

Now think about $A \cup (B \cap C)$. In Figure 1.10 the set A is shaded with horizontal lines, and $B \cap C$ is shaded with vertical lines. The union $A \cup (B \cap C)$ is represented by the totality of all shaded regions, as shown on the right.

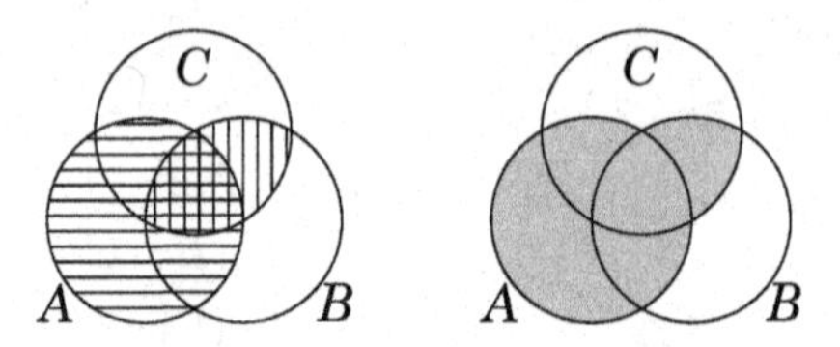

Figure 1.10. How to make a Venn diagram for $A \cup (B \cap C)$

Compare the diagrams for $(A \cup B) \cap C$ and $A \cup (B \cap C)$ in Figures 1.9 and 1.10. The fact that the diagrams are different indicates that $(A \cup B) \cap C \neq A \cup (B \cap C)$ in general. Thus an expression such as $A \cup B \cap C$ is absolutely meaningless because we can't tell whether it means $(A \cup B) \cap C$ or $A \cup (B \cap C)$. In summary, Venn diagrams have helped us understand the following.

Important Points:

- If an expression involving sets uses only $\cup$, then parentheses are optional.
- If an expression involving sets uses only $\cap$, then parentheses are optional.
- If an expression uses both $\cup$ and $\cap$, then parentheses are **essential**.

In the next section we will study types of expressions that use only $\cup$ or only $\cap$. These expressions will not require the use of parentheses.

Exercises for Section 1.7

1. Draw a Venn diagram for $\overline{A}$.
2. Draw a Venn diagram for $B - A$.
3. Draw a Venn diagram for $(A - B) \cap C$.
4. Draw a Venn diagram for $(A \cup B) - C$.
5. Draw Venn diagrams for $A \cup (B \cap C)$ and $(A \cup B) \cap (A \cup C)$. Based on your drawings, do you think $A \cup (B \cap C) = (A \cup B) \cap (A \cup C)$?
6. Draw Venn diagrams for $A \cap (B \cup C)$ and $(A \cap B) \cup (A \cap C)$. Based on your drawings, do you think $A \cap (B \cup C) = (A \cap B) \cup (A \cap C)$?
7. Suppose sets A and B are in a universal set U. Draw Venn diagrams for $\overline{A \cap B}$ and $\overline{A} \cup \overline{B}$. Based on your drawings, do you think it's true that $\overline{A \cap B} = \overline{A} \cup \overline{B}$?
8. Suppose sets A and B are in a universal set U. Draw Venn diagrams for $\overline{A \cup B}$ and $\overline{A} \cap \overline{B}$. Based on your drawings, do you think it's true that $\overline{A \cup B} = \overline{A} \cap \overline{B}$?
9. Draw a Venn diagram for $(A \cap B) - C$.
10. Draw a Venn diagram for $(A - B) \cup C$.

Following are Venn diagrams for expressions involving sets A, B and C. Write the corresponding expression.

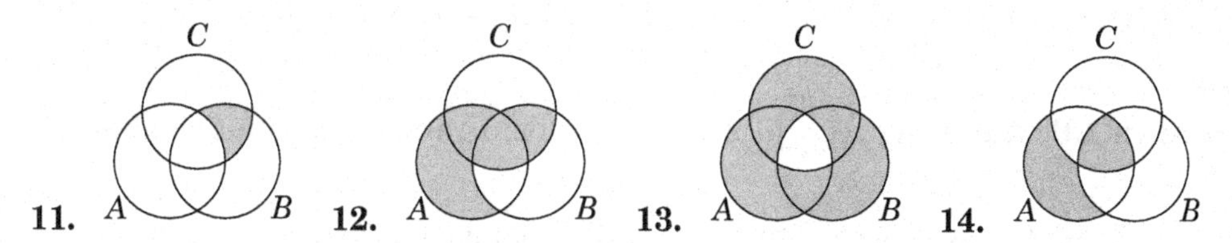

1.8 Indexed Sets

When a mathematical problem involves lots of sets, it is often convenient to keep track of them by using subscripts (also called indices). Thus instead of denoting three sets as A,B and C, we might instead write them as A_1,A_2 and A_3. These are called **indexed sets**.

Although we defined union and intersection to be operations that combine two sets, you by now have no difficulty forming unions and intersections of three or more sets. (For instance, in the previous section we drew Venn diagrams for the intersection and union of three sets.) But let's take a moment to write down careful definitions. Given sets $A_1,A_2,\ldots,A_n$, the set $A_1 \cup A_2 \cup A_3 \cup \cdots \cup A_n$ consists of everything that is in *at least one* of the sets A_i. Likewise $A_1 \cap A_2 \cap A_3 \cap \cdots \cap A_n$ consists of everything that is common to *all* of the sets A_i. Here is a careful definition.

Definition 1.7 Suppose $A_1,A_2,\ldots,A_n$ are sets. Then

$$\begin{aligned} A_1 \cup A_2 \cup A_3 \cup \cdots \cup A_n &= \{x : x \in A_i \text{ for } \textit{at least one set } A_i, \text{ for } 1 \le i \le n\}, \\ A_1 \cap A_2 \cap A_3 \cap \cdots \cap A_n &= \{x : x \in A_i \text{ for } \textit{every} \text{ set } A_i, \text{ for } 1 \le i \le n\}. \end{aligned}$$

But if the number n of sets is large, these expressions can get messy. To overcome this, we now develop some notation that is akin to sigma notation. You already know that sigma notation is a convenient symbolism for expressing sums of many numbers. Given numbers $a_1,a_2,a_3,\ldots,a_n$, then

$$\sum_{i=1}^{n} a_i = a_1 + a_2 + a_3 + \cdots + a_n.$$

Even if the list of numbers is infinite, the sum

$$\sum_{i=1}^{\infty} a_i = a_1 + a_2 + a_3 + \cdots + a_i + \cdots$$

is often still meaningful. The notation we are about to introduce is very similar to this. Given sets $A_1,A_2,A_3,\ \ldots,\ A_n$, we define

$$\bigcup_{i=1}^{n} A_i = A_1 \cup A_2 \cup A_3 \cup \cdots \cup A_n \qquad \text{and} \qquad \bigcap_{i=1}^{n} A_i = A_1 \cap A_2 \cap A_3 \cap \cdots \cap A_n.$$

Example 1.9 Suppose $A_1 = \{0,2,5\}$, $A_2 = \{1,2,5\}$ and $A_3 = \{2,5,7\}$. Then

$$\bigcup_{i=1}^{3} A_i = A_1 \cup A_2 \cup A_3 = \{0,1,2,5,7\} \qquad \text{and} \qquad \bigcap_{i=1}^{3} A_i = A_1 \cap A_2 \cap A_3 = \{2,5\}.$$

This notation is also used when the list of sets $A_1, A_2, A_3, \ldots$ is infinite:

$$\bigcup_{i=1}^{\infty} A_i = A_1 \cup A_2 \cup A_3 \cup \cdots = \{x : x \in A_i \text{ for at least one set } A_i \text{ with } 1 \le i\}.$$
$$\bigcap_{i=1}^{\infty} A_i = A_1 \cap A_2 \cap A_3 \cap \cdots = \{x : x \in A_i \text{ for every set } A_i \text{ with } 1 \le i\}.$$

Example 1.10 This example involves the following infinite list of sets.

$$A_1 = \{-1, 0, 1\}, \quad A_2 = \{-2, 0, 2\}, \quad A_3 = \{-3, 0, 3\}, \quad \cdots, \quad A_i = \{-i, 0, i\}, \quad \cdots$$

Observe that $\bigcup_{i=1}^{\infty} A_i = \mathbb{Z}$, and $\bigcap_{i=1}^{\infty} A_i = \{0\}$.

Here is a useful twist on our new notation. We can write

$$\bigcup_{i=1}^{3} A_i = \bigcup_{i \in \{1,2,3\}} A_i,$$

as this takes the union of the sets A_i for $i = 1, 2, 3$. Likewise:

$$\bigcap_{i=1}^{3} A_i = \bigcap_{i \in \{1,2,3\}} A_i$$
$$\bigcup_{i=1}^{\infty} A_i = \bigcup_{i \in \mathbb{N}} A_i$$
$$\bigcap_{i=1}^{\infty} A_i = \bigcap_{i \in \mathbb{N}} A_i$$

Here we are taking the union or intersection of a collection of sets A_i where i is an element of some set, be it $\{1, 2, 3\}$ or $\mathbb{N}$. In general, the way this works is that we will have a collection of sets A_i for $i \in I$, where I is the set of possible subscripts. The set I is called an **index set**.

It is important to realize that the set I need not even consist of integers. (We could subscript with letters or real numbers, etc.) Since we are programmed to think of i as an integer, let's make a slight notational change: We use α, not i, to stand for an element of I. Thus we are dealing with a collection of sets A_α for $\alpha \in I$. This leads to the following definition.

Definition 1.8 If we have a set A_α for every α in some index set I, then

$$\bigcup_{\alpha \in I} A_\alpha = \{x : x \in A_\alpha \text{ for at least one set } A_\alpha \text{ with } \alpha \in I\}$$
$$\bigcap_{\alpha \in I} A_\alpha = \{x : x \in A_\alpha \text{ for every set } A_\alpha \text{ with } \alpha \in I\}.$$

Example 1.11 Here the sets A_α will be subsets of $\mathbb{R}^2$. Let $I = [0,2] = \{x \in \mathbb{R} : 0 \le x \le 2\}$. For each number $\alpha \in I$, let $A_\alpha = \{(x,\alpha) : x \in \mathbb{R}, 1 \le x \le 2\}$. For instance, given $\alpha = 1 \in I$ the set $A_1 = \{(x,1) : x \in \mathbb{R}, 1 \le x \le 2\}$ is a horizontal line segment one unit above the x-axis and stretching between $x = 1$ and $x = 2$, as shown in Figure 1.11(a). Likewise $A_{\sqrt{2}} = \{(x,\sqrt{2}) : x \in \mathbb{R}, 1 \le x \le 2\}$ is a horizontal line segment $\sqrt{2}$ units above the x-axis and stretching between $x = 1$ and $x = 2$. A few other of the A_α are shown in Figure 1.11(a), but they can't all be drawn because there is one A_α for each of the infinitely many numbers $\alpha \in [0,2]$. The totality of them covers the shaded region in Figure 1.11(b), so this region is the union of all the A_α. Since the shaded region is the set $\{(x,y) \in \mathbb{R}^2 : 1 \le x \le 2, 0 \le y \le 2\} = [1,2] \times [0,2]$, it follows that

$$\bigcup_{\alpha \in [0,2]} A_\alpha = [1,2] \times [0,2].$$

Likewise, since there is no point (x,y) that is in every set A_α, we have

$$\bigcap_{\alpha \in [0,2]} A_\alpha = \emptyset.$$

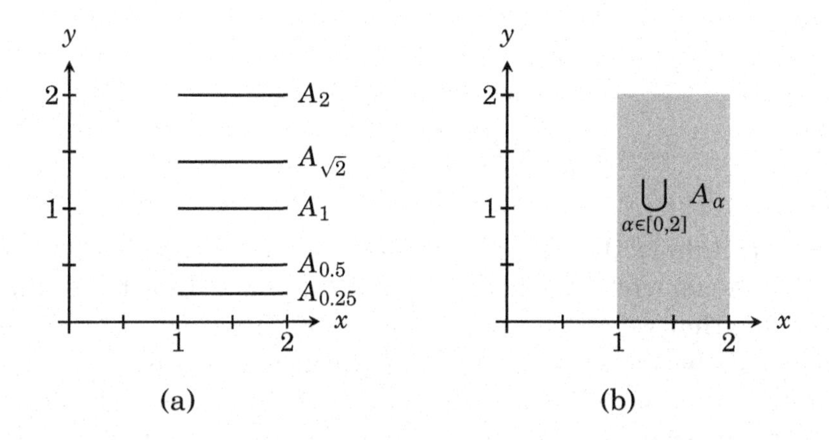

Figure 1.11. The union of an indexed collection of sets

One final comment. Observe that $A_\alpha = [1,2] \times \{\alpha\}$, so the above expressions can be written as

$$\bigcup_{\alpha \in [0,2]} [1,2] \times \{\alpha\} = [1,2] \times [0,2] \quad \text{and} \quad \bigcap_{\alpha \in [0,2]} [1,2] \times \{\alpha\} = \emptyset.$$

Example 1.12 In this example our sets are indexed by $\mathbb{R}^2$. For any $(a,b) \in \mathbb{R}^2$, let $P_{(a,b)}$ be the following subset of $\mathbb{R}^3$:

$$P_{(a,b)} = \{(x,y,z) \in \mathbb{R}^3 : ax + by = 0\}.$$

In words, given a point $(a,b) \in \mathbb{R}^2$, the corresponding set $P_{(a,b)}$ consists of all points (x,y,z) in $\mathbb{R}^3$ that satisfy the equation $ax + by = 0$. From previous math courses you will recognize this as a plane in $\mathbb{R}^3$, that is, $P_{(a,b)}$ is a plane in $\mathbb{R}^3$. Moreover, since any point $(0,0,z)$ on the z-axis automatically satisfies $ax + by = 0$, each $P_{(a,b)}$ contains the z-axis.

Figure 1.12 (left) shows the set $P_{(1,2)} = \{(x,y,z) \in \mathbb{R}^3 : x + 2y = 0\}$. It is the vertical plane that intersects the xy-plane at the line $x + 2y = 0$.

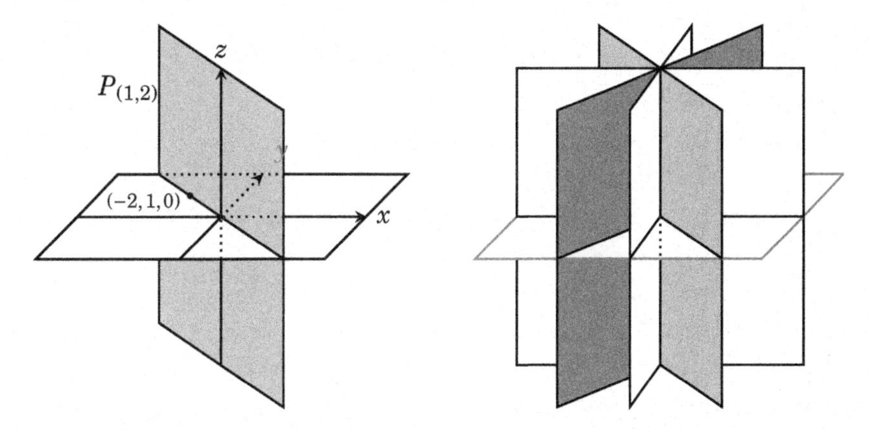

Figure 1.12. The sets $P_{(a,b)}$ are vertical planes containing the z-axis.

For any point $(a,b) \in \mathbb{R}^2$ with $(a,b) \neq (0,0)$, we can visualize $P_{(a,b)}$ as the vertical plane that cuts the xy-plane at the line $ax + by = 0$. Figure 1.12 (right) shows a few of the $P_{(a,b)}$. Since any two such planes intersect along the z-axis, and because the z-axis is a subset of every $P_{(a,b)}$, it is immediately clear that

$$\bigcap_{(a,b)\in\mathbb{R}^2} P_{(a,b)} = \{(0,0,z) : z \in \mathbb{R}\} = \text{“the } z\text{-axis”}.$$

For the union, note that any given point $(a,b,c) \in \mathbb{R}^3$ belongs to the set $P_{(-b,a)}$ because $(x,y,z) = (a,b,c)$ satisfies the equation $-bx + ay = 0$. (In fact, any (a,b,c) belongs to the special set $P_{(0,0)} = \mathbb{R}^3$, which is the only $P_{(a,b)}$ that is not a plane.) Since any point in $\mathbb{R}^3$ belongs to some $P_{(a,b)}$ we have

$$\bigcup_{(a,b)\in\mathbb{R}^2} P_{(a,b)} = \mathbb{R}^3.$$

Exercises for Section 1.8

1. Suppose $A_1 = \{a,b,d,e,g,f\}$, $A_2 = \{a,b,c,d\}$, $A_3 = \{b,d,a\}$ and $A_4 = \{a,b,h\}$.

(a) $\bigcup_{i=1}^{4} A_i =$ **(b)** $\bigcap_{i=1}^{4} A_i =$

2. Suppose $\begin{cases} A_1 &=& \{0,2,4,8,10,12,14,16,18,20,22,24\}, \\ A_2 &=& \{0,3,6,9,12,15,18,21,24\}, \\ A_3 &=& \{0,4,8,12,16,20,24\}. \end{cases}$

(a) $\bigcup_{i=1}^{3} A_i =$ **(b)** $\bigcap_{i=1}^{3} A_i =$

3. For each $n \in \mathbb{N}$, let $A_n = \{0,1,2,3,\ldots,n\}$.

(a) $\bigcup_{i\in\mathbb{N}} A_i =$ **(b)** $\bigcap_{i\in\mathbb{N}} A_i =$

4. For each $n \in \mathbb{N}$, let $A_n = \{-2n,0,2n\}$.

(a) $\bigcup_{i\in\mathbb{N}} A_i =$ **(b)** $\bigcap_{i\in\mathbb{N}} A_i =$

5. **(a)** $\bigcup_{i\in\mathbb{N}} [i,i+1] =$ **(b)** $\bigcap_{i\in\mathbb{N}} [i,i+1] =$

6. **(a)** $\bigcup_{i\in\mathbb{N}} [0,i+1] =$ **(b)** $\bigcap_{i\in\mathbb{N}} [0,i+1] =$

7. **(a)** $\bigcup_{i\in\mathbb{N}} \mathbb{R} \times [i,i+1] =$ **(b)** $\bigcap_{i\in\mathbb{N}} \mathbb{R} \times [i,i+1] =$

8. **(a)** $\bigcup_{\alpha\in\mathbb{R}} \{\alpha\} \times [0,1] =$ **(b)** $\bigcap_{\alpha\in\mathbb{R}} \{\alpha\} \times [0,1] =$

9. **(a)** $\bigcup_{X\in\mathscr{P}(\mathbb{N})} X =$ **(b)** $\bigcap_{X\in\mathscr{P}(\mathbb{N})} X =$

10. **(a)** $\bigcup_{x\in[0,1]} [x,1] \times [0,x^2] =$ **(b)** $\bigcap_{x\in[0,1]} [x,1] \times [0,x^2] =$

11. Is $\bigcap_{\alpha\in I} A_\alpha \subseteq \bigcup_{\alpha\in I} A_\alpha$ always true for any collection of sets A_α with index set I?

12. If $\bigcap_{\alpha\in I} A_\alpha = \bigcup_{\alpha\in I} A_\alpha$, what do you think can be said about the relationships between the sets A_α?

13. If $J \neq \emptyset$ and $J \subseteq I$, does it follow that $\bigcup_{\alpha\in J} A_\alpha \subseteq \bigcup_{\alpha\in I} A_\alpha$? What about $\bigcap_{\alpha\in J} A_\alpha \subseteq \bigcap_{\alpha\in I} A_\alpha$?

14. If $J \neq \emptyset$ and $J \subseteq I$, does it follow that $\bigcap_{\alpha\in I} A_\alpha \subseteq \bigcap_{\alpha\in J} A_\alpha$? Explain.

1.9 Sets that Are Number Systems

In practice, the sets we tend to be most interested in often have special properties and structures. For example, the sets $\mathbb{Z}$, $\mathbb{Q}$ and $\mathbb{R}$ are familiar number systems: Given such a set, any two of its elements can be added (or multiplied, etc.) together to produce another element in the set. These operations obey the familiar commutative, associative and distributive properties that we all have dealt with for years. Such properties lead to the standard algebraic techniques for solving equations. Even though we are concerned with the idea of proof, we will not find it necessary to define, prove or verify such properties and techniques; we will accept them as the ground rules upon which our further deductions are based.

We also accept as fact the natural ordering of the elements of $\mathbb{N}, \mathbb{Z}, \mathbb{Q}$ and $\mathbb{R}$, so that (for example) the meaning of "$5 < 7$" is understood and does not need to be justified or explained. Similarly, if $x \leq y$ and $a \neq 0$, we know that $ax \leq ay$ or $ax \geq ay$, depending on whether a is positive or negative.

Another thing that our ingrained understanding of the ordering of numbers tells us is that any non-empty subset of $\mathbb{N}$ has a smallest element. In other words, if $A \subseteq \mathbb{N}$ and $A \neq \emptyset$, then there is an element $x_0 \in A$ that is smaller than every other element of A. (To find it, start at 1, then move in increments to 2, 3, 4, etc., until you hit a number $x_0 \in A$; this is the smallest element of A.) Similarly, given an integer b, any non-empty subset $A \subseteq \{b, b+1, b+2, b+3, \ldots\}$ has a smallest element. This fact is sometimes called the **well-ordering principle**. There is no need to remember this term, but do be aware that we will use this simple, intuitive idea often in proofs, usually without a second thought.

The well-ordering principle seems innocent enough, but it actually says something very fundamental and special about the positive integers $\mathbb{N}$. In fact, the corresponding statement about the positive real numbers is false: The subset $A = \{\frac{1}{n} : n \in \mathbb{N}\}$ of the positive reals has no smallest element because for any $x_0 = \frac{1}{n} \in A$ that we might pick, there is always a smaller element $\frac{1}{n+1} \in A$.

One consequence of the well-ordering principle (as we will see below) is the familiar fact that any integer a can be divided by a non-zero integer b, resulting in a quotient q and remainder r. For example, $b = 3$ goes into $a = 17$ $q = 5$ times with remainder $r = 2$. In symbols, $17 = 5 \cdot 3 + 2$, or $a = qb + r$. This significant fact is called the **division algorithm**.

Fact 1.5 (The Division Algorithm) Given integers a and b with $b > 0$, there exist integers q and r for which $a = qb + r$ and $0 \leq r < b$.

Although there is no harm in accepting the division algorithm without proof, note that it does follow from the well-ordering principle. Here's how: Given integers a, b with $b > 0$, form the set

$$A = \{a - xb : x \in \mathbb{Z},\ 0 \le a - xb\} \subseteq \{0, 1, 2, 3, \ldots\}.$$

(For example, if $a = 17$ and $b = 3$ then $A = \{2, 5, 8, 11, 14, 17, 20, \ldots\}$ is the set of positive integers obtained by adding multiples of 3 to 17. Notice that the remainder $r = 2$ of $17 \div 3$ is the smallest element of this set.) In general, let r be the smallest element of the set $A = \{a - xb : x \in \mathbb{Z},\ 0 \le a - xb\}$. Then $r = a - qb$ for some $x = q \in \mathbb{Z}$, so $a = qb + r$. Moreover, $0 \le r < b$, as follows. The fact that $r \in A \subseteq \{0, 1, 2, 3 \ldots\}$ implies $0 \le r$. In addition, it cannot happen that $r \ge b$: If this were the case, then the non-negative number $r - b = (a - qb) - b = a - (q+1)b$ having form $a - xb$ would be a smaller element of A than r, and r was explicitly chosen as the smallest element of A. Since it is not the case that $r \ge b$, it must be that $r < b$. Therefore $0 \le r < b$. We've now produced a q and an r for which $a = qb + r$ and $0 \le r < b$.

Moving on, it is time to clarify a small issue. This chapter asserted that all of mathematics can be described with sets. But at the same time we maintained that some mathematical entities are not sets. (For instance, our approach was to say that an individual number, such as 5, is not itself a set, though it may be an *element* of a set.)

We have made this distinction because we need a place to stand as we explore sets: After all, it would appear suspiciously circular to declare that every mathematical entity is a set, and then go on to define a set as a collection whose members are sets!

But to most mathematicians, saying "The number 5 is not a set," is like saying "The number 5 is not a number."

The truth is that any number *can* itself be understood as a set. One way to do this is to begin with the identification $0 = \emptyset$. Then $1 = \{\emptyset\} = \{0\}$, and $2 = \{\emptyset, \{\emptyset\}\} = \{0, 1\}$, and $3 = \{\emptyset, \{\emptyset\}, \{\emptyset, \{\emptyset\}\}\} = \{0, 1, 2\}$. In general the natural number n is the set $n = \{0, 1, 2, \ldots, n-1\}$ of the n numbers (which are themselves sets) that come before it.

We will not undertake such a study here, but the elements of the number systems $\mathbb{Z}$, $\mathbb{Q}$ and $\mathbb{R}$ can all be defined in terms of sets. (Even the operations of addition, multiplication, etc., can be defined in set-theoretic terms.) In fact, mathematics itself can be regarded as the study of things that can be described as sets. Any mathematical entity is a set, whether or not we choose to think of it that way.

1.10 Russell's Paradox

This section contains some background information that may be interesting, but is not used in the remainder of the book.

The philosopher and mathematician Bertrand Russell (1872–1970) did groundbreaking work on the theory of sets and the foundations of mathematics. He was probably among the first to understand how the misuse of sets can lead to bizarre and paradoxical situations. He is famous for an idea that has come to be known as **Russell's paradox**.

Russell's paradox involves the following set of sets:

$$A = \{X : X \text{ is a set and } X \notin X\}. \tag{1.1}$$

In words, A is the set of all sets that do not include themselves as elements. Most sets we can think of are in A. The set $\mathbb{Z}$ of integers is not an integer (i.e., $\mathbb{Z} \notin \mathbb{Z}$) and therefore $\mathbb{Z} \in A$. Also $\emptyset \in A$ because $\emptyset$ is a set and $\emptyset \notin \emptyset$.

Is there a set that is not in A? Consider $B = \{\{\{\{\ldots\}\}\}\}$. Think of B as a box containing a box, containing a box, containing a box, and so on, forever. Or a set of Russian dolls, nested one inside the other, endlessly. The curious thing about B is that it has just one element, namely B itself:

$$B = \{\underbrace{\{\{\{\ldots\}\}\}}_{B}\}.$$

Thus $B \in B$. As B does not satisfy $B \notin B$, Equation (1.1) says $B \notin A$.

Russell's paradox arises from the question *"Is A an element of A?"*

For a set X, Equation (1.1) says $X \in A$ means the same thing as $X \notin X$. So for $X = A$, the previous line says $A \in A$ means the same thing as $A \notin A$. Conclusion: if $A \in A$ is true, then it is false; if $A \in A$ is false, then it is true. This is Russell's paradox.

Initially Russell's paradox sparked a crisis among mathematicians. How could a mathematical statement be both true and false? This seemed to be in opposition to the very essence of mathematics.

The paradox instigated a very careful examination of set theory and an evaluation of what can and cannot be regarded as a set. Eventually mathematicians settled upon a collection of axioms for set theory—the so-called **Zermelo-Fraenkel axioms**. One of these axioms is the well-ordering principle of the previous section. Another, the axiom of foundation, states that no non-empty set X is allowed to have the property $X \cap x \neq \emptyset$ for all its elements x. This rules out such circularly defined "sets" as $X = \{X\}$ introduced above. If we adhere to these axioms, then situations like Russell's paradox disappear. Most mathematicians accept all this on faith and happily ignore the Zermelo-Fraenkel axioms. Paradoxes like Russell's do not tend to come up in everyday mathematics—you have to go out of your way to construct them.

Still, Russell's paradox reminds us that precision of thought and language is an important part of doing mathematics. The next chapter deals with the topic of logic, a codification of thought and language.

Additional Reading on Sets. For a lively account of Bertrand Russell's life and work (including his paradox), see the graphic novel *Logicomix: An Epic Search For Truth*, by Apostolos Doxiadis and Christos Papadimitriou. Also see cartoonist Jessica Hagy's online strip *Indexed*—it is based largely on Venn diagrams.

CHAPTER 2

Logic

Logic is a systematic way of thinking that allows us to deduce new information from old information and to parse the meanings of sentences. You use logic informally in everyday life and certainly also in doing mathematics. For example, suppose you are working with a certain circle, call it "Circle X," and you have available the following two pieces of information.

1. Circle X has radius equal to 3.
2. If any circle has radius r, then its area is πr^2 square units.

You have no trouble putting these two facts together to get:

3. Circle X has area 9π square units.

In doing this you are using logic to combine existing information to produce new information. Because deducing new information is central to mathematics, logic plays a fundamental role. This chapter is intended to give you a sufficient mastery of it.

It is important to realize that logic is a process of deducing information correctly, *not* just deducing correct information. For example, suppose we were mistaken and Circle X actually had a radius of 4, not 3. Let's look at our exact same argument again.

1. Circle X has radius equal to 3.
2. If any circle has radius r, then its area is πr^2 square units.

---

3. Circle X has area 9π square units.

The sentence *"Circle X has radius equal to* 3.*"* is now untrue, and so is our conclusion *"Circle X has area* 9π *square units."* But the logic is perfectly correct; the information was combined correctly, even if some of it was false. This distinction between correct logic and correct information is significant because it is often important to follow the consequences of an incorrect assumption. Ideally, we want both our logic *and* our information to be correct, but the point is that they are different things.

In proving theorems, we apply logic to information that is considered obviously true (such as *"Any two points determine exactly one line."*) or is already known to be true (e.g., the Pythagorean theorem). If our logic is correct, then anything we deduce from such information will also be true (or at least as true as the "obviously true" information we began with).

2.1 Statements

The study of logic begins with statements. A **statement** is a sentence or a mathematical expression that is either definitely true or definitely false. You can think of statements as pieces of information that are either correct or incorrect. Thus statements are pieces of information that we might apply logic to in order to produce other pieces of information (which are also statements).

Example 2.1 Here are some examples of statements. They are all true.

If a circle has radius r, then its area is πr^2 square units.

Every even number is divisible by 2.

$2 \in \mathbb{Z}$

$\sqrt{2} \notin \mathbb{Z}$

$\mathbb{N} \subseteq \mathbb{Z}$

The set $\{0,1,2\}$ has three elements.

Some right triangles are isosceles.

Example 2.2 Here are some additional statements. They are all false.

All right triangles are isosceles.

$5 = 2$

$\sqrt{2} \notin \mathbb{R}$

$\mathbb{Z} \subseteq \mathbb{N}$

$\{0,1,2\} \cap \mathbb{N} = \emptyset$

Example 2.3 Here we pair sentences or expressions that are not statements with similar expressions that *are* statements.

NOT Statements:	Statements:
Add 5 to both sides.	Adding 5 to both sides of $x-5=37$ gives $x=42$.
$\mathbb{Z}$	$42 \in \mathbb{Z}$
42	42 is not a number.
What is the solution of $2x=84$?	The solution of $2x=84$ is 42.

Example 2.4 We will often use the letters P, Q, R and S to stand for specific statements. When more letters are needed we can use subscripts. Here are more statements, designated with letters. You decide which of them are true and which are false.

P : For every integer $n > 1$, the number $2^n - 1$ is prime.

Q : Every polynomial of degree n has at most n roots.

R : The function $f(x) = x^2$ is continuous.

$S_1 : \mathbb{Z} \subseteq \emptyset$

$S_2 : \{0, -1, -2\} \cap \mathbb{N} = \emptyset$

Designating statements with letters (as was done above) is a very useful shorthand. In discussing a particular statement, such as *"The function $f(x) = x^2$ is continuous,"* it is convenient to just refer to it as R to avoid having to write or say it many times.

Statements can contain variables. Here is an example.

P : If an integer x is a multiple of 6, then x is even.

This is a sentence that is true. (All multiples of 6 are even, so no matter which multiple of 6 the integer x happens to be, it is even.) Since the sentence P is definitely true, it is a statement. When a sentence or statement P contains a variable such as x, we sometimes denote it as $P(x)$ to indicate that it is saying something about x. Thus the above statement can be denoted as

$P(x)$: If an integer x is a multiple of 6, then x is even.

A statement or sentence involving two variables might be denoted $P(x, y)$, and so on.

It is quite possible for a sentence containing variables to not be a statement. Consider the following example.

$Q(x)$: The integer x is even.

Is this a statement? Whether it is true or false depends on just which integer x is. It is true if $x = 4$ and false if $x = 7$, etc. But without any stipulations on the value of x it is impossible to say whether $Q(x)$ is true or false. Since it is neither definitely true nor definitely false, $Q(x)$ cannot be a statement. A sentence such as this, whose truth depends on the value of one or more variables, is called an **open sentence**. The variables in an open sentence (or statement) can represent any type of entity, not just numbers. Here is an open sentence where the variables are functions:

$R(f,g)$: The function f is the derivative of the function g.

This open sentence is true if $f(x) = 2x$ and $g(x) = x^2$. It is false if $f(x) = x^3$ and $g(x) = x^2$, etc. We point out that a sentence such as $R(f,g)$ (that involves variables) can be denoted either as $R(f,g)$ or just R. We use the expression $R(f,g)$ when we want to emphasize that the sentence involves variables.

We will have more to say about open sentences later, but for now let's return to statements.

Statements are everywhere in mathematics. Any result or theorem that has been proved true is a statement. The quadratic formula and the Pythagorean theorem are both statements:

P: The solutions of the equation $ax^2 + bx + c = 0$ are $x = \dfrac{-b \pm \sqrt{b^2 - 4ac}}{2a}$.

Q: If a right triangle has legs of lengths a and b and hypotenuse of length c, then $a^2 + b^2 = c^2$.

Here is a very famous statement, so famous, in fact, that it has a name. It is called **Fermat's last theorem** after Pierre Fermat, a seventeenth-century French mathematician who scribbled it in the margin of a notebook.

R: For all numbers $a, b, c, n \in \mathbb{N}$ with $n > 2$, it is the case that $a^n + b^n \neq c^n$.

Fermat believed this statement was true. He noted that he could prove it was true, except his notebook's margin was too narrow to contain his proof. It is doubtful that he really had a correct proof in mind, for after his death generations of brilliant mathematicians tried unsuccessfully to prove that his statement was true (or false). Finally, in 1993, Andrew Wiles of Princeton University announced that he had devised a proof. Wiles had worked on the problem for over seven years, and his proof runs through hundreds of pages. The moral of this story is that some true statements are not obviously true.

Here is another statement famous enough to be named. It was first posed in the eighteenth century by the German mathematician Christian Goldbach, and thus is called the **Goldbach conjecture**:

S: Every even integer greater than 2 is a sum of two prime numbers.

You must agree that S is either true or false. It appears to be true, because when you examine even numbers that are bigger than 2, they seem to be sums of two primes: $4 = 2+2$, $6 = 3+3$, $8 = 3+5$, $10 = 5+5$, $12 = 5+7$,

$100 = 17 + 83$ and so on. But that's not to say there isn't some large even number that's not the sum of two primes. If such a number exists, then S is false. The thing is, in the over 260 years since Goldbach first posed this problem, no one has been able to determine whether it's true or false. But since it is clearly either true or false, S is a statement.

This book is about the methods that can be used to prove that S (or any other statement) is true or false. To prove that a statement is true, we start with obvious statements (or other statements that have been proven true) and use logic to deduce more and more complex statements until finally we obtain a statement such as S. Of course some statements are more difficult to prove than others, and S appears to be notoriously difficult; we will concentrate on statements that are easier to prove.

But the point is this: In proving that statements are true, we use logic to help us understand statements and to combine pieces of information to produce new pieces of information. In the next several sections we explore some standard ways that statements can be combined to form new statements, or broken down into simpler statements.

Exercises for Section 2.1

Decide whether or not the following are statements. In the case of a statement, say if it is true or false, if possible.

1. Every real number is an even integer.

2. Every even integer is a real number.

3. If x and y are real numbers and $5x = 5y$, then $x = y$.

4. Sets $\mathbb{Z}$ and $\mathbb{N}$.

5. Sets $\mathbb{Z}$ and $\mathbb{N}$ are infinite.

6. Some sets are finite.

7. The derivative of any polynomial of degree 5 is a polynomial of degree 6.

8. $\mathbb{N} \notin \mathscr{P}(\mathbb{N})$.

9. $\cos(x) = -1$

10. $(\mathbb{R} \times \mathbb{N}) \cap (\mathbb{N} \times \mathbb{R}) = \mathbb{N} \times \mathbb{N}$

11. The integer x is a multiple of 7.

12. If the integer x is a multiple of 7, then it is divisible by 7.

13. Either x is a multiple of 7, or it is not.

14. Call me Ishmael.

15. In the beginning, God created the heaven and the earth.

2.2 And, Or, Not

The word "and" can be used to combine two statements to form a new statement. Consider for example the following sentence.

R_1 : The number 2 is even **and** the number 3 is odd.

We recognize this as a true statement, based on our common-sense understanding of the meaning of the word "and." Notice that R_1 is made up of two simpler statements:

P : The number 2 is even.
Q : The number 3 is odd.

These are joined together by the word "and" to form the more complex statement R_1. The statement R_1 asserts that P and Q are both true. Since both P and Q are in fact true, the statement R_1 is also true.

Had one or both of P and Q been false, then R_1 would be false. For instance, each of the following statements is false.

R_2 : The number 1 is even **and** the number 3 is odd.
R_3 : The number 2 is even **and** the number 4 is odd.
R_4 : The number 3 is even **and** the number 2 is odd.

From these examples we see that any two statements P and Q can be combined to form a new statement "P **and** Q." In the spirit of using letters to denote statements, we now introduce the special symbol $\wedge$ to stand for the word "and." Thus if P and Q are statements, $P \wedge Q$ stands for the statement "P **and** Q." The statement $P \wedge Q$ is true if both P and Q are true; otherwise it is false. This is summarized in the following table, called a **truth table**.

P	Q	$P \wedge Q$
T	T	T
T	F	F
F	T	F
F	F	F

In this table, T stands for "True," and F stands for "False." (T and F are called **truth values**.) Each line lists one of the four possible combinations or truth values for P and Q, and the column headed by $P \wedge Q$ tells whether the statement $P \wedge Q$ is true or false in each case.

Statements can also be combined using the word "or." Consider the following four statements.

S_1 : The number 2 is even **or** the number 3 is odd.
S_2 : The number 1 is even **or** the number 3 is odd.
S_3 : The number 2 is even **or** the number 4 is odd.
S_4 : The number 3 is even **or** the number 2 is odd.

In mathematics, the assertion "P **or** Q" is always understood to mean that one *or both* of P and Q is true. Thus statements S_1, S_2, S_3 are all true, while S_4 is false. The symbol $\vee$ is used to stand for the word "or." So if P and Q are statements, $P \vee Q$ represents the statement "P **or** Q." Here is the truth table.

P	Q	$P \vee Q$
T	T	T
T	F	T
F	T	T
F	F	F

It is important to be aware that the meaning of "or" expressed in the above table differs from the way it is often used in everyday conversation. For example, suppose a university official makes the following threat:

You pay your tuition **or** you will be withdrawn from school.

You understand that this means that either you pay your tuition *or* you will be withdrawn from school, *but not both*. In mathematics we never use the word "or" in such a sense. For us "or" means exactly what is stated in the table for $\vee$. Thus $P \vee Q$ being true means *one or both* of P and Q is true. If we ever need to express the fact that exactly one of P and Q is true, we use one of the following constructions:

P **or** Q, **but not both**.

Either P **or** Q.

Exactly one of P **or** Q.

If the university official were a mathematician, he might have qualified his statement in one of the following ways.

Pay your tuition **or** you will be withdrawn from school, **but not both**.

Either you pay your tuition **or** you will be withdrawn from school.

To conclude this section, we mention another way of obtaining new statements from old ones. Given any statement P, we can form the new statement **"It is not true that** P." For example, consider the following statement.

The number 2 is even.

This statement is true. Now change it by inserting the words "It is not true that" at the beginning:

It is not true that the number 2 is even.

This new statement is false.

For another example, starting with the false statement "$2 \in \emptyset$," we get the true statement "It is not true that $2 \in \emptyset$."

We use the symbol $\sim$ to stand for the words "It's not true that," so $\sim P$ means **"It's not true that** P." We often read $\sim P$ simply as "not P." Unlike $\wedge$ and $\vee$, which combine two statements, the symbol $\sim$ just alters a single statement. Thus its truth table has just two lines, one for each possible truth value of P.

P	$\sim P$
T	F
F	T

The statement $\sim P$ is called the **negation** of P. The negation of a specific statement can be expressed in numerous ways. Consider

P : The number 2 is even.

Here are several ways of expressing its negation.

$\sim P$: It's not true that the number 2 is even.
$\sim P$: It is false that the number 2 is even.
$\sim P$: The number 2 is not even.

In this section we've learned how to combine or modify statements with the operations $\wedge$, $\vee$ and $\sim$. Of course we can also apply these operations to open sentences or a mixture of open sentences and statements. For example, (x is an even integer) $\wedge$ (3 is an odd integer) is an open sentence that is a combination of an open sentence and a statement.

Exercises for Section 2.2

Express each statement or open sentence in one of the forms $P \wedge Q$, $P \vee Q$, or $\sim P$. Be sure to also state exactly what statements P and Q stand for.

1. The number 8 is both even and a power of 2.
2. The matrix A is not invertible.
3. $x \neq y$
4. $x < y$
5. $y \geq x$
6. There is a quiz scheduled for Wednesday or Friday.
7. The number x equals zero, but the number y does not.
8. At least one of the numbers x and y equals 0.
9. $x \in A - B$
10. $x \in A \cup B$
11. $A \in \{X \in \mathscr{P}(\mathbb{N}) : |\overline{X}| < \infty\}$
12. Happy families are all alike, but each unhappy family is unhappy in its own way. (Leo Tolstoy, *Anna Karenina*)
13. Human beings want to be good, but not too good, and not all the time. (George Orwell)
14. A man should look for what is, and not for what he thinks should be. (Albert Einstein)

2.3 Conditional Statements

There is yet another way to combine two statements. Suppose we have in mind a specific integer a. Consider the following statement about a.

R : If the integer a is a multiple of 6, then a is divisible by 2.

We immediately spot this as a true statement based on our knowledge of integers and the meanings of the words "if" and "then." If integer a is a multiple of 6, then a is even, so therefore a is divisible by 2. Notice that R is built up from two simpler statements:

P : The integer a is a multiple of 6.
Q : The integer a is divisible by 2.
R : If P, then Q.

In general, given any two statements P and Q whatsoever, we can form the new statement *"If P, then Q."* This is written symbolically as $P \Rightarrow Q$ which we read as *"If P, then Q,"* or *"P implies Q."* Like $\wedge$ and $\vee$, the symbol $\Rightarrow$ has a very specific meaning. When we assert that the statement $P \Rightarrow Q$ is true, we mean that *if* P is true *then* Q must also be true. (In other words we mean that the condition P being true forces Q to be true.) A statement of form $P \Rightarrow Q$ is called a **conditional** statement because it means Q will be true *under the condition* that P is true.

You can think of $P \Rightarrow Q$ as being a promise that whenever P is true, Q will be true also. There is only one way this promise can be broken (i.e. be false) and that is if P is true but Q is false. Thus the truth table for the promise $P \Rightarrow Q$ is as follows:

P	Q	$P \Rightarrow Q$
T	T	T
T	F	F
F	T	T
F	F	T

Perhaps you are bothered by the fact that $P \Rightarrow Q$ is true in the last two lines of this table. Here's an example to convince you that the table is correct. Suppose your professor makes the following promise:

If you pass the final exam, **then** you will pass the course.

Your professor is making the promise

(You pass the exam) $\Rightarrow$ (You pass the course).

Under what circumstances did she lie? There are four possible scenarios, depending on whether or not you passed the exam and whether or not you passed the course. These scenarios are tallied in the following table.

You pass exam	You pass course	(You pass exam) $\Rightarrow$ (You pass course)
T	T	T
T	F	F
F	T	T
F	F	T

The first line describes the scenario where you pass the exam and you pass the course. Clearly the professor kept her promise, so we put a T in the third column to indicate that she told the truth. In the second line, you passed the exam, but your professor gave you a failing grade in the course. In this case she broke her promise, and the F in the third column indicates that what she said was untrue.

Now consider the third row. In this scenario you failed the exam but still passed the course. How could that happen? Maybe your professor felt sorry for you. But that doesn't make her a liar. Her only promise was that if you passed the exam then you would pass the course. She did not say passing the exam was the *only way* to pass the course. Since she didn't lie, then she told the truth, so there is a T in the third column.

Finally look at the fourth row. In that scenario you failed the exam and you failed the course. Your professor did not lie; she did exactly what she said she would do. Hence the T in the third column.

In mathematics, whenever we encounter the construction *"If P, then Q"* it means exactly what the truth table for $\Rightarrow$ expresses. But of course there are other grammatical constructions that also mean $P \Rightarrow Q$. Here is a summary of the main ones.

$$\left.\begin{array}{l} \text{If } P \text{, then } Q. \\ Q \text{ if } P. \\ Q \text{ whenever } P. \\ Q \text{, provided that } P. \\ \text{Whenever } P \text{, then also } Q. \\ P \text{ is a sufficient condition for } Q. \\ \text{For } Q \text{, it is sufficient that } P. \\ Q \text{ is a necessary condition for } P. \\ \text{For } P \text{, it is necessary that } Q. \\ P \text{ only if } Q. \end{array}\right\} \; P \Rightarrow Q$$

These can all be used in the place of (and mean exactly the same thing as) *"If P, then Q."* You should analyze the meaning of each one and convince yourself that it captures the meaning of $P \Rightarrow Q$. For example, $P \Rightarrow Q$ means the condition of P being true is enough (i.e., sufficient) to make Q true; hence *"P is a sufficient condition for Q."*

The wording can be tricky. Often an everyday situation involving a conditional statement can help clarify it. For example, consider your professor's promise:

(You pass the exam) $\Rightarrow$ (You pass the course)

This means that your passing the exam is a sufficient (though perhaps not necessary) condition for your passing the course. Thus your professor might just as well have phrased her promise in one of the following ways.

Passing the exam is a sufficient condition for passing the course.

For you to pass the course, it is sufficient that you pass the exam.

However, when we want to say *"If P, then Q"* in everyday conversation, we do not normally express this as *"Q is a necessary condition for P"* or *"P only if Q."* But such constructions are not uncommon in mathematics. To understand why they make sense, notice that $P \Rightarrow Q$ being true means that it's impossible that P is true but Q is false, so in order for P to be true it is necessary that Q is true; hence "*Q is a necessary condition for P.*" And this means that P can only be true if Q is true, i.e., "*P only if Q.*"

Exercises for Section 2.3

Without changing their meanings, convert each of the following sentences into a sentence having the form "*If P, then Q.*"

1. A matrix is invertible provided that its determinant is not zero.
2. For a function to be continuous, it is sufficient that it is differentiable.
3. For a function to be integrable, it is necessary that it is continuous.
4. A function is rational if it is a polynomial.
5. An integer is divisible by 8 only if it is divisible by 4.
6. Whenever a surface has only one side, it is non-orientable.
7. A series converges whenever it converges absolutely.
8. A geometric series with ratio r converges if $|r| < 1$.
9. A function is integrable provided the function is continuous.
10. The discriminant is negative only if the quadratic equation has no real solutions.
11. You fail only if you stop writing. (Ray Bradbury)
12. People will generally accept facts as truth only if the facts agree with what they already believe. (Andy Rooney)
13. Whenever people agree with me I feel I must be wrong. (Oscar Wilde)

2.4 Biconditional Statements

It is important to understand that $P \Rightarrow Q$ is not the same as $Q \Rightarrow P$. To see why, suppose that a is some integer and consider the statements

$$(a \text{ is a multiple of } 6) \quad \Rightarrow \quad (a \text{ is divisible by } 2),$$
$$(a \text{ is divisible by } 2) \quad \Rightarrow \quad (a \text{ is a multiple of } 6).$$

The first statement asserts that if a is a multiple of 6 then a is divisible by 2. This is clearly true, for any multiple of 6 is even and therefore divisible by 2. The second statement asserts that if a is divisible by 2 then it is a multiple of 6. This is not necessarily true, for $a = 4$ (for instance) is divisible by 2, yet not a multiple of 6. Therefore the meanings of $P \Rightarrow Q$ and $Q \Rightarrow P$ are in general quite different. The conditional statement $Q \Rightarrow P$ is called the **converse** of $P \Rightarrow Q$, so a conditional statement and its converse express entirely different things.

But sometimes, if P and Q are just the right statements, it can happen that $P \Rightarrow Q$ and $Q \Rightarrow P$ are both necessarily true. For example, consider the statements

$$\begin{aligned} (a \text{ is even}) &\Rightarrow (a \text{ is divisible by } 2), \\ (a \text{ is divisible by } 2) &\Rightarrow (a \text{ is even}). \end{aligned}$$

No matter what value a has, both of these statements are true. Since both $P \Rightarrow Q$ and $Q \Rightarrow P$ are true, it follows that $(P \Rightarrow Q) \wedge (Q \Rightarrow P)$ is true.

We now introduce a new symbol $\Leftrightarrow$ to express the meaning of the statement $(P \Rightarrow Q) \wedge (Q \Rightarrow P)$. The expression $P \Leftrightarrow Q$ is understood to have exactly the same meaning as $(P \Rightarrow Q) \wedge (Q \Rightarrow P)$. According to the previous section, $Q \Rightarrow P$ is read as "P *if* Q," and $P \Rightarrow Q$ can be read as "P *only if* Q." Therefore we pronounce $P \Leftrightarrow Q$ as "P *if and only if* Q." For example, given an integer a, we have the true statement

$$(a \text{ is even}) \Leftrightarrow (a \text{ is divisible by } 2),$$

which we can read as *"Integer a is even if and only if a is divisible by 2."*

The truth table for $\Leftrightarrow$ is shown below. Notice that in the first and last rows, both $P \Rightarrow Q$ and $Q \Rightarrow P$ are true (according to the truth table for $\Rightarrow$), so $(P \Rightarrow Q) \wedge (Q \Rightarrow P)$ is true, and hence $P \Leftrightarrow Q$ is true. However, in the middle two rows one of $P \Rightarrow Q$ or $Q \Rightarrow P$ is false, so $(P \Rightarrow Q) \wedge (Q \Rightarrow P)$ is false, making $P \Leftrightarrow Q$ false.

P	Q	$P \Leftrightarrow Q$
T	T	T
T	F	F
F	T	F
F	F	T

Compare the statement $R : (a \text{ is even}) \Leftrightarrow (a \text{ is divisible by } 2)$ with this truth table. If a is even then the two statements on either side of $\Leftrightarrow$ are true, so according to the table R is true. If a is odd then the two statements on either side of $\Leftrightarrow$ are false, and again according to the table R is true. Thus R is true no matter what value a has. In general, $P \Leftrightarrow Q$ being true means P and Q are both true or both false.

Not surprisingly, there are many ways of saying $P \Leftrightarrow Q$ in English. The following constructions all mean $P \Leftrightarrow Q$:

$$\left.\begin{array}{l} P \text{ if and only if } Q. \\ P \text{ is a necessary and sufficient condition for } Q. \\ \text{For } P \text{ it is necessary and sufficient that } Q. \\ \text{If } P\text{, then } Q\text{, and conversely.} \end{array}\right\} P \Leftrightarrow Q$$

The first three of these just combine constructions from the previous section to express that $P \Rightarrow Q$ and $Q \Rightarrow P$. In the last one, the words "*...and conversely*" mean that in addition to "*If P, then Q*" being true, the converse statement "*If Q, then P*" is also true.

Exercises for Section 2.4

Without changing their meanings, convert each of the following sentences into a sentence having the form "*P if and only if Q.*"

1. For matrix A to be invertible, it is necessary and sufficient that $\det(A) \neq 0$.

2. If a function has a constant derivative then it is linear, and conversely.

3. If $xy = 0$ then $x = 0$ or $y = 0$, and conversely.

4. If $a \in \mathbb{Q}$ then $5a \in \mathbb{Q}$, and if $5a \in \mathbb{Q}$ then $a \in \mathbb{Q}$.

5. For an occurrence to become an adventure, it is necessary and sufficient for one to recount it. (Jean-Paul Sartre)

2.5 Truth Tables for Statements

You should now know the truth tables for $\wedge$, $\vee$, $\sim$, $\Rightarrow$ and $\Leftrightarrow$. They should be *internalized* as well as memorized. You must understand the symbols thoroughly, for we now combine them to form more complex statements.

For example, suppose we want to convey that one or the other of P and Q is true but they are not both true. No single symbol expresses this, but we could combine them as

$$(P \vee Q) \wedge \sim (P \wedge Q),$$

which literally means:

P or Q is true, and it is not the case that both P and Q are true.

This statement will be true or false depending on the truth values of P and Q. In fact we can make a truth table for the entire statement. Begin as usual by listing the possible true/false combinations of P and Q on four lines. The statement $(P \vee Q) \wedge \sim (P \wedge Q)$ contains the individual statements $(P \vee Q)$ and $(P \wedge Q)$, so we next tally their truth values in the third and fourth columns. The fifth column lists values for $\sim (P \wedge Q)$, and these are just the opposites of the corresponding entries in the fourth column. Finally, combining the third and fifth columns with $\wedge$, we get the values for $(P \vee Q) \wedge \sim (P \wedge Q)$ in the sixth column.

P	Q	$(P \vee Q)$	$(P \wedge Q)$	$\sim (P \wedge Q)$	$(P \vee Q) \wedge \sim (P \wedge Q)$
T	T	T	T	F	**F**
T	F	T	F	T	**T**
F	T	T	F	T	**T**
F	F	F	F	T	**F**

This truth table tells us that $(P \vee Q) \wedge \sim (P \wedge Q)$ is true precisely when one but not both of P and Q are true, so it has the meaning we intended. (Notice that the middle three columns of our truth table are just "helper columns" and are not necessary parts of the table. In writing truth tables, you may choose to omit such columns if you are confident about your work.)

For another example, consider the following familiar statement concerning two real numbers x and y:

The product xy equals zero if and only if $x = 0$ or $y = 0$.

This can be modeled as $(xy = 0) \Leftrightarrow (x = 0 \ \vee \ y = 0)$. If we introduce letters P, Q and R for the statements $xy = 0$, $x = 0$ and $y = 0$, it becomes $P \Leftrightarrow (Q \vee R)$. Notice that the parentheses are necessary here, for without them we wouldn't know whether to read the statement as $P \Leftrightarrow (Q \vee R)$ or $(P \Leftrightarrow Q) \vee R$.

Making a truth table for $P \Leftrightarrow (Q \vee R)$ entails a line for each T/F combination for the three statements P, Q and R. The eight possible combinations are tallied in the first three columns of the following table.

P	Q	R	$Q \vee R$	$P \Leftrightarrow (Q \vee R)$
T	T	T	T	**T**
T	T	F	T	**T**
T	F	T	T	**T**
T	F	F	F	**F**
F	T	T	T	**F**
F	T	F	T	**F**
F	F	T	T	**F**
F	F	F	F	**T**

We fill in the fourth column using our knowledge of the truth table for $\vee$. Finally the fifth column is filled in by combining the first and fourth columns with our understanding of the truth table for $\Leftrightarrow$. The resulting table gives the true/false values of $P \Leftrightarrow (Q \vee R)$ for all values of P, Q and R.

Notice that when we plug in various values for x and y, the statements $P: xy = 0$, $Q: x = 0$ and $R: y = 0$ have various truth values, but the statement $P \Leftrightarrow (Q \vee R)$ is always true. For example, if $x = 2$ and $y = 3$, then P, Q and R are all false. This scenario is described in the last row of the table, and there we see that $P \Leftrightarrow (Q \vee R)$ is true. Likewise if $x = 0$ and $y = 7$, then P and Q are true and R is false, a scenario described in the second line of the table, where again $P \Leftrightarrow (Q \vee R)$ is true. There is a simple reason why $P \Leftrightarrow (Q \vee R)$ is true for any values of x and y: It is that $P \Leftrightarrow (Q \vee R)$ represents $(xy = 0) \Leftrightarrow (x = 0 \;\vee\; y = 0)$, which is a *true mathematical statement*. It is absolutely impossible for it to be false.

This may make you wonder about the lines in the table where $P \Leftrightarrow (Q \vee R)$ is false. Why are they there? The reason is that $P \Leftrightarrow (Q \vee R)$ can also represent a false statement. To see how, imagine that at the end of the semester your professor makes the following promise.

> You pass the class if and only if you get an "A" on the final or you get a "B" on the final.

This promise has the form $P \Leftrightarrow (Q \vee R)$, so its truth values are tabulated in the above table. Imagine it turned out that you got an "A" on the exam but failed the course. Then surely your professor lied to you. In fact, P is false, Q is true and R is false. This scenario is reflected in the sixth line of the table, and indeed $P \Leftrightarrow (Q \vee R)$ is false (i.e., it is a lie).

The moral of this example is that people can lie, but true mathematical statements *never* lie.

We close this section with a word about the use of parentheses. The symbol $\sim$ is analogous to the minus sign in algebra. It negates the expression it precedes. Thus $\sim P \vee Q$ means $(\sim P) \vee Q$, not $\sim (P \vee Q)$. In $\sim (P \vee Q)$, the value of the entire expression $P \vee Q$ is negated.

Exercises for Section 2.5

Write a truth table for the logical statements in problems 1–9:

1. $P \vee (Q \Rightarrow R)$
2. $(Q \vee R) \Leftrightarrow (R \wedge Q)$
3. $\sim (P \Rightarrow Q)$
4. $\sim (P \vee Q) \vee (\sim P)$
5. $(P \wedge \sim P) \vee Q$
6. $(P \wedge \sim P) \wedge Q$
7. $(P \wedge \sim P) \Rightarrow Q$
8. $P \vee (Q \wedge \sim R)$
9. $\sim (\sim P \vee \sim Q)$
10. Suppose the statement $((P \wedge Q) \vee R) \Rightarrow (R \vee S)$ is false. Find the truth values of P, Q, R and S. (This can be done without a truth table.)
11. Suppose P is false and that the statement $(R \Rightarrow S) \Leftrightarrow (P \wedge Q)$ is true. Find the truth values of R and S. (This can be done without a truth table.)

2.6 Logical Equivalence

In contemplating the truth table for $P \Leftrightarrow Q$, you probably noticed that $P \Leftrightarrow Q$ is true exactly when P and Q are both true or both false. In other words, $P \Leftrightarrow Q$ is true precisely when at least one of the statements $P \wedge Q$ or $\sim P \wedge \sim Q$ is true. This may tempt us to say that $P \Leftrightarrow Q$ means the same thing as $(P \wedge Q) \vee (\sim P \wedge \sim Q)$.

To see if this is really so, we can write truth tables for $P \Leftrightarrow Q$ and $(P \wedge Q) \vee (\sim P \wedge \sim Q)$. In doing this, it is more efficient to put these two statements into the same table, as follows. (This table has helper columns for the intermediate expressions $\sim P$, $\sim Q$, $(P \wedge Q)$ and $(\sim P \wedge \sim Q)$.)

P	Q	$\sim P$	$\sim Q$	$(P \wedge Q)$	$(\sim P \wedge \sim Q)$	$(P \wedge Q) \vee (\sim P \wedge \sim Q)$	$P \Leftrightarrow Q$
T	*T*	*F*	*F*	*T*	*F*	**T**	**T**
T	*F*	*F*	*T*	*F*	*F*	**F**	**F**
F	*T*	*T*	*F*	*F*	*F*	**F**	**F**
F	*F*	*T*	*T*	*F*	*T*	**T**	**T**

The table shows that $P \Leftrightarrow Q$ and $(P \wedge Q) \vee (\sim P \wedge \sim Q)$ have the same truth value, no matter the values P and Q. It is as if $P \Leftrightarrow Q$ and $(P \wedge Q) \vee (\sim P \wedge \sim Q)$ are algebraic expressions that are equal no matter what is "plugged into" variables P and Q. We express this state of affairs by writing

$$P \Leftrightarrow Q \quad = \quad (P \wedge Q) \vee (\sim P \wedge \sim Q)$$

and saying that $P \Leftrightarrow Q$ and $(P \wedge Q) \vee (\sim P \wedge \sim Q)$ are **logically equivalent**.

In general, two statements are **logically equivalent** if their truth values match up line-for-line in a truth table.

Logical equivalence is important because it can give us different (and potentially useful) ways of looking at the same thing. As an example, the following table shows that $P \Rightarrow Q$ is logically equivalent to $(\sim Q) \Rightarrow (\sim P)$.

P	Q	$\sim P$	$\sim Q$	$(\sim Q) \Rightarrow (\sim P)$	$P \Rightarrow Q$
T	*T*	*F*	*F*	**T**	**T**
T	*F*	*F*	*T*	**F**	**F**
F	*T*	*T*	*F*	**T**	**T**
F	*F*	*T*	*T*	**T**	**T**

The fact that $P \Rightarrow Q \;=\; (\sim Q) \Rightarrow (\sim P)$ is useful because so many theorems have the form $P \Rightarrow Q$. As we will see in Chapter 5, proving such a theorem may be easier if we express it in the logically equivalent form $(\sim Q) \Rightarrow (\sim P)$.

There are two pairs of logically equivalent statements that come up again and again throughout this book and beyond. They are prevalent enough to be dignified by a special name: **DeMorgan's laws.**

Fact 2.1 (DeMorgan's Laws)

1. $\sim(P \wedge Q) = (\sim P) \vee (\sim Q)$
2. $\sim(P \vee Q) = (\sim P) \wedge (\sim Q)$

The first of DeMorgan's laws is verified by the following table. You are asked to verify the second in one of the exercises.

P	Q	$\sim P$	$\sim Q$	$P \wedge Q$	$\sim(P \wedge Q)$	$(\sim P) \vee (\sim Q)$
T	T	F	F	T	**F**	**F**
T	F	F	T	F	**T**	**T**
F	T	T	F	F	**T**	**T**
F	F	T	T	F	**T**	**T**

DeMorgan's laws are actually very natural and intuitive. Consider the statement $\sim(P \wedge Q)$, which we can interpret as meaning that *it is not the case that both P and Q are true.* If it is not the case that both P and Q are true, then at least one of P or Q is false, in which case $(\sim P) \vee (\sim Q)$ is true. Thus $\sim(P \wedge Q)$ means the same thing as $(\sim P) \vee (\sim Q)$.

DeMorgan's laws can be very useful. Suppose we happen to know that some statement having form $\sim(P \vee Q)$ is true. The second of DeMorgan's laws tells us that $(\sim Q) \wedge (\sim P)$ is also true, hence $\sim P$ and $\sim Q$ are both true as well. Being able to quickly obtain such additional pieces of information can be extremely useful.

Here is a summary of some significant logical equivalences. Those that are not immediately obvious can be verified with a truth table.

$$P \Rightarrow Q = (\sim Q) \Rightarrow (\sim P) \qquad \text{Contrapositive law} \qquad (2.1)$$

$$\left.\begin{array}{r} \sim(P \wedge Q) = \sim P \vee \sim Q \\ \sim(P \vee Q) = \sim P \wedge \sim Q \end{array}\right\} \qquad \text{DeMorgan's laws} \qquad (2.2)$$

$$\left.\begin{array}{r} P \wedge Q = Q \wedge P \\ P \vee Q = Q \vee P \end{array}\right\} \qquad \text{Commutative laws} \qquad (2.3)$$

$$\left.\begin{array}{r} P \wedge (Q \vee R) = (P \wedge Q) \vee (P \wedge R) \\ P \vee (Q \wedge R) = (P \vee Q) \wedge (P \vee R) \end{array}\right\} \qquad \text{Distributive laws} \qquad (2.4)$$

$$\left.\begin{array}{r} P \wedge (Q \wedge R) = (P \wedge Q) \wedge R \\ P \vee (Q \vee R) = (P \vee Q) \vee R \end{array}\right\} \qquad \text{Associative laws} \qquad (2.5)$$

Notice how the distributive law $P \wedge (Q \vee R) = (P \wedge Q) \vee (P \wedge R)$ has the same structure as the distributive law $p \cdot (q + r) = p \cdot q + p \cdot r$ from algebra.

Concerning the associative laws, the fact that $P \wedge (Q \wedge R) = (P \wedge Q) \wedge R$ means that the position of the parentheses is irrelevant, and we can write this as $P \wedge Q \wedge R$ without ambiguity. Similarly, we may drop the parentheses in an expression such as $P \vee (Q \vee R)$.

But parentheses are essential when there is a mix of $\wedge$ and $\vee$, as in $P \vee (Q \wedge R)$. Indeed, $P \vee (Q \wedge R)$ and $(P \vee Q) \wedge R$ are **not** logically equivalent. (See Exercise 13 for Section 2.6, below.)

Exercises for Section 2.6

A. Use truth tables to show that the following statements are logically equivalent.

1. $P \wedge (Q \vee R) = (P \wedge Q) \vee (P \wedge R)$

2. $P \vee (Q \wedge R) = (P \vee Q) \wedge (P \vee R)$

3. $P \Rightarrow Q = (\sim P) \vee Q$

4. $\sim (P \vee Q) = (\sim P) \wedge (\sim Q)$

5. $\sim (P \vee Q \vee R) = (\sim P) \wedge (\sim Q) \wedge (\sim R)$

6. $\sim (P \wedge Q \wedge R) = (\sim P) \vee (\sim Q) \vee (\sim R)$

7. $P \Rightarrow Q = (P \wedge \sim Q) \Rightarrow (Q \wedge \sim Q)$

8. $\sim P \Leftrightarrow Q = (P \Rightarrow \sim Q) \wedge (\sim Q \Rightarrow P)$

B. Decide whether or not the following pairs of statements are logically equivalent.

9. $P \wedge Q$ and $\sim (\sim P \vee \sim Q)$

10. $(P \Rightarrow Q) \vee R$ and $\sim ((P \wedge \sim Q) \wedge \sim R)$

11. $(\sim P) \wedge (P \Rightarrow Q)$ and $\sim (Q \Rightarrow P)$

12. $\sim (P \Rightarrow Q)$ and $P \wedge \sim Q$

13. $P \vee (Q \wedge R)$ and $(P \vee Q) \wedge R$

14. $P \wedge (Q \vee \sim Q)$ and $(\sim P) \Rightarrow (Q \wedge \sim Q)$

2.7 Quantifiers

Using symbols $\wedge$, $\vee$, $\sim$, $\Rightarrow$ and $\Leftrightarrow$, we can deconstruct many English sentences into a symbolic form. As we have seen, this symbolic form can help us understand the logical structure of sentences and how different sentences may actually have the same meaning (as in logical equivalence).

But these symbols alone are not powerful enough to capture the full meaning of every statement. To help overcome this defect, we introduce two new symbols that correspond to common mathematical phrases. The symbol "$\forall$" stands for the phrase *"For all"* or *"For every."* The symbol "$\exists$" stands for the phrase *"There exists a"* or *"There is a."* Thus the statement

For every $n \in \mathbb{Z}$, $2n$ is even,

can be expressed in either of the following ways:

$\forall n \in \mathbb{Z}, 2n$ is even,

$\forall n \in \mathbb{Z}, E(2n)$.

Likewise, a statement such as

There exists a subset X of $\mathbb{N}$ for which $|X| = 5$.

can be translated as

$$\exists X, (X \subseteq \mathbb{N}) \wedge (|X| = 5) \quad \text{or} \quad \exists X \subseteq \mathbb{N}, |X| = 5 \quad \text{or} \quad \exists X \in \mathscr{P}(\mathbb{N}), |X| = 5.$$

The symbols $\forall$ and $\exists$ are called **quantifiers** because they refer in some sense to the quantity (i.e., all or some) of the variable that follows them. Symbol $\forall$ is called the **universal quantifier** and $\exists$ is called the **existential quantifier**. Statements which contain them are called **quantified** statements. A statement beginning with $\forall$ is called a **universally quantified** statement, and one beginning with $\exists$ is called an **existentially quantified** statement.

Example 2.5 The following English statements are paired with their translations into symbolic form.

Every integer that is not odd is even.
$\forall n \in \mathbb{Z}, \sim (n \text{ is odd }) \Rightarrow (n \text{ is even}),$ or $\forall n \in \mathbb{Z}, \sim O(n) \Rightarrow E(n).$

There is an integer that is not even.
$\exists n \in \mathbb{Z}, \sim E(n).$

For every real number x, there is a real number y for which $y^3 = x$.
$\forall x \in \mathbb{R}, \exists y \in \mathbb{R}, y^3 = x.$

Given any two rational numbers a and b, it follows that ab is rational.
$\forall a, b \in \mathbb{Q}, ab \in \mathbb{Q}.$

Given a set S (such as, but not limited to, $\mathbb{N}$, $\mathbb{Z}$, $\mathbb{Q}$ etc.), a quantified statement of form $\forall x \in S, P(x)$ is understood to be true if $P(x)$ is true for every $x \in S$. If there is at least one $x \in S$ for which $P(x)$ is false, then $\forall x \in S, P(x)$ is a false statement. Similarly, $\exists x \in S, P(x)$ is true provided that $P(x)$ is true for at least one element $x \in S$; otherwise it is false. Thus each statement in Example 2.5 is true. Here are some examples of quantified statements that are false:

Example 2.6 The following false quantified statements are paired with their translations.

Every integer is even.
$\forall n \in \mathbb{Z}, E(n).$

There is an integer n for which $n^2 = 2$.
$\exists n \in \mathbb{Z}, n^2 = 2$.

For every real number x, there is a real number y for which $y^2 = x$.
$\forall x \in \mathbb{R}, \exists y \in \mathbb{R}, y^2 = x$.

Given any two rational numbers a and b, it follows that $\sqrt{ab}$ is rational.
$\forall a, b \in \mathbb{Q}, \sqrt{ab} \in \mathbb{Q}$.

Example 2.7 When a statement contains two quantifiers you must be very alert to their order, for reversing the order can change the meaning. Consider the following statement from Example 2.5.

$\forall x \in \mathbb{R}, \exists y \in \mathbb{R}, y^3 = x$.

This statement is true, for no matter what number x is there exists a number $y = \sqrt[3]{x}$ for which $y^3 = x$. Now reverse the order of the quantifiers to get the new statement

$\exists y \in \mathbb{R}, \forall x \in \mathbb{R}, y^3 = x$.

This new statement says that there exists a particular number y with the property that $y^3 = x$ for *every* real number x. Since no number y can have this property, the statement is false. The two statements above have entirely different meanings.

Quantified statements are often misused in casual conversation. Maybe you've heard someone say *"All students do not pay full tuition."* when they mean *"Not all students pay full tuition."* While the mistake is perhaps marginally forgivable in casual conversation, it must never be made in a mathematical context. Do not say *"All integers are not even."* because that means there are no even integers. Instead, say *"Not all integers are even."*

Exercises for Section 2.7

Write the following as English sentences. Say whether they are true or false.

1. $\forall x \in \mathbb{R}, x^2 > 0$

2. $\forall x \in \mathbb{R}, \exists n \in \mathbb{N}, x^n \geq 0$

3. $\exists a \in \mathbb{R}, \forall x \in \mathbb{R}, ax = x$

4. $\forall X \in \mathscr{P}(\mathbb{N}), X \subseteq \mathbb{R}$

5. $\forall n \in \mathbb{N}, \exists X \in \mathscr{P}(\mathbb{N}), |X| < n$

6. $\exists n \in \mathbb{N}, \forall X \in \mathscr{P}(\mathbb{N}), |X| < n$

7. $\forall X \subseteq \mathbb{N}, \exists n \in \mathbb{Z}, |X| = n$

8. $\forall n \in \mathbb{Z}, \exists X \subseteq \mathbb{N}, |X| = n$

9. $\forall n \in \mathbb{Z}, \exists m \in \mathbb{Z}, m = n + 5$

10. $\exists m \in \mathbb{Z}, \forall n \in \mathbb{Z}, m = n + 5$

2.8 More on Conditional Statements

It is time to address a very important point about conditional statements that contain variables. To motivate this, let's return to the following example concerning integers x:

$$(x \text{ is a multiple of } 6) \Rightarrow (x \text{ is even}).$$

As noted earlier, since every multiple of 6 is even, this is a true statement no matter what integer x is. We could even underscore this fact by writing this true statement as

$$\forall x \in \mathbb{Z}, (x \text{ is a multiple of } 6) \Rightarrow (x \text{ is even}).$$

But now switch things around to get the different statement

$$(x \text{ is even}) \Rightarrow (x \text{ is a multiple of } 6).$$

This is true for some values of x such as -6, 12, 18, etc., but false for others (such as 2, 4, etc.). Thus we do not have a statement, but rather an open sentence. (Recall from Section 2.1 that an *open sentence* is a sentence whose truth value depends on the value of a certain variable or variables.) However, by putting a universal quantifier in front we get

$$\forall x \in \mathbb{Z}, (x \text{ is even}) \Rightarrow (x \text{ is a multiple of } 6),$$

which is definitely false, so this new expression is a statement, *not an open sentence*. In general, given any two open sentences $P(x)$ and $Q(x)$ about integers x, the expression $\forall x \in \mathbb{Z}, P(x) \Rightarrow Q(x)$ is either true or false, so it is a statement, not an open sentence.

Now we come to the very important point. In mathematics, whenever $P(x)$ and $Q(x)$ are open sentences concerning elements x in some set S (depending on context), an expression of form $P(x) \Rightarrow Q(x)$ is understood to be the *statement* $\forall x \in S, P(x) \Rightarrow Q(x)$. In other words, if a conditional statement is not explicitly quantified then there is an implied universal quantifier in front of it. This is done because statements of the form $\forall x \in S, P(x) \Rightarrow Q(x)$ are so common in mathematics that we would get tired of putting the $\forall x \in S$ in front of them.

Thus the following sentence is a true statement (as it is true for all x).

If x is a multiple of 6, then x is even.

Likewise, the next sentence is a false statement (as it is not true for all x).

If x is even, then x is a multiple of 6.

This leads to the following significant interpretation of a conditional statement, which is more general than (but consistent with) the interpretation from Section 2.3.

Definition 2.1 If P and Q are statements or open sentences, then

"If P, then Q,"

is a statement. This statement is true if it's impossible for P to be true while Q is false. It is false if there is at least one instance in which P is true but Q is false.

Thus the following are **true** statements:

If $x \in \mathbb{R}$, then $x^2 + 1 > 0$.

If a function f is differentiable on $\mathbb{R}$, then f is continuous on $\mathbb{R}$.

Likewise, the following are **false** statements:

If p is a prime number, then p is odd. (2 is prime.)

If f is a rational function, then f has an asymptote. (x^2 is rational.)

2.9 Translating English to Symbolic Logic

In writing (and reading) proofs of theorems, we must always be alert to the logical structure and meanings of the sentences. Sometimes it is necessary or helpful to parse them into expressions involving logic symbols. This may be done mentally or on scratch paper, or occasionally even explicitly within the body of a proof. The purpose of this section is to give you sufficient practice in translating English sentences into symbolic form so that you can better understand their logical structure. Here are some examples:

Example 2.8 Consider the Mean Value Theorem from Calculus:

If f is continuous on the interval $[a,b]$ and differentiable on (a,b), then there is a number $c \in (a,b)$ for which $f'(c) = \frac{f(b)-f(a)}{b-a}$.

Here is a translation to symbolic form:

$$\Big(\big(f \text{ cont. on } [a,b]\big) \wedge \big(f \text{ is diff. on } (a,b)\big) \Big) \Rightarrow \Big(\exists\, c \in (a,b), f'(c) = \frac{f(b)-f(a)}{b-a} \Big).$$

Example 2.9 Consider Goldbach's conjecture, from Section 2.1:

Every even integer greater than 2 is the sum of two primes.

This can be translated in the following ways, where P is the set of prime numbers and $S = \{4, 6, 8, 10, \dots\}$ is the set of even integers greater than 2.

$$(n \in S) \Rightarrow (\exists\, p, q \in P, n = p + q)$$
$$\forall\, n \in S, \exists\, p, q \in P, n = p + q$$

These translations of Goldbach's conjecture illustrate an important point. The first has the basic structure $(n \in S) \Rightarrow Q(n)$ and the second has structure $\forall\, n \in S, Q(n)$, yet they have exactly the same meaning. This is significant. Every universally quantified statement can be expressed as a conditional statement.

Fact 2.2 Suppose S is a set and $Q(x)$ is a statement about x for each $x \in S$. The following statements mean the same thing:

$$\forall\, x \in S, Q(x)$$
$$(x \in S) \Rightarrow Q(x).$$

This fact is significant because so many theorems have the form of a conditional statement. (The Mean Value Theorem is an example!) In proving a theorem we have to think carefully about what it says. Sometimes a theorem will be expressed as a universally quantified statement but it will be more convenient to think of it as a conditional statement. Understanding the above fact allows us to switch between the two forms.

We close this section with some final points. In translating a statement, be attentive to its intended meaning. Don't jump into, for example, automatically replacing every "and" with $\wedge$ and "or" with $\vee$. An example:

At least one of the integers x and y is even.

Don't be led astray by the presence of the word "and." The meaning of the statement is that one or both of the numbers is even, so it should be translated with "or," not "and":

$(x \text{ is even}) \vee (y \text{ is even})$.

Finally, the logical meaning of "but" can be captured by "and." The sentence "*The integer x is even, but the integer y is odd*," is translated as

$(x \text{ is even}) \wedge (y \text{ is odd})$.

Exercises for Section 2.9

Translate each of the following sentences into symbolic logic.

1. If f is a polynomial and its degree is greater than 2, then f' is not constant.
2. The number x is positive but the number y is not positive.
3. If x is prime then $\sqrt{x}$ is not a rational number.
4. For every prime number p there is another prime number q with $q > p$.
5. For every positive number ε, there is a positive number δ for which $|x-a| < \delta$ implies $|f(x) - f(a)| < \varepsilon$.
6. For every positive number ε there is a positive number M for which $|f(x)-b| < \varepsilon$, whenever $x > M$.
7. There exists a real number a for which $a + x = x$ for every real number x.
8. I don't eat anything that has a face.
9. If x is a rational number and $x \neq 0$, then $\tan(x)$ is not a rational number.
10. If $\sin(x) < 0$, then it is not the case that $0 \leq x \leq \pi$.
11. There is a Providence that protects idiots, drunkards, children and the United States of America. (Otto von Bismarck)
12. You can fool some of the people all of the time, and you can fool all of the people some of the time, but you can't fool all of the people all of the time. (Abraham Lincoln)
13. Everything is funny as long as it is happening to somebody else. (Will Rogers)

2.10 Negating Statements

Given a statement R, the statement $\sim R$ is called the **negation** of R. If R is a complex statement, then it is often the case that its negation $\sim R$ can be written in a simpler or more useful form. The process of finding this form is called **negating** R. In proving theorems it is often necessary to negate certain statements. We now investigate how to do this.

We have already examined part of this topic. **DeMorgan's laws**

$$\sim(P \wedge Q) = (\sim P) \vee (\sim Q) \tag{2.6}$$

$$\sim(P \vee Q) = (\sim P) \wedge (\sim Q) \tag{2.7}$$

(from Section 2.6) can be viewed as rules that tell us how to negate the statements $P \wedge Q$ and $P \vee Q$. Here are some examples that illustrate how DeMorgan's laws are used to negate statements involving "and" or "or."

Example 2.10 Consider negating the following statement.

R : You can solve it by factoring or with the quadratic formula.

Now, R means (You can solve it by factoring) $\vee$ (You can solve it with Q.F.), which we will denote as $P \vee Q$. The negation of this is

$$\sim(P \vee Q) \;=\; (\sim P) \wedge (\sim Q).$$

Therefore, in words, the negation of R is

$\sim R$: You can't solve it by factoring and you can't solve it with the quadratic formula.

Maybe you can find $\sim R$ without invoking DeMorgan's laws. That is good; you have internalized DeMorgan's laws and are using them unconsciously.

Example 2.11 We will negate the following sentence.

R : The numbers x and y are both odd.

This statement means (x is odd) $\wedge$ (y is odd), so its negation is

$$\begin{aligned} \sim\big((x \text{ is odd}) \wedge (y \text{ is odd})\big) &= \sim(x \text{ is odd}) \vee \sim(y \text{ is odd}) \\ &= (x \text{ is even}) \vee (y \text{ is even}). \end{aligned}$$

Therefore the negation of R can be expressed in the following ways:

$\sim R$: The number x is even or the number y is even.
$\sim R$: At least one of x and y is even.

Now let's move on to a slightly different kind of problem. It's often necessary to find the negations of quantified statements. For example, consider $\sim(\forall x \in \mathbb{N}, P(x))$. Reading this in words, we have the following:

It is not the case that $P(x)$ is true for all natural numbers x.

This means $P(x)$ is false for at least one x. In symbols, this is $\exists x \in \mathbb{N}, \sim P(x)$. Thus $\sim(\forall x \in \mathbb{N}, P(x)) = \exists x \in \mathbb{N}, \sim P(x)$. Similarly, you can reason out that $\sim(\exists x \in \mathbb{N}, P(x)) = \forall x \in \mathbb{N}, \sim P(x)$. In general:

$$\sim(\forall x \in S, P(x)) \;=\; \exists x \in S, \sim P(x), \tag{2.8}$$

$$\sim(\exists x \in S, P(x)) \;=\; \forall x \in S, \sim P(x). \tag{2.9}$$

Example 2.12 Consider negating the following statement.

R : The square of every real number is non-negative.

Symbolically, R can be expressed as $\forall x \in \mathbb{R}, x^2 \geq 0$, and thus its negation is $\sim(\forall x \in \mathbb{R}, x^2 \geq 0) = \exists x \in \mathbb{R}, \sim(x^2 \geq 0) = \exists x \in \mathbb{R}, x^2 < 0$. In words, this is

$\sim R$: There exists a real number whose square is negative.

Observe that R is true and $\sim R$ is false. You may be able to get $\sim R$ immediately, without using Equation (2.8) as we did above. If so, that is good; if not, you will probably be there soon.

If a statement has multiple quantifiers, negating it will involve several iterations of Equations (2.8) and (2.9). Consider the following:

S : For every real number x there is a real number y for which $y^3 = x$.

This statement asserts any real number x has a cube root y, so it's true. Symbolically S can be expressed as

$$\forall x \in \mathbb{R}, \exists y \in \mathbb{R}, y^3 = x.$$

Let's work out the negation of this statement.

$$\begin{aligned} \sim(\forall x \in \mathbb{R}, \exists y \in \mathbb{R}, y^3 = x) &= \exists x \in \mathbb{R}, \sim(\exists y \in \mathbb{R}, y^3 = x) \\ &= \exists x \in \mathbb{R}, \forall y \in \mathbb{R}, \sim(y^3 = x) \\ &= \exists x \in \mathbb{R}, \forall y \in \mathbb{R}, y^3 \neq x. \end{aligned}$$

Therefore the negation is the following (false) statement.

$\sim S$: There is a real number x for which $y^3 \neq x$ for all real numbers y.

In writing proofs you will sometimes have to negate a conditional statement $P \Rightarrow Q$. The remainder of this section describes how to do this. To begin, look at the expression $\sim(P \Rightarrow Q)$, which literally says *"$P \Rightarrow Q$ is false."* You know from the truth table for $\Rightarrow$ that the only way that $P \Rightarrow Q$ can be false is if P is true and Q is false. Therefore $\sim(P \Rightarrow Q) = P \wedge \sim Q$.

$$\sim(P \Rightarrow Q) = P \wedge \sim Q \tag{2.10}$$

(In fact, in Exercise 12 of Section 2.6, you used a truth table to verify that these two statements are indeed logically equivalent.)

Example 2.13 Negate the following statement about a particular (i.e., constant) number a.

R : If a is odd then a^2 is odd.

Using Equation (2.10), we get the following negation.

$\sim R$: a is odd and a^2 is not odd.

Example 2.14 This example is like the previous one, but the constant a is replaced by a variable x. We will negate the following statement.

R : If x is odd then x^2 is odd.

As discussed in Section 2.8, we interpret this as the universally quantified statement

$R : \forall x \in \mathbb{Z}, (x \text{ odd}) \Rightarrow (x^2 \text{ odd}).$

By Equations (2.8) and (2.10), we get the following negation for R.

$$\begin{aligned} \sim\big(\forall x \in \mathbb{Z}, (x \text{ odd}) \Rightarrow (x^2 \text{ odd})\big) &= \exists x \in \mathbb{Z}, \sim\big((x \text{ odd}) \Rightarrow (x^2 \text{ odd})\big) \\ &= \exists x \in \mathbb{Z}, (x \text{ odd}) \wedge \sim (x^2 \text{ odd}). \end{aligned}$$

Translating back into words, we have

$\sim R$: There is an odd integer x whose square is not odd.

Notice that R is true and $\sim R$ is false.

The above Example 2.14 showed how to negate a conditional statement $P(x) \Rightarrow Q(x)$. This type of problem can sometimes be embedded in more complex negation. See Exercise 5 below (and its solution).

Exercises for Section 2.10

Negate the following sentences.

1. The number x is positive, but the number y is not positive.
2. If x is prime, then $\sqrt{x}$ is not a rational number.
3. For every prime number p, there is another prime number q with $q > p$.
4. For every positive number ε, there is a positive number δ such that $|x-a| < \delta$ implies $|f(x)-f(a)| < \varepsilon$.
5. For every positive number ε, there is a positive number M for which $|f(x)-b| < \varepsilon$ whenever $x > M$.
6. There exists a real number a for which $a + x = x$ for every real number x.

7. I don't eat anything that has a face.

8. If x is a rational number and $x \neq 0$, then $\tan(x)$ is not a rational number.

9. If $\sin(x) < 0$, then it is not the case that $0 \leq x \leq \pi$.

10. If f is a polynomial and its degree is greater than 2, then f' is not constant.

11. You can fool all of the people all of the time.

12. Whenever I have to choose between two evils, I choose the one I haven't tried yet. (Mae West)

2.11 Logical Inference

Suppose we know that a statement of form $P \Rightarrow Q$ is true. This tells us that whenever P is true, Q will also be true. By itself, $P \Rightarrow Q$ being true does not tell us that either P or Q is true (they could both be false, or P could be false and Q true). However if in addition we happen to know that P is true then it must be that Q is true. This is called a **logical inference**: Given two true statements we can infer that a third statement is true. In this instance true statements $P \Rightarrow Q$ and P are "added together" to get Q. This is described below with $P \Rightarrow Q$ and P stacked one atop the other with a line separating them from Q. The intended meaning is that $P \Rightarrow Q$ combined with P produces Q.

$$\frac{\begin{array}{l} P \Rightarrow Q \\ P \end{array}}{Q} \qquad \frac{\begin{array}{l} P \Rightarrow Q \\ \sim Q \end{array}}{\sim P} \qquad \frac{\begin{array}{l} P \vee Q \\ \sim P \end{array}}{Q}$$

Two other logical inferences are listed above. In each case you should convince yourself (based on your knowledge of the relevant truth tables) that the truth of the statements above the line forces the statement below the line to be true.

Following are some additional useful logical inferences. The first expresses the obvious fact that if P and Q are both true then the statement $P \wedge Q$ will be true. On the other hand, $P \wedge Q$ being true forces P (also Q) to be true. Finally, if P is true, then $P \vee Q$ must be true, no matter what statement Q is.

$$\frac{\begin{array}{l} P \\ Q \end{array}}{P \wedge Q} \qquad \frac{P \wedge Q}{P} \qquad \frac{P}{P \vee Q}$$

These inferences are so intuitively obvious that they scarcely need to be mentioned. However, they represent certain patterns of reasoning that we will frequently apply to sentences in proofs, so we should be cognizant of the fact that we are using them.

2.12 An Important Note

It is important to be aware of the reasons that we study logic. There are three very significant reasons. First, the truth tables we studied tell us the exact meanings of the words such as "and," "or," "not" and so on. For instance, whenever we use or read the *"If..., then"* construction in a mathematical context, logic tells us exactly what is meant. Second, the rules of inference provide a system in which we can produce new information (statements) from known information. Finally, logical rules such as DeMorgan's laws help us correctly change certain statements into (potentially more useful) statements with the same meaning. Thus logic helps us understand the meanings of statements and it also produces new meaningful statements.

Logic is the glue that holds strings of statements together and pins down the exact meaning of certain key phrases such as the *"If..., then"* or *"For all"* constructions. Logic is the common language that all mathematicians use, so we must have a firm grip on it in order to write and understand mathematics.

But despite its fundamental role, logic's place is in the background of what we do, not the forefront. From here on, the beautiful symbols $\wedge$, $\vee$, $\Rightarrow$, $\Leftrightarrow$, $\sim$, $\forall$ and $\exists$ are rarely written. But we are aware of their meanings constantly. When reading or writing a sentence involving mathematics we parse it with these symbols, either mentally or on scratch paper, so as to understand the true and unambiguous meaning.

CHAPTER 3

Counting

It may seem peculiar that a college-level text has a chapter on counting. At its most basic level, counting is a process of pointing to each object in a collection and calling off *"one, two, three,..."* until the quantity of objects is determined. How complex could that be? Actually, counting can become quite subtle, and in this chapter we explore some of its more sophisticated aspects. Our goal is still to answer the question *"How many?"* but we introduce mathematical techniques that bypass the actual process of counting individual objects.

Almost every branch of mathematics uses some form of this "sophisticated counting." Many such counting problems can be modeled with the idea of a *list*, so we start there.

3.1 Counting Lists

A **list** is an ordered sequence of objects. A list is denoted by an opening parenthesis, followed by the objects, separated by commas, followed by a closing parenthesis. For example (a,b,c,d,e) is a list consisting of the first five letters of the English alphabet, in order. The objects a,b,c,d,e are called the **entries** of the list; the first entry is a, the second is b, and so on. If the entries are rearranged we get a different list, so, for instance,

$$(a,b,c,d,e) \neq (b,a,c,d,e).$$

A list is somewhat like a set, but instead of being a mere collection of objects, the entries of a list have a definite *order*. Note that for sets we have

$$\{a,b,c,d,e\} = \{b,a,c,d,e\},$$

but—as noted above—the analogous equality for lists does not hold.

Unlike sets, lists are allowed to have repeated entries. For example $(5,3,5,4,3,3)$ is a perfectly acceptable list, as is (S,O,S). The number of entries in a list is called its **length**. Thus $(5,3,5,4,3,3)$ has length six, and (S,O,S) has length three.

Occasionally we may get sloppy and write lists without parentheses and commas; for instance, we may express (S,O,S) as SOS if there is no danger of confusion. But be alert that doing this can lead to ambiguity. Is it reasonable that $(9,10,11)$ should be the same as 91011? If so, then $(9,10,11) = 91011 = (9,1,0,1,1)$, which makes no sense. We will thus almost always adhere to the parenthesis/comma notation for lists.

Lists are important because many real-world phenomena can be described and understood in terms of them. For example, your phone number (with area code) can be identified as a list of ten digits. Order is essential, for rearranging the digits can produce a different phone number. A *byte* is another important example of a list. A byte is simply a length-eight list of 0's and 1's. The world of information technology revolves around bytes.

To continue our examples of lists, $(a,15)$ is a list of length two. Likewise $(0,(0,1,1))$ is a list of length two whose second entry is a list of length three. The list $(\mathbb{N},\mathbb{Z},\mathbb{R})$ has length three, and each of its entries is a set. We emphasize that for two lists to be equal, they must have exactly the same entries in exactly the same order. Consequently if two lists are equal, then they must have the same length. Said differently, if two lists have different lengths, then they are not equal. For example, $(0,0,0,0,0,0) \neq (0,0,0,0,0)$. For another example note that

$$(g,r,o,c,e,r,y,\ l,i,s,t) \qquad \neq \qquad \left(\boxed{\begin{array}{l}\text{bread}\\ \text{milk}\\ \text{eggs}\\ \text{mustard}\\ \text{coffee}\end{array}} \right)$$

because the list on the left has length eleven but the list on the right has just one entry (a piece of paper with some words on it).

There is one very special list which has no entries at all. It is called the **empty list**, and is denoted $()$. It is the only list whose length is zero.

One often needs to count up the number of possible lists that satisfy some condition or property. For example, suppose we need to make a list of length three having the property that the first entry must be an element of the set $\{a,b,c\}$, the second entry must be in $\{5,7\}$ and the third entry must be in $\{a,x\}$. Thus $(a,5,a)$ and $(b,5,a)$ are two such lists. How many such lists are there all together? To answer this question, imagine making the list by selecting the first element, then the second and finally the third. This is described in Figure 3.1. The choices for the first list entry are a,b or c, and the left of the diagram branches out in three directions, one for each choice. Once this choice is made there are two choices (5 or 7) for the second entry, and this is described graphically by two branches from each of the three choices for the first entry. This pattern continues for the choice for the third entry, which is either a or x. Thus, in the diagram there are $3\cdot 2\cdot 2 = 12$ paths from left to right, each corresponding to a particular choice for each entry in the list. The corresponding lists are tallied at the far-right end of each path. So, to answer our original question, there are 12 possible lists with the stated properties.

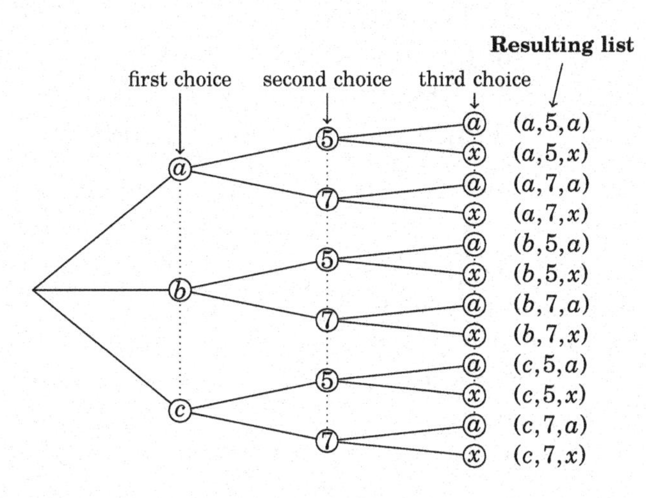

Figure 3.1. Constructing lists of length 3

We summarize the type of reasoning used above in an important fact called the **multiplication principle**.

Fact 3.1 (Multiplication Principle) Suppose in making a list of length n there are a_1 possible choices for the first entry, a_2 possible choices for the second entry, a_3 possible choices for the third entry and so on. Then the total number of different lists that can be made this way is the product $a_1\cdot a_2\cdot a_3\cdot\cdots\cdot a_n$.

So, for instance, in the above example we had $a_1 = 3, a_2 = 2$ and $a_3 = 2$, so the total number of lists was $a_1\cdot a_2\cdot a_3 = 3\cdot 2\cdot 2 = 12$. Now let's look at some additional examples of how the multiplication principle can be used.

Example 3.1 A standard license plate consists of three letters followed by four numbers. For example, *JRB-4412* and *MMX-8901* are two standard license plates. (Vanity plates such as *LV2COUNT* are not included among the standard plates.) How many different standard license plates are possible?

To answer this question, note that any standard license plate such as *JRB-4412* corresponds to a length-7 list *(J,R,B,4,4,1,2)*, so the question can be answered by counting how many such lists are possible. We use the multiplication principle. There are $a_1 = 26$ possibilities (one for each letter of the alphabet) for the first entry of the list. Similarly, there are $a_2 = 26$ possibilities for the second entry and $a_3 = 26$ possibilities for the third entry. There are $a_4 = 10$ possibilities for the fourth entry, and likewise $a_5 = a_6 = a_7 = 10$. Therefore there are a total of $a_1 \cdot a_2 \cdot a_3 \cdot a_4 \cdot a_5 \cdot a_6 \cdot a_7 = 26 \cdot 26 \cdot 26 \cdot 10 \cdot 10 \cdot 10 \cdot 10 =$ **175,760,000 possible standard license plates.**

There are two types of list-counting problems. On one hand, there are situations in which the same symbol or symbols may appear multiple times in different entries of the list. For example, license plates or telephone numbers can have repeated symbols. The sequence *CCX-4144* is a perfectly valid license plate in which the symbols *C* and *4* appear more than once. On the other hand, for some lists repeated symbols do not make sense or are not allowed. For instance, imagine drawing 5 cards from a standard 52-card deck and laying them in a row. Since no 2 cards in the deck are identical, this list has no repeated entries. We say that *repetition is allowed* in the first type of list and *repetition is not allowed* in the second kind of list. (Often we call a list in which repetition is not allowed a **non-repetitive list**.) The following example illustrates the difference.

Example 3.2 Consider making lists from symbols *A, B, C, D, E, F, G*.

(a) How many length-4 lists are possible if repetition is allowed?

(b) How many length-4 lists are possible if repetition is **not** allowed?

(c) How many length-4 lists are possible if repetition is **not** allowed and the list must contain an *E*?

(d) How many length-4 lists are possible if repetition is allowed and the list must contain an *E*?

Solutions:

(a) Imagine the list as containing four boxes that we fill with selections from the letters *A,B,C,D,E,F* and *G*, as illustrated below.

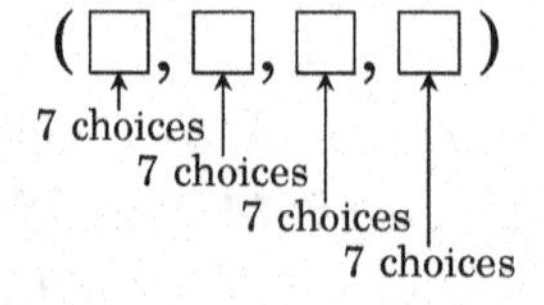

There are seven possibilities for the contents of each box, so the total number of lists that can be made this way is $7 \cdot 7 \cdot 7 \cdot 7 =$ **2401**.

(b) This problem is the same as the previous one except that repetition is not allowed. We have seven choices for the first box, but once it is filled we can no longer use the symbol that was placed in it. Hence there are only six possibilities for the second box. Once the second box has been filled we have used up two of our letters, and there are only five left to choose from in filling the third box. Finally, when the third box is filled we have only four possible letters for the last box.

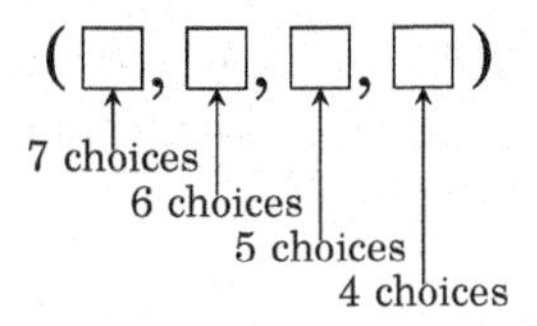

Thus the answer to our question is that there are $7 \cdot 6 \cdot 5 \cdot 4 = \mathbf{840}$ lists in which repetition does not occur.

(c) We are asked to count the length-4 lists in which repetition is not allowed and the symbol E must appear somewhere in the list. Thus E occurs once and only once in each such list. Let us divide these lists into four categories depending on whether the E occurs as the first, second, third or fourth entry. These four types of lists are illustrated below.

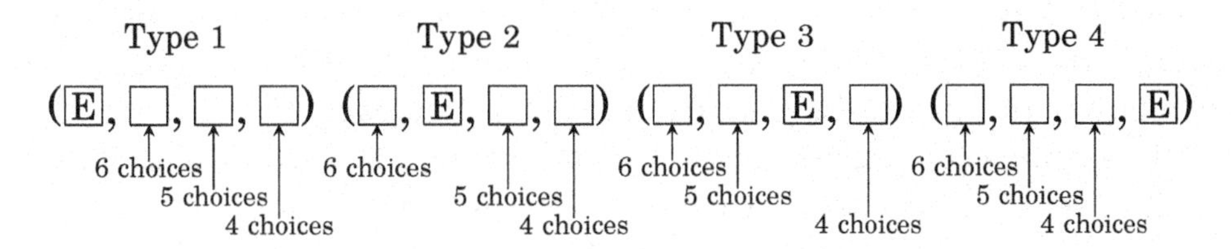

Consider lists of the first type, in which the E appears in the first entry. We have six remaining choices (*A,B,C,D,F* or *G*) for the second entry, five choices for the third entry and four choices for the fourth entry. Hence there are $6 \cdot 5 \cdot 4 = 120$ lists having an E in the first entry. As indicated in the above diagram, there are also $6 \cdot 5 \cdot 4 = 120$ lists having an E in the second, third or fourth entry. Thus there are $120 + 120 + 120 + 120 = \mathbf{480}$ such lists all together.

(d) Now we must find the number of length-four lists where repetition is allowed and the list must contain an E. Our strategy is as follows. By Part (a) of this exercise there are $7 \cdot 7 \cdot 7 \cdot 7 = 7^4 = 2401$ lists where repetition is allowed. Obviously this is not the answer to our current question, for many of these lists contain no E. We will subtract from 2401 the number of lists that **do not** contain an E. In making a list that does not contain an E, we have six choices for each list entry (because we can choose any one of the six letters *A,B,C,D,F* or *G*). Thus there are $6 \cdot 6 \cdot 6 \cdot 6 = 6^4 = 1296$ lists that do not have an E. Therefore the final answer to our question is that there are $2401 - 1296 = \mathbf{1105}$ lists with repetition allowed that contain at least one E.

Perhaps you wondered if Part (d) of Example 3.2 could be solved with a setup similar to that of Part (c). Let's try doing it that way. We want to count the length-4 lists (with repetition allowed) that contain at least one E. The following diagram is adapted from Part (c), the only difference being that there are now seven choices in each slot because we are allowed to repeat any of the seven letters.

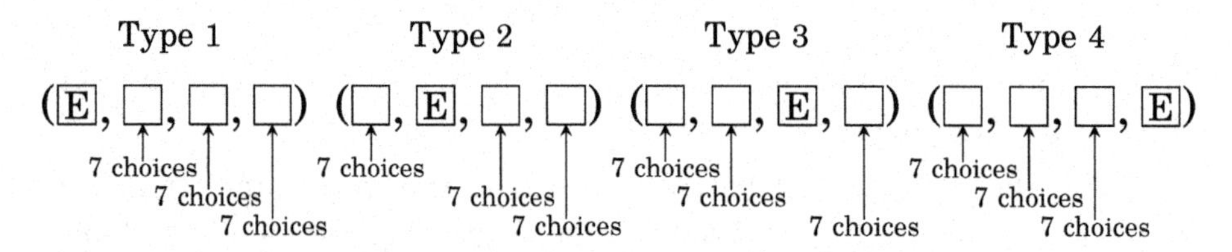

This gives a total of $7^3 + 7^3 + 7^3 + 7^3 = 1372$ lists, an answer that is substantially larger than the (correct) value of 1105 that we got in our solution to Part (d) above. It is not hard to see what went wrong. The list (E,E,A,B) is of type 1 *and* type 2, so it got counted *twice*. Similarly (E,E,C,E) is of type 1, 3 and 4, so it got counted three times. In fact, you can find many similar lists that were counted multiple times.

In solving counting problems, we must always be careful to avoid this kind of double-counting or triple-counting, or worse.

Exercises for Section 3.1

Note: A calculator may be helpful for some of the exercises in this chapter. This is the only chapter for which a calculator may be helpful. (As for the exercises in the other chapters, a calculator makes them harder.)

1. Consider lists made from the letters *T,H,E,O,R,Y*, with repetition allowed.

(a) How many length-4 lists are there?

(b) How many length-4 lists are there that begin with T?

(c) How many length-4 lists are there that do not begin with T?

2. Airports are identified with 3-letter codes. For example, the Richmond, Virginia airport has the code *RIC*, and Portland, Oregon has *PDX*. How many different 3-letter codes are possible?

3. How many lists of length 3 can be made from the symbols *A,B,C,D,E,F* if...

(**a**) ... repetition is allowed.

(**b**) ... repetition is not allowed.

(**c**) ... repetition is not allowed and the list must contain the letter A.

(**d**) ... repetition is allowed and the list must contain the letter A.

4. Five cards are dealt off of a standard 52-card deck and lined up in a row. How many such line-ups are there in which all 5 cards are of the same suit?

5. Five cards are dealt off of a standard 52-card deck and lined up in a row. How many such line-ups are there in which all 5 cards are of the same color (i.e., all black or all red)?

6. Five cards are dealt off of a standard 52-card deck and lined up in a row. How many such line-ups are there in which exactly one of the 5 cards is a queen?

7. This problem involves 8-digit binary strings such as 10011011 or 00001010 (i.e., 8-digit numbers composed of 0's and 1's).

(**a**) How many such strings are there?

(**b**) How many such strings end in 0?

(**c**) How many such strings have the property that their second and fourth digits are 1's?

(**d**) How many such strings have the property that their second **or** fourth digits are 1's?

8. This problem concerns lists made from the symbols A,B,C,D,E.

(**a**) How many such length-5 lists have at least one letter repeated?

(**b**) How many such length-6 lists have at least one letter repeated?

9. This problem concerns 4-letter codes made from the letters $A,B,C,D,...,Z$.

(**a**) How many such codes can be made?

(**b**) How many such codes have no two consecutive letters the same?

10. This problem concerns lists made from the letters A,B,C,D,E,F,G,H,I,J.

(**a**) How many length-5 lists can be made from these letters if repetition is not allowed and the list must begin with a vowel?

(**b**) How many length-5 lists can be made from these letters if repetition is not allowed and the list must begin and end with a vowel?

(**c**) How many length-5 lists can be made from these letters if repetition is not allowed and the list must contain exactly one A?

11. This problem concerns lists of length 6 made from the letters A,B,C,D,E,F,G,H. How many such lists are possible if repetition is not allowed and the list contains two consecutive vowels?

12. Consider the lists of length six made with the symbols P, R, O, F, S, where repetition is allowed. (For example, the following is such a list: (P,R,O,O,F,S).) How many such lists can be made if the list must end in an S and the symbol O is used more than once?

3.2 Factorials

In working the examples from Section 3.1, you may have noticed that often we need to count the number of non-repetitive lists of length n that are made from n symbols. In fact, this particular problem occurs with such frequency that a special idea, called a *factorial*, is introduced to handle it.

The table below motivates this idea. The first column lists successive integer values n (beginning with 0) and the second column contains a set $\{A,B,\cdots\}$ of n symbols. The third column contains all the possible non-repetitive lists of length n which can be made from these symbols. Finally, the last column tallies up how many lists there are of that type. Notice that when $n = 0$ there is only one list of length 0 that can be made from 0 symbols, namely the empty list $()$. Thus the value 1 is entered in the last column of that row.

n	Symbols	Non-repetitive lists of length n made from the symbols	$n!$
0	$\{\}$	$()$	1
1	$\{A\}$	(A)	1
2	$\{A,B\}$	$(A,B),(B,A)$	2
3	$\{A,B,C\}$	$(A,B,C),(A,C,B),(B,C,A),(B,A,C),(C,A,B),(C,B,A)$	6
4	$\{A,B,C,D\}$	$(A,B,C,D),(A,B,D,C),(A,C,B,D),(A,C,D,B),(A,D,B,C),(A,D,C,B)$ $(B,A,C,D),(B,A,D,C),(B,C,A,D),(B,C,D,A),(B,D,A,C),(B,D,C,A)$ $(C,A,B,D),(C,A,D,B),(C,B,A,D),(C,B,D,A),(C,D,A,B),(C,D,B,A)$ $(D,A,B,C),(D,A,C,B),(D,B,A,C),(D,B,C,A),(D,C,A,B),(D,C,B,A)$	24
$\vdots$	$\vdots$	$\vdots$	$\vdots$

For $n > 0$, the number that appears in the last column can be computed using the multiplication principle. The number of non-repetitive lists of length n that can be made from n symbols is $n(n-1)(n-2)\cdots 3\cdot 2\cdot 1$. Thus, for instance, the number in the last column of the row for $n = 4$ is $4\cdot 3\cdot 2\cdot 1 = 24$.

The number that appears in the last column of Row n is called the **factorial** of n. It is denoted as $n!$ (read "n factorial"). Here is the definition:

Definition 3.1 If n is a non-negative integer, then the **factorial** of n, denoted $n!$, is the number of non-repetitive lists of length n that can be made from n symbols. Thus $0! = 1$ and $1! = 1$. If $n > 1$, then $n! = n(n-1)(n-2)\cdots 3\cdot 2\cdot 1$.

It follows that

$$\begin{aligned} 0! &= 1 \\ 1! &= 1 \\ 2! &= 2 \cdot 1 = 2 \\ 3! &= 3 \cdot 2 \cdot 1 = 6 \\ 4! &= 4 \cdot 3 \cdot 2 \cdot 1 = 24 \\ 5! &= 5 \cdot 4 \cdot 3 \cdot 2 \cdot 1 = 120 \\ 6! &= 6 \cdot 5 \cdot 4 \cdot 3 \cdot 2 \cdot 1 = 720, \quad \text{and so on.} \end{aligned}$$

Students are often tempted to say $0! = 0$, but this is wrong. The correct value is $0! = 1$, as the above definition and table tell us. Here is another way to see that $0!$ must equal 1: Notice that $5! = 5 \cdot 4 \cdot 3 \cdot 2 \cdot 1 = 5 \cdot (4 \cdot 3 \cdot 2 \cdot 1) = 5 \cdot 4!$. Also $4! = 4 \cdot 3 \cdot 2 \cdot 1 = 4 \cdot (3 \cdot 2 \cdot 1) = 4 \cdot 3!$. Generalizing this reasoning, we have the following formula.

$$n! = n \cdot (n-1)! \tag{3.1}$$

Plugging in $n = 1$ gives $1! = 1 \cdot (1-1)! = 1 \cdot 0!$, that is, $1! = 1 \cdot 0!$. If we mistakenly thought $0!$ were 0, this would give the incorrect result $1! = 0$.

We round out our discussion of factorials with an example.

Example 3.3 This problem involves making lists of length seven from the symbols $0, 1, 2, 3, 4, 5$ and 6.

(a) How many such lists are there if repetition is not allowed?

(b) How many such lists are there if repetition is not allowed *and* the first three entries must be odd?

(c) How many such lists are there in which repetition is allowed, and the list must contain at least one repeated number?

To answer the first question, note that there are seven symbols, so the number of lists is $7! = \mathbf{5040}$. To answer the second question, notice that the set $\{0, 1, 2, 3, 4, 5, 6\}$ contains three odd numbers and four even numbers. Thus in making the list the first three entries must be filled by odd numbers and the final four must be filled with even numbers. By the multiplication principle, the number of such lists is $3 \cdot 2 \cdot 1 \cdot 4 \cdot 3 \cdot 2 \cdot 1 = 3!4! = \mathbf{144}$.

To answer the third question, notice that there are $7^7 = 823{,}543$ lists in which repetition is allowed. The set of all such lists includes lists that are non-repetitive (e.g., $(0, 6, 1, 2, 4, 3, 5)$) as well as lists that have some repetition (e.g., $(6, 3, 6, 2, 0, 0, 0)$). We want to compute the number of lists that have at least one repeated number. To find the answer we can subtract the number of non-repetitive lists of length seven from the total number of possible lists of length seven. Therefore the answer is $7^7 - 7! = 823{,}543 - 5040 = \mathbf{818{,}503}$.

We close this section with a formula that combines the ideas of the first and second sections of the present chapter. One of the main problems of Section 3.1 was as follows: Given n symbols, how many non-repetitive lists of length k can be made from the n symbols? We learned how to apply the multiplication principle to obtain the answer

$$n(n-1)(n-2)\cdots(n-k+1).$$

Notice that by cancellation this value can also be written as

$$\frac{n(n-1)(n-2)\cdots(n-k+1)(n-k)(n-k-1)\cdots 3\cdot 2\cdot 1}{(n-k)(n-k-1)\cdots 3\cdot 2\cdot 1} = \frac{n!}{(n-k)!}.$$

We summarize this as follows:

Fact 3.2 The number of non-repetitive lists of length k whose entries are chosen from a set of n possible entries is $\frac{n!}{(n-k)!}$.

For example, consider finding the number of non-repetitive lists of length five that can be made from the symbols $1,2,3,4,5,6,7,8$. We will do this two ways. By the multiplication principle, the answer is $8\cdot 7\cdot 6\cdot 5\cdot 4 = 6720$. Using the formula from Fact 3.2, the answer is $\frac{8!}{(8-5)!} = \frac{8!}{3!} = \frac{40{,}320}{6} = 6720$.

The new formula isn't really necessary, but it is a nice repackaging of an old idea and will prove convenient in the next section.

Exercises for Section 3.2

1. What is the smallest n for which $n!$ has more than 10 digits?
2. For which values of n does $n!$ have n or fewer digits?
3. How many 5-digit positive integers are there in which there are no repeated digits and all digits are odd?
4. Using only pencil and paper, find the value of $\frac{100!}{95!}$.
5. Using only pencil and paper, find the value of $\frac{120!}{118!}$.
6. There are two 0's at the end of $10! = 3{,}628{,}800$. Using only pencil and paper, determine how many 0's are at the end of the number $100!$.
7. Compute how many 9-digit numbers can be made from the digits $1,2,3,4,5,6,7,8,9$ if repetition is not allowed and all the odd digits occur first (on the left) followed by all the even digits (i.e. as in 137598264, but not 123456789).
8. Compute how many 7-digit numbers can be made from the digits $1,2,3,4,5,6,7$ if there is no repetition and the odd digits must appear in an unbroken sequence. (Examples: 3571264 or 2413576 or 2467531, etc., but **not** 7234615.)

9. There is a very interesting function $\Gamma : [0,\infty) \to \mathbb{R}$ called the **gamma function**. It is defined as $\Gamma(x) = \int_0^\infty t^{x-1}e^{-t}dt$. It has the remarkable property that if $x \in \mathbb{N}$, then $\Gamma(x) = (x-1)!$. Check that this is true for $x = 1,2,3,4$.
   Notice that this function provides a way of extending factorials to numbers other than integers. Since $\Gamma(n) = (n-1)!$ for all $n \in \mathbb{N}$, we have the formula $n! = \Gamma(n+1)$. But Γ can be evaluated at any number in $[0,\infty)$, not just at integers, so we have a formula for $n!$ for any $n \in [0,\infty)$. Extra credit: Compute $\pi!$.
10. There is another significant function called **Stirling's formula** that provides an approximation to factorials. It states that $n! \approx \sqrt{2\pi n}\left(\frac{n}{e}\right)^n$. It is an approximation to $n!$ in the sense that $\frac{n!}{\sqrt{2\pi n}\left(\frac{n}{e}\right)^n}$ approaches 1 as n approaches ∞. Use Stirling's formula to find approximations to 5!, 10!, 20! and 50!.

3.3 Counting Subsets

The previous two sections were concerned with counting the number of lists that can be made by selecting k entries from a set of n possible entries. We turn now to a related question: How many *subsets* can be made by selecting k elements from a set with n elements?

To highlight the differences between these two problems, look at the set $A = \{a,b,c,d,e\}$. First, think of the non-repetitive lists that can be made from selecting two entries from A. By Fact 3.2 (on the previous page), there are $\frac{5!}{(5-2)!} = \frac{5!}{3!} = \frac{120}{6} = 20$ such lists. They are as follows.

$$\begin{array}{cccccccccc} (a,b), & (a,c), & (a,d), & (a,e), & (b,c), & (b,d), & (b,e), & (c,d), & (c,e) & (d,e) \\ (b,a), & (c,a), & (d,a), & (e,a), & (c,b), & (d,b), & (e,b), & (d,c), & (e,c) & (e,d) \end{array}$$

Next consider the *subsets* of A that can made from selecting two elements from A. There are only ten such subsets, as follows.

$$\{a,b\},\ \{a,c\},\ \{a,d\},\ \{a,e\},\ \{b,c\},\ \{b,d\},\ \{b,e\},\ \{c,d\},\ \{c,e\},\ \{d,e\}.$$

The reason that there are more lists than subsets is that changing the order of the entries of a list produces a different list, but changing the order of the elements of a set does not change the set. Using elements $a,b \in A$, we can make two lists (a,b) and (b,a), but only one subset $\{a,b\}$.

In this section we are concerned not with counting lists, but with counting subsets. As was noted above, the basic question is this: How many subsets can be made by choosing k elements from an n-element set? We begin with some notation that gives a name to the answer to this question.

Definition 3.2 If n and k are integers, then $\binom{n}{k}$ denotes the number of subsets that can be made by choosing k elements from a set with n elements. The symbol $\binom{n}{k}$ is read "n choose k." (Some textbooks write $C(n,k)$ instead of $\binom{n}{k}$.)

To illustrate this definition, the following table computes the values of $\binom{4}{k}$ for various values of k by actually listing all the subsets of the 4-element set $A = \{a,b,c,d\}$ that have cardinality k. The values of k appear in the far-left column. To the right of each k are all of the subsets (if any) of A of size k. For example, when $k = 1$, set A has four subsets of size k, namely $\{a\}$, $\{b\}$, $\{c\}$ and $\{d\}$. Therefore $\binom{4}{1} = 4$. Similarly, when $k = 2$ there are six subsets of size k so $\binom{4}{2} = 6$.

k	k-element subsets of $\{a,b,c,d\}$	$\binom{4}{k}$
-1		$\binom{4}{-1} = 0$
0	$\emptyset$	$\binom{4}{0} = 1$
1	$\{a\},\{b\},\{c\},\{d\}$	$\binom{4}{1} = 4$
2	$\{a,b\},\{a,c\},\{a,d\},\{b,c\},\{b,d\},\{c,d\}$	$\binom{4}{2} = 6$
3	$\{a,b,c\},\{a,b,d\},\{a,c,d\},\{b,c,d\}$	$\binom{4}{3} = 4$
4	$\{a,b,c,d\}$	$\binom{4}{4} = 1$
5		$\binom{4}{5} = 0$
6		$\binom{4}{6} = 0$

When $k = 0$, there is only one subset of A that has cardinality k, namely the empty set, $\emptyset$. Therefore $\binom{4}{0} = 1$.

Notice that if k is negative or greater than $|A|$, then A has no subsets of cardinality k, so $\binom{4}{k} = 0$ in these cases. In general $\binom{n}{k} = 0$ whenever $k < 0$ or $k > n$. In particular this means $\binom{n}{k} = 0$ if n is negative.

Although it was not hard to work out the values of $\binom{4}{k}$ by writing out subsets in the above table, this method of actually listing sets would not be practical for computing $\binom{n}{k}$ when n and k are large. We need a formula. To find one, we will now carefully work out the value of $\binom{5}{3}$ in such a way that a pattern will emerge that points the way to a formula for any $\binom{n}{k}$.

To begin, note that $\binom{5}{3}$ is the number of 3-element subsets of $\{a,b,c,d,e\}$. These are listed in the following table. We see that in fact $\binom{5}{3} = 10$.

$\longleftarrow \binom{5}{3} \longrightarrow$

$\{a,b,c\}$	$\{a,b,d\}$	$\{a,b,e\}$	$\{a,c,d\}$	$\{a,c,e\}$	$\{a,d,e\}$	$\{b,c,d\}$	$\{b,c,e\}$	$\{b,d,e\}$	$\{c,d,e\}$

The formula will emerge when we expand this table as follows. Taking any one of the ten 3-element sets above, we can make 3! different non-repetitive lists from its elements. For example, consider the first set $\{a,b,c\}$. The first column of the following table tallies the $3! = 6$ different lists that can be the letters $\{a,b,c\}$. The second column tallies the lists that can be made from $\{a,b,d\}$, and so on.

$\longleftarrow \binom{5}{3} \longrightarrow$

↑	*abc*	*abd*	*abe*	*acd*	*ace*	*ade*	*bcd*	*bce*	*bde*	*cde*
	acb	*adb*	*aeb*	*adc*	*aec*	*aed*	*bdc*	*bec*	*bed*	*ced*
	bac	*bad*	*bae*	*cad*	*cae*	*dae*	*cbd*	*cbe*	*dbe*	*dce*
3!	*bca*	*bda*	*bea*	*cda*	*cea*	*dea*	*cdb*	*ceb*	*deb*	*dec*
	cba	*dba*	*eba*	*dca*	*eca*	*eda*	*dcb*	*ecb*	*edb*	*edc*
↓	*cab*	*dab*	*eab*	*dac*	*eac*	*ead*	*dbc*	*ebc*	*ebd*	*ecd*

This table has $\binom{5}{3}$ columns and 3! rows, so it has a total of $3!\binom{5}{3}$ lists. But notice also that the table consists of every non-repetitive length-3 list that can be made from the symbols $\{a,b,c,d,e\}$. We know from Fact 3.2 that there are $\frac{5!}{(5-3)!}$ such lists. Thus the total number of lists in the table is $3!\binom{5}{3} = \frac{5!}{(5-3)!}$. Dividing both sides of this equation by 3!, we get

$$\binom{5}{3} = \frac{5!}{3!(5-3)!}.$$

Working this out, you will find that it does give the correct value of 10.

But there was nothing special about the values 5 and 3. We could do the above analysis for any $\binom{n}{k}$ instead of $\binom{5}{3}$. The table would have $\binom{n}{k}$ columns and $k!$ rows. We would get

$$\binom{n}{k} = \frac{n!}{k!(n-k)!}.$$

We summarize this as follows:

Fact 3.3 If $n, k \in \mathbb{Z}$ and $0 \le k \le n$, then $\binom{n}{k} = \frac{n!}{k!(n-k)!}$. Otherwise $\binom{n}{k} = 0$.

Let's now use our new knowledge to work some exercises.

Example 3.4 How many 4-element subsets does $\{1,2,3,4,5,6,7,8,9\}$ have?
The answer is $\binom{9}{4} = \frac{9!}{4!(9-4)!} = \frac{9!}{4!5!} = \frac{9\cdot8\cdot7\cdot6\cdot5!}{4!5!} = \frac{9\cdot8\cdot7\cdot6}{4!} = \frac{9\cdot8\cdot7\cdot6}{24} = \mathbf{126}$.

Example 3.5 A single 5-card hand is dealt off of a standard 52-card deck. How many different 5-card hands are possible?

To answer this, think of the deck as being a set D of 52 cards. Then a 5-card hand is just a 5-element subset of D. For example, here is one of many different 5-card hands that might be dealt from the deck.

$$\left\{ 7\clubsuit,\ 2\clubsuit,\ 3\heartsuit,\ A\spadesuit,\ 5\diamondsuit \right\}$$

The total number of possible hands equals the number of 5-element subsets of D, that is

$$\binom{52}{5} = \frac{52!}{5!\cdot 47!} = \frac{52\cdot51\cdot50\cdot49\cdot48\cdot47!}{5!\cdot47!} = \frac{52\cdot51\cdot50\cdot49\cdot48}{5!} = 2{,}598{,}960.$$

Thus the answer to our question is that there are 2,598,960 different five-card hands that can be dealt from a deck of 52 cards.

Example 3.6 This problem concerns 5-card hands that can be dealt off of a 52-card deck. How many such hands are there in which two of the cards are clubs and three are hearts?

Solution: Think of such a hand as being described by a list of length two of the form

$$\left(\left\{ {*}\clubsuit,\ {*}\clubsuit \right\},\ \left\{ {*}\heartsuit,\ {*}\heartsuit,\ {*}\heartsuit \right\} \right),$$

where the first entry is a 2-element subset of the set of 13 club cards, and the second entry is a 3-element subset of the set of 13 heart cards. There are $\binom{13}{2}$ choices for the first entry and $\binom{13}{3}$ choices for the second entry, so by the multiplication principle there are $\binom{13}{2}\binom{13}{3} = \frac{13!}{2!11!}\frac{13!}{3!10!} = 22{,}308$ such lists. Answer: There are **22,308 possible 5-card hands with two clubs and three hearts.**

Example 3.7 Imagine a lottery that works as follows. A bucket contains 36 balls numbered $1,2,3,4,\ldots,36$. Six of these balls will be drawn randomly. For \$1 you buy a ticket that has six blanks: $\boxed{\square\square\square\square\square\square}$. You fill in the blanks with six different numbers between 1 and 36. You win \$1,000,000 if you chose the same numbers that are drawn, regardless of order. What are your chances of winning?

Solution: In filling out the ticket you are choosing six numbers from a set of 36 numbers. Thus there are $\binom{36}{6} = \frac{36!}{6!(36-6)!} = 1,947,792$ different combinations of numbers you might write. Only one of these will be a winner. **Your chances of winning are one in 1,947,792.**

Exercises for Section 3.3

1. Suppose a set A has 37 elements. How many subsets of A have 10 elements? How many subsets have 30 elements? How many have 0 elements?
2. Suppose A is a set for which $|A| = 100$. How many subsets of A have 5 elements? How many subsets have 10 elements? How many have 99 elements?
3. A set X has exactly 56 subsets with 3 elements. What is the cardinality of X?
4. Suppose a set B has the property that $\left|\{X : X \in \mathscr{P}(B), |X| = 6\}\right| = 28$. Find $|B|$.
5. How many 16-digit binary strings contain exactly seven 1's? (Examples of such strings include 0111000011110000 and 0011001100110010, etc.)
6. $\left|\{X \in \mathscr{P}(\{0,1,2,3,4,5,6,7,8,9\}) : |X| = 4\}\right| =$
7. $\left|\{X \in \mathscr{P}(\{0,1,2,3,4,5,6,7,8,9\}) : |X| < 4\}\right| =$
8. This problem concerns lists made from the symbols A,B,C,D,E,F,G,H,I.
   (a) How many length-5 lists can be made if repetition is not allowed and the list is in alphabetical order? (Example: *BDEFI* or *ABCGH*, but not *BACGH*.)
   (b) How many length-5 lists can be made if repetition is not allowed and the list is **not** in alphabetical order?
9. This problem concerns lists of length 6 made from the letters A,B,C,D,E,F, without repetition. How many such lists have the property that the D occurs before the A?
10. A department consists of 5 men and 7 women. From this department you select a committee with 3 men and 2 women. In how many ways can you do this?
11. How many positive 10-digit integers contain no 0's and exactly three 6's?
12. Twenty-one people are to be divided into two teams, the Red Team and the Blue Team. There will be 10 people on Red Team and 11 people on Blue Team. In how many ways can this be done?
13. Suppose n and k are integers for which $0 \le k \le n$. Use the formula $\binom{n}{k} = \frac{n!}{k!(n-k)!}$ to show that $\binom{n}{k} = \binom{n}{n-k}$.
14. Suppose $n,k \in \mathbb{Z}$, and $0 \le k \le n$. Use Definition 3.2 alone (without using Fact 3.3) to show that $\binom{n}{k} = \binom{n}{n-k}$.

3.4 Pascal's Triangle and the Binomial Theorem

There are some beautiful and significant patterns among the numbers $\binom{n}{k}$. This section investigates a pattern based on one equation in particular. It happens that

$$\binom{n+1}{k} = \binom{n}{k-1} + \binom{n}{k} \tag{3.2}$$

for any integers n and k with $1 \le k \le n$.

To see why this is true, recall that $\binom{n+1}{k}$ equals the number of k-element subsets of a set with $n+1$ elements. Now, the set $A = \{0,1,2,3,\ldots,n\}$ has $n+1$ elements, so $\binom{n+1}{k}$ equals the number of k-element subsets of A. Such subsets can be divided into two types: those that contain 0 and those that do not contain 0. To make a k-element subset that contains 0 we can start with $\{0\}$ and then append to this set an additional $k-1$ numbers selected from $\{1,2,3,\ldots,n\}$. There are $\binom{n}{k-1}$ ways to make this selection, so there are $\binom{n}{k-1}$ k-element subsets of A that contain 0. Concerning the k-element subsets of A that do not contain 0, there are $\binom{n}{k}$ of these sets, for we can form them by selecting k elements from the n-element set $\{1,2,3,\ldots,n\}$. In light of all this, Equation (3.2) just expresses the obvious fact that the number of k-element subsets of A equals the number of k-element subsets that contain 0 plus the number of k-element subsets that do not contain 0.

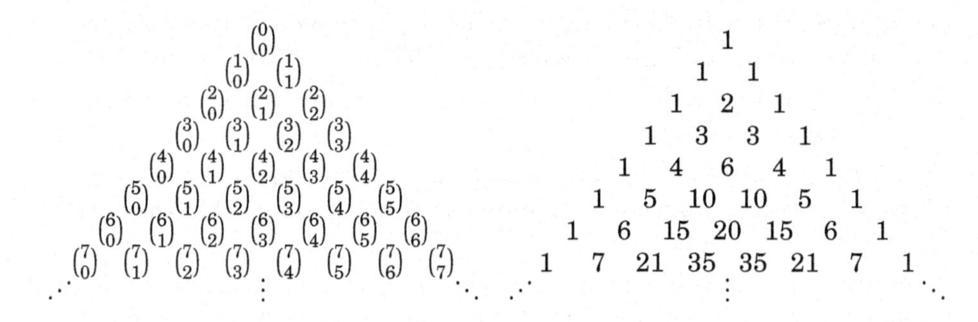

Figure 3.2. Pascal's triangle

Now that we have seen why Equation (3.2) is true, we are going to arrange the numbers $\binom{n}{k}$ in a triangular pattern that highlights various relationships among them. The left-hand side of Figure 3.2 shows numbers $\binom{n}{k}$ arranged in a pyramid with $\binom{0}{0}$ at the apex, just above a row containing $\binom{1}{k}$ with $k=0$ and $k=1$. Below *this* is a row listing the values of $\binom{2}{k}$ for $k=0,1,2$. In general, each row listing the numbers $\binom{n}{k}$ is just above a row listing the numbers $\binom{n+1}{k}$.

Any number $\binom{n+1}{k}$ for $0 < k < n$ in this pyramid is immediately below and between the the two numbers $\binom{n}{k-1}$ and $\binom{n}{k}$ in the previous row. But Equation 3.2 says $\binom{n+1}{k} = \binom{n}{k-1} + \binom{n}{k}$, and therefore any number (other than 1) in the pyramid is the sum of the two numbers immediately above it.

This pattern is especially evident on the right of Figure 3.2, where each $\binom{n}{k}$ is worked out. Notice how 21 is the sum of the numbers 6 and 15 above it. Similarly, 5 is the sum of the 1 and 4 above *it* and so on.

The arrangement on the right of Figure 3.2 is called **Pascal's triangle.** (It is named after Blaise Pascal, 1623–1662, a French mathematician and philosopher who discovered many of its properties.) Although we have written only the first eight rows of Pascal's triangle (beginning with Row 0 at the apex), it obviously could be extended downward indefinitely. We could add an additional row at the bottom by placing a 1 at each end and obtaining each remaining number by adding the two numbers above its position. Doing this would give the following row:

$$1 \quad 8 \quad 28 \quad 56 \quad 70 \quad 56 \quad 28 \quad 8 \quad 1$$

This row consists of the numbers $\binom{8}{k}$ for $0 \le k \le 8$, and we have computed them without the formula $\binom{8}{k} = \frac{8!}{k!(8-k)!}$. Any $\binom{n}{k}$ can be computed this way.

The very top row (containing only 1) is called *Row* 0. Row 1 is the next down, followed by Row 2, then Row 3, etc. With this labeling, Row n consists of the numbers $\binom{n}{k}$ for $0 \le k \le n$.

Notice that Row n appears to be a list of the coefficients of $(x+y)^n$. For example $(x+y)^2 = \mathbf{1}x^2 + \mathbf{2}xy + \mathbf{1}y^2$, and Row 2 lists the coefficients 1 2 1. Similarly $(x+y)^3 = \mathbf{1}x^3 + \mathbf{3}x^2y + \mathbf{3}xy^2 + \mathbf{1}y^3$, and Row 3 is 1 3 3 1. Pascal's triangle is shown on the left of Figure 3.3 and on the right are the expansions of $(x+y)^n$ for $0 \le n \le 5$. In every case (at least as far as you care to check) the numbers in Row n match up with the coefficients of $(x+y)^n$.

$$
\begin{array}{ccccccccccc}
 & & & & & 1 & & & & & \\
 & & & & 1 & & 1 & & & & \\
 & & & 1 & & 2 & & 1 & & & \\
 & & 1 & & 3 & & 3 & & 1 & & \\
 & 1 & & 4 & & 6 & & 4 & & 1 & \\
1 & & 5 & & 10 & & 10 & & 5 & & 1 \\
\cdot^{\cdot^{\cdot}} & & & & & \vdots & & & & & \ddots
\end{array}
\qquad
\begin{array}{c}
1 \\
1x + 1y \\
1x^2 + 2xy + 1y^2 \\
1x^3 + 3x^2y + 3xy^2 + 1y^3 \\
1x^4 + 4x^3y + 6x^2y^2 + 4xy^3 + 1y^4 \\
1x^5 + 5x^4y + 10x^3y^2 + 10x^2y^3 + 5xy^4 + 1y^5 \\
\cdot^{\cdot^{\cdot}} \qquad \vdots \qquad \ddots
\end{array}
$$

Figure 3.3. The nth row of Pascal's triangle lists the coefficients of $(x+y)^n$

In fact this turns out to be true for every n. This result is known as the binomial theorem, and it is worth mentioning here. It tells how to raise a binomial $x+y$ to a non-negative integer power n.

Theorem 3.1 **(Binomial Theorem)** If n is a non-negative integer, then $(x+y)^n = \binom{n}{0}x^n + \binom{n}{1}x^{n-1}y + \binom{n}{2}x^{n-2}y^2 + \binom{n}{3}x^{n-3}y^3 + \cdots + \binom{n}{n-1}xy^{n-1} + \binom{n}{n}y^n$.

For now we will be content to accept the binomial theorem without proof. (You will be asked to prove it in an exercise in Chapter 10.) You may find it useful from time to time. For instance, you can apply it if you ever need to expand an expression such as $(x+y)^7$. To do this, look at Row 7 of Pascal's triangle in Figure 3.2 and apply the binomial theorem to get

$$(x+y)^7 = x^7 + 7x^6y + 21x^5y^2 + 35x^4y^3 + 35x^3y^4 + 21x^2y^5 + 7xy^6 + y^7.$$

For another example,

$$\begin{aligned}(2a-b)^4 &= ((2a)+(-b))^4 \\ &= (2a)^4 + 4(2a)^3(-b) + 6(2a)^2(-b)^2 + 4(2a)(-b)^3 + (-b)^4 \\ &= 16a^4 - 32a^3b + 24a^2b^2 - 8ab^3 + b^4.\end{aligned}$$

Exercises for Section 3.4

1. Write out Row 11 of Pascal's triangle.
2. Use the binomial theorem to find the coefficient of x^8y^5 in $(x+y)^{13}$.
3. Use the binomial theorem to find the coefficient of x^8 in $(x+2)^{13}$.
4. Use the binomial theorem to find the coefficient of x^6y^3 in $(3x-2y)^9$.
5. Use the binomial theorem to show $\sum_{k=0}^{n}\binom{n}{k} = 2^n$.
6. Use Definition 3.2 (page 74) and Fact 1.3 (page 12) to show $\sum_{k=0}^{n}\binom{n}{k} = 2^n$.
7. Use the binomial theorem to show $\sum_{k=0}^{n} 3^k\binom{n}{k} = 4^n$.
8. Use Fact 3.3 (page 76) to derive Equation 3.2 (page 78).
9. Use the binomial theorem to show $\binom{n}{0} - \binom{n}{1} + \binom{n}{2} - \binom{n}{3} + \binom{n}{4} - \cdots + (-1)^n\binom{n}{n} = 0$.
10. Show that the formula $k\binom{n}{k} = n\binom{n-1}{k-1}$ is true for all integers n, k with $0 \le k \le n$.
11. Use the binomial theorem to show $9^n = \sum_{k=0}^{n}(-1)^k\binom{n}{k}10^{n-k}$.
12. Show that $\binom{n}{k}\binom{k}{m} = \binom{n}{m}\binom{n-m}{k-m}$.
13. Show that $\binom{n}{3} = \binom{2}{2} + \binom{3}{2} + \binom{4}{2} + \binom{5}{2} + \cdots + \binom{n-1}{2}$.
14. The first five rows of Pascal's triangle appear in the digits of powers of 11: $11^0 = 1$, $11^1 = 11$, $11^2 = 121$, $11^3 = 1331$ and $11^4 = 14641$. Why is this so? Why does the pattern not continue with 11^5?

3.5 Inclusion-Exclusion

Many counting problems involve computing the cardinality of a union $A \cup B$ of two finite sets. We examine this kind of problem now.

First we develop a formula for $|A \cup B|$. It is tempting to say that $|A \cup B|$ must equal $|A| + |B|$, but that is not quite right. If we count the elements of A and then count the elements of B and add the two figures together, we get $|A| + |B|$. But if A and B have some elements in common, then we have counted each element in $A \cap B$ *twice*.

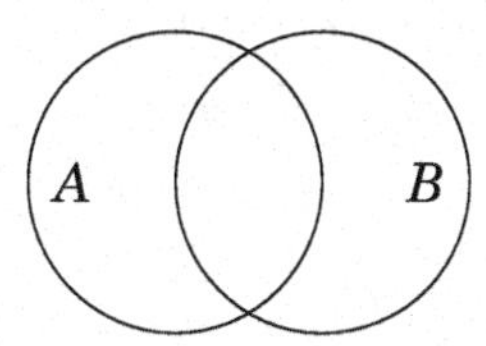

Therefore $|A| + |B|$ exceeds $|A \cup B|$ by $|A \cap B|$, and consequently $|A \cup B| = |A| + |B| - |A \cap B|$. This can be a useful equation.

$$|A \cup B| = |A| + |B| - |A \cap B| \tag{3.3}$$

Notice that the sets A, B and $A \cap B$ are all generally smaller than $A \cup B$, so Equation (3.3) has the potential of reducing the problem of determining $|A \cup B|$ to three simpler counting problems. It is sometimes called an *inclusion-exclusion* formula because elements in $A \cap B$ are included (twice) in $|A|+|B|$, then excluded when $|A \cap B|$ is subtracted. Notice that if $A \cap B = \emptyset$, then we do in fact get $|A \cup B| = |A| + |B|$; conversely if $|A \cup B| = |A| + |B|$, then it must be that $A \cap B = \emptyset$.

Example 3.8 A 3-card hand is dealt off of a standard 52-card deck. How many different such hands are there for which all 3 cards are red or all three cards are face cards?

Solution: Let A be the set of 3-card hands where all three cards are red (i.e., either $\heartsuit$ or $\diamondsuit$). Let B be the set of 3-card hands in which all three cards are face cards (i.e., J,K or Q of any suit). These sets are illustrated below.

$$A = \left\{ \left\{ 5\heartsuit, K\diamondsuit, 2\heartsuit \right\}, \left\{ K\heartsuit, J\heartsuit, Q\heartsuit \right\}, \left\{ A\diamondsuit, 6\diamondsuit, 6\heartsuit \right\}, \ldots \right\} \quad \text{(Red cards)}$$

$$B = \left\{ \left\{ K\spadesuit, K\diamondsuit, J\clubsuit \right\}, \left\{ K\heartsuit, J\heartsuit, Q\heartsuit \right\}, \left\{ Q\diamondsuit, Q\clubsuit, Q\heartsuit \right\}, \ldots \right\} \quad \text{(Face cards)}$$

We seek the number of 3-card hands that are all red or all face cards, and this number is $|A \cup B|$. By Formula (3.3), $|A \cup B| = |A| + |B| - |A \cap B|$. Let's examine $|A|, |B|$ and $|A \cap B|$ separately. Any hand in A is formed by selecting three cards from the 26 red cards in the deck, so $|A| = \binom{26}{3}$. Similarly, any hand in B is formed by selecting three cards from the 12 face cards in the deck, so $|B| = \binom{12}{3}$. Now think about $A \cap B$. It contains all the 3-card hands made up of cards that are red face cards.

$$A \cap B \;=\; \left\{ \left\{ K\heartsuit, K\diamondsuit, J\heartsuit \right\}, \left\{ K\heartsuit, J\heartsuit, Q\heartsuit \right\}, \left\{ Q\diamondsuit, J\diamondsuit, Q\heartsuit \right\}, \ldots \right\} \quad \text{(Red face cards)}$$

The deck has only 6 red face cards, so $|A \cap B| = \binom{6}{3}$.

Now we can answer our question. The number of 3-card hands that are all red or all face cards is $|A \cup B| = |A| + |B| - |A \cap B| = \binom{26}{3} + \binom{12}{3} - \binom{6}{3} = 2600 + 220 - 20 = \mathbf{2800}$.

There is an analogue to Equation (3.3) that involves three sets. Consider three sets A, B and C, as represented in the following Venn Diagram.

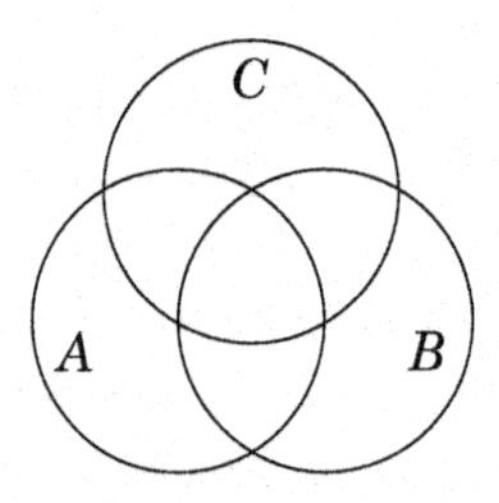

Using the same kind of reasoning that resulted in Equation (3.3), you can convince yourself that

$$|A \cup B \cup C| = |A| + |B| + |C| - |A \cap B| - |A \cap C| - |B \cap C| + |A \cap B \cap C|. \tag{3.4}$$

There's probably not much harm in ignoring this one for now, but if you find this kind of thing intriguing you should definitely take a course in combinatorics. (Ask your instructor!)

As we've noted, Equation (3.3) becomes $|A \cup B| = |A| + |B|$ if it happens that $A \cap B = \emptyset$. Also, in Equation (3.4), note that if $A \cap B = \emptyset$, $A \cap C = \emptyset$ and $B \cap C = \emptyset$, we get the simple formula $|A \cup B \cup C| = |A| + |B| + |C|$. In general, we have the following formula for n sets, none of which overlap. It is sometimes called the **addition principle.**

Fact 3.4 (Addition Principle) If $A_1, A_2, \ldots, A_n$ are sets with $A_i \cap A_j = \emptyset$ whenever $i \neq j$, then $|A_1 \cup A_2 \cup \cdots \cup A_n| = |A_1| + |A_2| + \cdots + |A_n|$.

Example 3.9 How many 7-digit binary strings (0010100, 1101011, etc.) have an odd number of 1's?

Solution: Let A be the set of all 7-digit binary strings with an odd number of 1's, so the answer to the question will be $|A|$. To compute $|A|$, we break A up into smaller parts. Notice any string in A will have either one, three, five or seven 1's. Let A_1 be the set of 7-digit binary strings with only one 1. Let A_3 be the set of 7-digit binary strings with three 1's. Let A_5 be the set of 7-digit binary strings with five 1's, and let A_7 be the set of 7-digit binary strings with seven 1's. Therefore $A = A_1 \cup A_3 \cup A_5 \cup A_7$. Notice that any two of the sets A_i have empty intersection, so Fact 3.4 gives $|A| = |A_1| + |A_3| + |A_5| + |A_7|$.

Now the problem is to find the values of the individual terms of this sum. For instance take A_3, the set of 7-digit binary strings with three 1's. Such a string can be formed by selecting three out of seven positions for the 1's and putting 0's in the other spaces. Therefore $|A_3| = \binom{7}{3}$. Similarly $|A_1| = \binom{7}{1}$, $|A_5| = \binom{7}{5}$, and $|A_7| = \binom{7}{7}$. Finally the answer to our question is $|A| = |A_1| + |A_3| + |A_5| + |A_7| = \binom{7}{1} + \binom{7}{3} + \binom{7}{5} + \binom{7}{7} = 7 + 35 + 21 + 1 = 64$. **There are 64 seven-digit binary strings with an odd number of 1's.**

You may already have been using the Addition Principle intuitively, without thinking of it as a free-standing result. For instance, we used it in Example 3.2(c) when we divided lists into four types and computed the number of lists of each type.

Exercises for Section 3.5

1. At a certain university 523 of the seniors are history majors or math majors (or both). There are 100 senior math majors, and 33 seniors are majoring in both history and math. How many seniors are majoring in history?
2. How many 4-digit positive integers are there for which there are no repeated digits, or for which there may be repeated digits, but all are odd?
3. How many 4-digit positive integers are there that are even or contain no 0's?
4. This problem involves lists made from the letters *T,H,E,O,R,Y*, with repetition allowed.

   (a) How many 4-letter lists are there that don't begin with *T*, or don't end in *Y*?

   (b) How many 4-letter lists are there in which the sequence of letters *T,H,E* appears consecutively?

   (c) How many 5-letter lists are there in which the sequence of letters *T,H,E* appears consecutively?

5. How many 7-digit binary strings begin in 1 or end in 1 or have exactly four 1's?

6. Is the following statement true or false? Explain. If $A_1 \cap A_2 \cap A_3 = \emptyset$, then $|A_1 \cup A_2 \cup A_3| = |A_1| + |A_2| + |A_3|$.

7. This problem concerns 4-card hands dealt off of a standard 52-card deck. How many 4-card hands are there for which all 4 cards are of the same suit or all 4 cards are red?

8. This problem concerns 4-card hands dealt off of a standard 52-card deck. How many 4-card hands are there for which all 4 cards are of different suits or all 4 cards are red?

9. A 4-letter list is made from the letters L,I,S,T,E,D according to the following rule: Repetition is allowed, and the first two letters on the list are vowels or the list ends in D. How many such lists are possible?

10. A 5-card poker hand is called a *flush* if all cards are the same suit. How many different flushes are there?

Part II

How to Prove Conditional Statements

CHAPTER 4

Direct Proof

It is time to prove some theorems. There are various strategies for doing this; we now examine the most straightforward approach, a technique called *direct proof.* As we begin, it is important to keep in mind the meanings of three key terms: *Theorem*, *proof* and *definition*.

A **theorem** is a mathematical statement that is true and can be (and has been) verified as true. A **proof** of a theorem is a written verification that shows that the theorem is definitely and unequivocally true. A proof should be understandable and convincing to anyone who has the requisite background and knowledge. This knowledge includes an understanding of the meanings of the mathematical words, phrases and symbols that occur in the theorem and its proof. It is crucial that both the writer of the proof and the readers of the proof agree on the exact meanings of all the words, for otherwise there is an intolerable level of ambiguity. A **definition** is an exact, unambiguous explanation of the meaning of a mathematical word or phrase. We will elaborate on the terms *theorem* and *definition* in the next two sections, and then finally we will be ready to begin writing proofs.

4.1 Theorems

A **theorem** is a statement that is true and has been proved to be true. You have encountered many theorems in your mathematical education. Here are some theorems taken from an undergraduate calculus text. They will be familiar to you, though you may not have read all the proofs.

Theorem: Let f be differentiable on an open interval I and let $c \in I$. If $f(c)$ is the maximum or minimum value of f on I, then $f'(c) = 0$.

Theorem: If $\sum_{k=1}^{\infty} a_k$ converges, then $\lim_{k \to \infty} a_k = 0$.

Theorem: Suppose f is continuous on the interval $[a,b]$. Then f is integrable on $[a,b]$.

Theorem: Every absolutely convergent series converges.

Observe that each of these theorems either has the conditional form *"If P, then Q,"* or can be put into that form. The first theorem has an initial sentence *"Let f be differentiable on an open interval I, and let $c \in I$,"* which sets up some notation, but a conditional statement follows it. The third theorem has form *"Suppose P. Then Q,"* but this means the same thing as *"If P, then Q."* The last theorem can be re-expressed as *"If a series is absolutely convergent, then it is convergent."*

A theorem of form *"If P, then Q,"* can be regarded as a device that produces new information from P. Whenever we are dealing with a situation in which P is true, then the theorem guarantees that, in addition, Q is true. Since this kind of expansion of information is useful, theorems of form *"If P, then Q,"* are very common.

But not *every* theorem is a conditional statement. Some have the form of the biconditional $P \Leftrightarrow Q$, but, as we know, that can be expressed as *two* conditional statements. Other theorems simply state facts about specific things. For example, here is another theorem from your study of calculus.

Theorem: The series $1+\frac{1}{2}+\frac{1}{3}+\frac{1}{4}+\frac{1}{5}+\cdots$ diverges.

It would be difficult (or at least awkward) to restate this as a conditional statement. Still, it is true that most theorems are conditional statements, so much of this book will concentrate on that type of theorem.

It is important to be aware that there are a number of words that mean essentially the same thing as the word "theorem," but are used in slightly different ways. In general the word "theorem" is reserved for a statement that is considered important or significant (the Pythagorean theorem, for example). A statement that is true but not as significant is sometimes called a **proposition**. A **lemma** is a theorem whose main purpose is to help prove another theorem. A **corollary** is a result that is an immediate consequence of a theorem or proposition. It is not important that you remember all these words now, for their meanings will become clear with usage.

Our main task is to learn how to prove theorems. As the above examples suggest, proving theorems requires a clear understanding of the meaning of the conditional statement, and that is the primary reason we studied it so extensively in Chapter 2. In addition, it is also crucial to understand the role of definitions.

4.2 Definitions

A proof of a theorem should be absolutely convincing. Ambiguity must be avoided. Everyone must agree on the exact meaning of each mathematical term. In Chapter 1 we defined the meanings of the sets $\mathbb{N}$, $\mathbb{Z}$, $\mathbb{R}$, $\mathbb{Q}$ and $\emptyset$, as well as the meanings of the symbols $\in$ and $\subseteq$, and we shall make frequent use of these things. Here is another definition that we use often.

Definition 4.1 An integer n is **even** if $n = 2a$ for some integer $a \in \mathbb{Z}$.

Thus, for example, 10 is even because $10 = 2 \cdot 5$. Also, according to the definition, 7 is not even because there is no integer a for which $7 = 2a$. While there would be nothing wrong with defining an integer to be odd if it's not even, the following definition is more concrete.

Definition 4.2 An integer n is **odd** if $n = 2a+1$ for some integer $a \in \mathbb{Z}$.

Thus 7 is odd because $7 = 2 \cdot 3 + 1$. We will use these definitions whenever the concept of even or odd numbers arises. If in a proof a certain number turns out to be even, the definition allows us to write it as $2a$ for an appropriate integer a. If some quantity has form $2b+1$ where b is an integer, then the definition tells us the quantity is odd.

Definition 4.3 Two integers have the **same parity** if they are both even or they are both odd. Otherwise they have **opposite parity**.

Thus 5 and -17 have the same parity, as do 8 and 0; but 3 and 4 have opposite parity.

Two points about definitions are in order. First, in this book the word or term being defined appears in boldface type. Second, it is common to express definitions as conditional statements even though the biconditional would more appropriately convey the meaning. Consider the definition of an even integer. You understand full well that if n is even then $n = 2a$ (for $a \in \mathbb{Z}$), *and* if $n = 2a$, then n is even. Thus, technically the definition should read *"An integer n is even if and only if $n = 2a$ for some $a \in \mathbb{Z}$."* However, it is an almost-universal convention that definitions are phrased in the conditional form, even though they are interpreted as being in the biconditional form. There is really no good reason for this, other than economy of words. It is the standard way of writing definitions, and we have to get used to it.

Here is another definition that we will use often.

Definition 4.4 Suppose a and b are integers. We say that a **divides** b, written $a \mid b$, if $b = ac$ for some $c \in \mathbb{Z}$. In this case we also say that a is a **divisor** of b, and that b is a **multiple** of a.

For example, 5 divides 15 because $15 = 5 \cdot 3$. We write this as $5 \mid 15$. Similarly $8 \mid 32$ because $32 = 8 \cdot 4$, and $-6 \mid 6$ because $6 = -6 \cdot -1$. However, 6 does not divide 9 because there is no integer c for which $9 = 6 \cdot c$. We express this as $6 \nmid 9$, which we read as "6 *does not divide* 9."

Be careful of your interpretation of the symbols. There is a big difference between the expressions $a \mid b$ and a/b. The expression $a \mid b$ is a *statement*, while a/b is a fraction. For example, $8 \mid 16$ is true and $8 \mid 20$ is false. By contrast, $8/16 = 0.5$ and $8/20 = 0.4$ are numbers, not statements. Be careful not to write one when you mean the other.

Every integer has a set of integers that divide it. For example, the set of divisors of 6 is $\{a \in \mathbb{Z} : a \mid 6\} = \{-6,-3,-2,-1,1,2,3,6\}$. The set of divisors of 5 is $\{-5,-1,1,5\}$. The set of divisors of 0 is $\mathbb{Z}$. This brings us to the following definition, with which you are already familiar.

Definition 4.5 A natural number n is **prime** if it has exactly two positive divisors, 1 and n.

For example, 2 is prime, as are 5 and 17. The definition implies that 1 is not prime, as it only has one (not two) positive divisor, namely 1. An integer n is **composite** if it factors as $n = ab$ where $a, b > 1$.

Definition 4.6 The **greatest common divisor** of integers a and b, denoted $\gcd(a,b)$, is the largest integer that divides both a and b. The **least common multiple** of non-zero integers a and b, denoted $\text{lcm}(a,b)$, is smallest positive integer that is a multiple of both a and b.

So $\gcd(18,24) = 6$, $\gcd(5,5) = 5$ and $\gcd(32,-8) = 8$. Also $\gcd(50,18) = 2$, but $\gcd(50,9) = 1$. Note that $\gcd(0,6) = 6$, because, although every integer divides 0, the largest divisor of 6 is 6.

The expression $\gcd(0,0)$ is problematic. Every integer divides 0, so the only conclusion is that $\gcd(0,0) = \infty$. We circumvent this irregularity by simply agreeing to consider $\gcd(a,b)$ only when a and b are not both zero.

Continuing our examples, $\text{lcm}(4,6) = 12$, and $\text{lcm}(7,7) = 7$.

Of course not all terms can be defined. If every word in a definition were defined, there would be separate definitions for the words that appeared in those definitions, and so on, until the chain of defined terms became circular. Thus we accept some ideas as being so intuitively clear that they require no definitions or verifications. For example, we will not find it necessary to define what an integer (or a real number) is. Nor will we define addition, multiplication, subtraction and division, though we will use these operations freely. We accept and use such things as the distributive and commutative properties of addition and multiplication, as well as other standard properties of arithmetic and algebra.

As mentioned in Section 1.9, we accept as fact the natural ordering of the elements of $\mathbb{N}, \mathbb{Z}, \mathbb{Q}$ and $\mathbb{R}$, so that (for example) statements such as "$5 < 7$," and "$x < y$ implies $-x > -y$," do not need to be justified.

In addition, we accept the following fact without justification or proof.

Fact 4.1 Suppose a and b are integers. Then:

- $a+b \in \mathbb{Z}$
- $a-b \in \mathbb{Z}$
- $ab \in \mathbb{Z}$

These three statements can be combined. For example, we see that if a, b and c are integers, then $a^2b - ca + b$ is also an integer.

We will also accept as obvious the fact that any integer a can be divided by a non-zero integer b, resulting in a unique quotient q and remainder r. For example, $b = 3$ goes into $a = 17$ $q = 5$ times with remainder $r = 2$. In symbols, $17 = 5 \cdot 3 + 2$, or $a = qb + r$. This fact, called the *division algorithm*, was mentioned on page 29.

(The Division Algorithm) Given integers a and b with $b > 0$, there exist unique integers q and r for which $a = qb + r$ and $0 \le r < b$.

Another fact that we will accept without proof (at least for now) is that every natural number greater than 1 has a unique factorization into primes. For example, the number 1176 can be factored into primes as $1176 = 2 \cdot 2 \cdot 2 \cdot 3 \cdot 7 \cdot 7 = 2^3 \cdot 3 \cdot 7^2$. By *unique* we mean that *any* factorization of 1176 into primes will have exactly the same factors (i.e., three 2's, one 3 and two 7's). Thus, for example, there is no valid factorization of 1176 that has a factor of 5. You may be so used to factoring numbers into primes that it seems obvious that there cannot be different prime factorizations of the same number, but in fact this is a fundamental result whose proof is not transparent. Nonetheless, we will be content to assume that every natural number greater than 1 has a unique factorization into primes. (We will revisit the issue of a proof in Section 10.2.)

We will introduce other accepted facts, as well as definitions, as needed.

4.3 Direct Proof

This section explains a simple way to prove theorems or propositions that have the form of conditional statements. The technique is called **direct proof**. To simplify the discussion, our first examples will involve proving statements that are almost obviously true. Thus we will call the statements *propositions* rather than theorems. (Remember, a proposition is a statement that, although true, is not as significant as a theorem.)

To understand how the technique of direct proof works, suppose we have some proposition of the following form.

Proposition If P, then Q.

This proposition is a conditional statement of form $P \Rightarrow Q$. Our goal is to show that this conditional statement is true. To see how to proceed, look at the truth table.

P	Q	$P \Rightarrow Q$
T	T	T
T	F	**F**
F	T	T
F	F	T

The table shows that if P is false, the statement $P \Rightarrow Q$ is automatically true. This means that if we are concerned with showing $P \Rightarrow Q$ is true, we don't have to worry about the situations where P is false (as in the last two lines of the table) because the statement $P \Rightarrow Q$ will be automatically true in those cases. But we must be very careful about the situations where P is true (as in the first two lines of the table). We must show that the condition of P being true forces Q to be true also, for that means the second line of the table cannot happen.

This gives a fundamental outline for proving statements of the form $P \Rightarrow Q$. Begin by assuming that P is true (remember, we don't need to worry about P being false) and show this forces Q to be true. We summarize this as follows.

Outline for Direct Proof

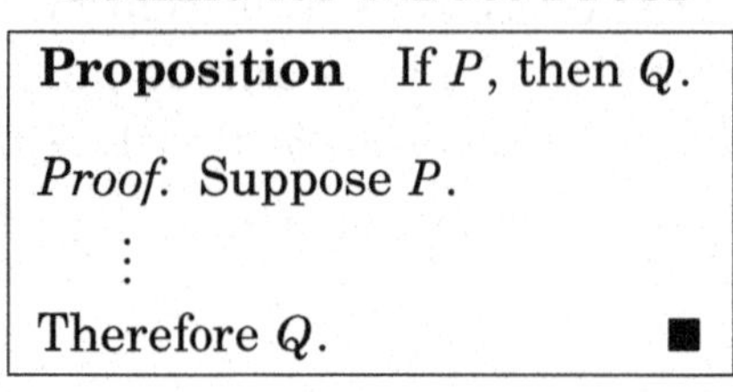

Proposition If P, then Q.

Proof. Suppose P.

$\vdots$

Therefore Q. ■

So the setup for direct proof is remarkably simple. The first line of the proof is the sentence *"Suppose P."* The last line is the sentence *"Therefore Q."* Between the first and last line we use logic, definitions and standard math facts to transform the statement P to the statement Q. It is common to use the word *"Proof"* to indicate the beginning of a proof, and the symbol ■ to indicate the end.

As our first example, let's prove that if x is odd then x^2 is also odd. (Granted, this is not a terribly impressive result, but we will move on to more significant things in due time.) The first step in the proof is to fill in the outline for direct proof. This is a lot like painting a picture, where the basic structure is sketched in first. We leave some space between the first and last line of the proof. The following series of frames indicates the steps you might take to fill in this space with a logical chain of reasoning.

Proposition If x is odd, then x^2 is odd.

Proof. Suppose x is odd.

Therefore x^2 is odd. ■

Now that we have written the first and last lines, we need to fill in the space with a chain of reasoning that shows that x being odd forces x^2 to be odd.

In doing this it's always advisable to use any definitions that apply. The first line says x is odd, and by Definition 4.2 it must be that $x = 2a+1$ for some $a \in \mathbb{Z}$, so we write this in as our second line.

Proposition If x is odd, then x^2 is odd.

Proof. Suppose x is odd.
Then $x = 2a+1$ for some $a \in \mathbb{Z}$, by definition of an odd number.

Therefore x^2 is odd. ■

Now jump down to the last line, which says x^2 is odd. Think about what the line immediately above it would have to be in order for us to conclude that x^2 is odd. By the definition of an odd number, we would have to have $x^2 = 2a+1$ for some $a \in \mathbb{Z}$. However, the symbol a now appears earlier in the proof in a different context, so we should use a different symbol, say b.

Proposition If x is odd, then x^2 is odd.

Proof. Suppose x is odd.
Then $x = 2a+1$ for some $a \in \mathbb{Z}$, by definition of an odd number.

Thus $x^2 = 2b+1$ *for an integer* b.
Therefore x^2 is odd, *by definition of an odd number.* ■

We are almost there. We can bridge the gap as follows.

Proposition If x is odd, then x^2 is odd.

Proof. Suppose x is odd.
Then $x = 2a+1$ for some $a \in \mathbb{Z}$, by definition of an odd number.
Thus $x^2 = (2a+1)^2 = 4a^2+4a+1 = 2(2a^2+2a)+1$.
So $x^2 = 2b+1$ *where* b *is the integer* $b = 2a^2+2a$.
Thus $x^2 = 2b+1$ for an integer b.
Therefore x^2 is odd, by definition of an odd number. ■

Finally, we may wish to clean up our work and write the proof in paragraph form. Here is our final version.

Proposition If x is odd, then x^2 is odd.

Proof. Suppose x is odd. Then $x = 2a+1$ for some $a \in \mathbb{Z}$, by definition of an odd number. Thus $x^2 = (2a+1)^2 = 4a^2+4a+1 = 2(2a^2+2a)+1$, so $x^2 = 2b+1$ where $b = 2a^2+2a \in \mathbb{Z}$. Therefore x^2 is odd, by definition of an odd number. ■

At least initially, it's generally a good idea to write the first and last line of your proof first, and then fill in the gap, sometimes jumping alternately between top and bottom until you meet in the middle, as we did above. This way you are constantly reminded that you are aiming for the statement at the bottom. Sometimes you will leave too much space, sometimes not enough. Sometimes you will get stuck before figuring out what to do. This is normal. Mathematicians do scratch work just as artists do sketches for their paintings.

Here is another example. Consider proving following proposition.

Proposition Let a, b and c be integers. If $a \mid b$ and $b \mid c$, then $a \mid c$.

Let's apply the basic outline for direct proof. To clarify the procedure we will write the proof in stages again.

> **Proposition** Let a, b and c be integers. If $a \mid b$ and $b \mid c$, then $a \mid c$.
>
> *Proof.* Suppose $a \mid b$ and $b \mid c$.
>
> Therefore $a \mid c$. ■

Our first step is to apply Definition 4.4 to the first line. The definition says $a \mid b$ means $b = ac$ for some integer c, but since c already appears in a different context on the first line, we must use a different letter, say d. Similarly let's use a new letter e in the definition of $b \mid c$.

> **Proposition** Let a, b and c be integers. If $a \mid b$ and $b \mid c$, then $a \mid c$.
>
> *Proof.* Suppose $a \mid b$ and $b \mid c$.
> *By Definition 4.4, we know $a \mid b$ means there is an integer d with $b = ad$.*
> *Likewise, $b \mid c$ means there is an integer e for which $c = be$.*
>
> Therefore $a \mid c$. ■

We have almost bridged the gap. The line immediately above the last line should show that $a \mid c$. According to Definition 4.4, this line should say that $c = ax$ for some integer x. We can get this equation from the lines at the top, as follows.

> **Proposition** Let a, b and c be integers. If $a \mid b$ and $b \mid c$, then $a \mid c$.
>
> *Proof.* Suppose $a \mid b$ and $b \mid c$.
> By Definition 4.4, we know $a \mid b$ means there is an integer d with $b = ad$.
> Likewise, $b \mid c$ means there is an integer e for which $c = be$.
> *Thus $c = be = (ad)e = a(de)$, so $c = ax$ for the integer $x = de$.*
> Therefore $a \mid c$. ■

The next example is presented all at once rather than in stages.

Proposition If x is an even integer, then x^2-6x+5 is odd.

Proof. Suppose x is an even integer.
Then $x = 2a$ for some $a \in \mathbb{Z}$, by definition of an even integer.
So $x^2-6x+5 = (2a)^2-6(2a)+5 = 4a^2-12a+5 = 4a^2-12a+4+1 = 2(2a^2-6a+2)+1$.
Therefore we have $x^2-6x+5 = 2b+1$, where $b = 2a^2-6a+2 \in \mathbb{Z}$.
Consequently x^2-6x+5 is odd, by definition of an odd number. ■

One doesn't normally use a separate line for each sentence in a proof, but for clarity we will often do this in the first few chapters of this book.

Our next example illustrates a standard technique for showing two quantities are equal. If we can show $m \le n$ and $n \le m$ then it follows that $m = n$. In general, the reasoning involved in showing $m \le n$ can be quite different from that of showing $n \le m$.

Recall Definition 4.6 of a least common multiple on page 90.

Proposition If $a, b, c \in \mathbb{N}$, then $\text{lcm}(ca, cb) = c \cdot \text{lcm}(a, b)$.

Proof. Assume $a, b, c \in \mathbb{N}$. Let $m = \text{lcm}(ca, cb)$ and $n = c \cdot \text{lcm}(a, b)$. We will show $m = n$. By definition, $\text{lcm}(a, b)$ is a multiple of both a and b, so $\text{lcm}(a, b) = ax = by$ for some $x, y \in \mathbb{Z}$. From this we see that $n = c \cdot \text{lcm}(a, b) = cax = cby$ is a multiple of both ca and cb. But $m = \text{lcm}(ca, cb)$ is the *smallest* multiple of both ca and cb. Thus $m \le n$.

On the other hand, as $m = \text{lcm}(ca, cb)$ is a multiple of both ca and cb, we have $m = cax = cby$ for some $x, y \in \mathbb{Z}$. Then $\frac{1}{c}m = ax = by$ is a multiple of both a and b. Therefore $\text{lcm}(a, b) \le \frac{1}{c}m$, so $c \cdot \text{lcm}(a, b) \le m$, that is, $n \le m$.

We've shown $m \le n$ and $n \le m$, so $m = n$. The proof is complete. ■

The examples we've looked at so far have all been proofs of statements about integers. In our next example, we are going to prove that if x and y are positive real numbers for which $x \le y$, then $\sqrt{x} \le \sqrt{y}$. You may feel that the proof is not as "automatic" as the proofs we have done so far. Finding the right steps in a proof can be challenging, and that is part of the fun.

Proposition Let x and y be positive numbers. If $x \le y$, then $\sqrt{x} \le \sqrt{y}$.

Proof. Suppose $x \le y$. Subtracting y from both sides gives $x - y \le 0$.
This can be written as $\sqrt{x}^2 - \sqrt{y}^2 \le 0$.
Factor this to get $(\sqrt{x} - \sqrt{y})(\sqrt{x} + \sqrt{y}) \le 0$.
Dividing both sides by the positive number $\sqrt{x} + \sqrt{y}$ produces $\sqrt{x} - \sqrt{y} \le 0$.
Adding $\sqrt{y}$ to both sides gives $\sqrt{x} \le \sqrt{y}$. ■

This proposition tells us that whenever $x \leq y$, we can take the square root of both sides and be assured that $\sqrt{x} \leq \sqrt{y}$. This can be useful, as we will see in our next proposition.

That proposition will concern the expression $2\sqrt{xy} \leq x+y$. Notice when you substitute random positive values for the variables, the expression is true. For example, for $x=6$ and $y=4$, the left side is $2\sqrt{6\cdot 4} = 4\sqrt{6} \approx 9.79$, which is less than the right side $6+4=10$. Is it true that $2\sqrt{xy} \leq x+y$ for any positive x and y? How could we prove it?

To see how, let's first cast this into the form of a conditional statement: If x and y are positive real numbers, then $2\sqrt{xy} \leq x+y$. The proof begins with the assumption that x and y are positive, and ends with $2\sqrt{xy} \leq x+y$. In mapping out a strategy, it can be helpful to work backwards, working from $2\sqrt{xy} \leq x+y$ to something that is obviously true. Then the steps can be reversed in the proof. In this case, squaring both sides of $2\sqrt{xy} \leq x+y$ gives us

$$4xy \leq x^2 + 2xy + y^2.$$

Now subtract $4xy$ from both sides and factor.

$$\begin{aligned} 0 &\leq x^2 - 2xy + y^2 \\ 0 &\leq (x-y)^2 \end{aligned}$$

But this last line is clearly true, since the square of $x-y$ cannot be negative! This gives us a strategy for the proof, which follows.

Proposition If x and y are positive real numbers, then $2\sqrt{xy} \leq x+y$.

Proof. Suppose x and y are positive real numbers.
Then $0 \leq (x-y)^2$, that is, $0 \leq x^2 - 2xy + y^2$.
Adding $4xy$ to both sides gives $4xy \leq x^2 + 2xy + y^2$.
Factoring yields $4xy \leq (x+y)^2$.
Previously we proved that such an inequality still holds after taking the square root of both sides; doing so produces $2\sqrt{xy} \leq x+y$. ■

Notice that in the last step of the proof we took the square root of both sides of $4xy \leq (x+y)^2$ and got $\sqrt{4xy} \leq \sqrt{(x+y)^2}$, and the fact that this did not reverse the symbol $\leq$ followed from our previous proposition. This is an important point. Often the proof of a proposition or theorem uses another proposition or theorem (that has already been proved).

4.4 Using Cases

In proving a statement is true, we sometimes have to examine multiple cases before showing the statement is true in all possible scenarios. This section illustrates a few examples.

Our examples will concern the expression $1+(-1)^n(2n-1)$. Here is a table showing its value for various integers for n. Notice that $1+(-1)^n(2n-1)$ is a multiple of 4 in every line.

n	$1+(-1)^n(2n-1)$
1	0
2	4
3	−4
4	8
5	−8
6	12

Is $1+(-1)^n(2n-1)$ always a multiple of 4? We prove the answer is "yes" in our next example. Notice, however, that the expression $1+(-1)^n(2n-1)$ behaves differently depending on whether n is even or odd, for in the first case $(-1)^n = 1$, and in the second $(-1)^n = -1$. Thus the proof must examine these two possibilities separately.

Proposition If $n \in \mathbb{N}$, then $1+(-1)^n(2n-1)$ is a multiple of 4.

Proof. Suppose $n \in \mathbb{N}$.
Then n is either even or odd. Let's consider these two cases separately.

Case 1. Suppose n is even. Then $n = 2k$ for some $k \in \mathbb{Z}$, and $(-1)^n = 1$. Thus $1+(-1)^n(2n-1) = 1+(1)(2\cdot 2k-1) = 4k$, which is a multiple of 4.

Case 2. Suppose n is odd. Then $n = 2k+1$ for some $k \in \mathbb{Z}$, and $(-1)^n = -1$. Thus $1+(-1)^n(2n-1) = 1-(2(2k+1)-1) = -4k$, which is a multiple of 4.

These cases show that $1+(-1)^n(2n-1)$ is always a multiple of 4. ■

Now let's examine the flip side of the question. We just proved that $1+(-1)^n(2n-1)$ is always a multiple of 4, but can we get *every* multiple of 4 this way? The following proposition and proof give an affirmative answer.

Proposition Every multiple of 4 equals $1+(-1)^n(2n-1)$ for some $n \in \mathbb{N}$.

Proof. In conditional form, the proposition is as follows:
If k is a multiple of 4, then there is an $n \in \mathbb{N}$ for which $1+(-1)^n(2n-1)=k$.
What follows is a proof of this conditional statement.
Suppose k is a multiple of 4.
This means $k=4a$ for some integer a.
We must produce an $n \in \mathbb{N}$ for which $1+(-1)^n(2n-1)=k$.
This is done by cases, depending on whether a is zero, positive or negative.
Case 1. Suppose $a=0$. Let $n=1$. Then $1+(-1)^n(2n-1)=1+(-1)^1(2-1)=0$ $=4\cdot 0=4a=k$.
Case 2. Suppose $a>0$. Let $n=2a$, which is in $\mathbb{N}$ because a is positive. Also n is even, so $(-1)^n=1$. Thus $1+(-1)^n(2n-1)=1+(2n-1)=2n=2(2a)=4a=k$.
Case 3. Suppose $a<0$. Let $n=1-2a$, which is an element of $\mathbb{N}$ because a is negative, making $1-2a$ positive. Also n is odd, so $(-1)^n=-1$. Thus $1+(-1)^n(2n-1)=1-(2n-1)=1-(2(1-2a)-1)=4a=k$.

The above cases show that no matter whether a multiple $k=4a$ of 4 is zero, positive or negative, $k=1+(-1)^n(2n-1)$ for some $n \in \mathbb{N}$. ■

4.5 Treating Similar Cases

Occasionally two or more cases in a proof will be so similar that writing them separately seems tedious or unnecessary. Here is an example.

Proposition If two integers have opposite parity, then their sum is odd.

Proof. Suppose m and n are two integers with opposite parity.
We need to show that $m+n$ is odd. This is done in two cases, as follows.
Case 1. Suppose m is even and n is odd. Thus $m=2a$ and $n=2b+1$ for some integers a and b. Therefore $m+n=2a+2b+1=2(a+b)+1$, which is odd (by Definition 4.2).
Case 2. Suppose m is odd and n is even. Thus $m=2a+1$ and $n=2b$ for some integers a and b. Therefore $m+n=2a+1+2b=2(a+b)+1$, which is odd (by Definition 4.2).

In either case, $m+n$ is odd. ■

The two cases in this proof are entirely alike except for the order in which the even and odd terms occur. It is entirely appropriate to just do one case and indicate that the other case is nearly identical. The phrase "*Without loss of generality...*" is a common way of signaling that the proof is treating just one of several nearly identical cases. Here is a second version of the above example.

Proposition If two integers have opposite parity, then their sum is odd.

Proof. Suppose m and n are two integers with opposite parity.
We need to show that $m+n$ is odd.
Without loss of generality, suppose m is even and n is odd.
Thus $m = 2a$ and $n = 2b+1$ for some integers a and b.
Therefore $m+n = 2a+2b+1 = 2(a+b)+1$, which is odd (by Definition 4.2). ■

In reading proofs in other texts, you may sometimes see the phrase "Without loss of generality" abbreviated as "WLOG." However, in the interest of transparency we will avoid writing it this way. In a similar spirit, it is advisable—at least until you become more experienced in proof writing—that you write out all cases, no matter how similar they appear to be.

Please check your understanding by doing the following exercises. The odd numbered problems have complete proofs in the Solutions section in the back of the text.

Exercises for Chapter 4

Use the method of direct proof to prove the following statements.

1. If x is an even integer, then x^2 is even.
2. If x is an odd integer, then x^3 is odd.
3. If a is an odd integer, then a^2+3a+5 is odd.
4. Suppose $x,y \in \mathbb{Z}$. If x and y are odd, then xy is odd.
5. Suppose $x,y \in \mathbb{Z}$. If x is even, then xy is even.
6. Suppose $a,b,c \in \mathbb{Z}$. If $a \mid b$ and $a \mid c$, then $a \mid (b+c)$.
7. Suppose $a,b \in \mathbb{Z}$. If $a \mid b$, then $a^2 \mid b^2$.
8. Suppose a is an integer. If $5 \mid 2a$, then $5 \mid a$.
9. Suppose a is an integer. If $7 \mid 4a$, then $7 \mid a$.
10. Suppose a and b are integers. If $a \mid b$, then $a \mid (3b^3 - b^2 + 5b)$.
11. Suppose $a,b,c,d \in \mathbb{Z}$. If $a \mid b$ and $c \mid d$, then $ac \mid bd$.
12. If $x \in \mathbb{R}$ and $0 < x < 4$, then $\frac{4}{x(4-x)} \geq 1$.
13. Suppose $x,y \in \mathbb{R}$. If $x^2+5y = y^2+5x$, then $x = y$ or $x+y = 5$.
14. If $n \in \mathbb{Z}$, then $5n^2+3n+7$ is odd. (Try cases.)
15. If $n \in \mathbb{Z}$, then n^2+3n+4 is even. (Try cases.)
16. If two integers have the same parity, then their sum is even. (Try cases.)
17. If two integers have opposite parity, then their product is even.
18. Suppose x and y are positive real numbers. If $x < y$, then $x^2 < y^2$.

19. Suppose a, b and c are integers. If $a^2 \mid b$ and $b^3 \mid c$, then $a^6 \mid c$.

20. If a is an integer and $a^2 \mid a$, then $a \in \{-1, 0, 1\}$.

21. If p is prime and k is an integer for which $0 < k < p$, then p divides $\binom{p}{k}$.

22. If $n \in \mathbb{N}$, then $n^2 = 2\binom{n}{2} + \binom{n}{1}$. (You may need a separate case for $n = 1$.)

23. If $n \in \mathbb{N}$, then $\binom{2n}{n}$ is even.

24. If $n \in \mathbb{N}$ and $n \geq 2$, then the numbers $n!+2$, $n!+3$, $n!+4$, $n!+5$, $\ldots$, $n!+n$ are all composite. (Thus for any $n \geq 2$, one can find n consecutive composite numbers. This means there are arbitrarily large "gaps" between prime numbers.)

25. If $a, b, c \in \mathbb{N}$ and $c \leq b \leq a$, then $\binom{a}{b}\binom{b}{c} = \binom{a}{b-c}\binom{a-b+c}{c}$.

26. Every odd integer is a difference of two squares. (Example $7 = 4^2 - 3^2$, etc.)

27. Suppose $a, b \in \mathbb{N}$. If $\gcd(a, b) > 1$, then $b \mid a$ or b is not prime.

28. If $a, b, c \in \mathbb{Z}$, then $c \cdot \gcd(a, b) \leq \gcd(ca, cb)$.

CHAPTER 5

Contrapositive Proof

We now examine an alternative to direct proof called **contrapositive proof**. Like direct proof, the technique of contrapositive proof is used to prove conditional statements of the form *"If P, then Q."* Although it is possible to use direct proof exclusively, there are occasions where contrapositive proof is much easier.

5.1 Contrapositive Proof

To understand how contrapositive proof works, imagine that you need to prove a proposition of the following form.

Proposition If P, then Q.

This is a conditional statement of form $P \Rightarrow Q$. Our goal is to show that this conditional statement is true. Recall that in Section 2.6 we observed that $P \Rightarrow Q$ is logically equivalent to $\sim Q \Rightarrow \sim P$. For convenience, we duplicate the truth table that verifies this fact.

P	Q	$\sim Q$	$\sim P$	$P \Rightarrow Q$	$\sim Q \Rightarrow \sim P$
T	T	F	F	**T**	**T**
T	F	T	F	**F**	**F**
F	T	F	T	**T**	**T**
F	F	T	T	**T**	**T**

According to the table, statements $P \Rightarrow Q$ and $\sim Q \Rightarrow \sim P$ are different ways of expressing exactly the same thing. The expression $\sim Q \Rightarrow \sim P$ is called the **contrapositive form** of $P \Rightarrow Q$.[1]

[1] Do not confuse the words *contrapositive* and *converse*. Recall from Section 2.4 that the *converse* of $P \Rightarrow Q$ is the statement $Q \Rightarrow P$, which is not logically equivalent to $P \Rightarrow Q$.

Since $P \Rightarrow Q$ is logically equivalent to $\sim Q \Rightarrow \sim P$, it follows that to prove $P \Rightarrow Q$ is true, it suffices to instead prove that $\sim Q \Rightarrow \sim P$ is true. If we were to use direct proof to show $\sim Q \Rightarrow \sim P$ is true, we would assume $\sim Q$ is true use this to deduce that $\sim P$ is true. This in fact is the basic approach of contrapositive proof, summarized as follows.

Outline for Contrapositive Proof

Proposition If P, then Q.
Proof. Suppose $\sim Q$.
$\vdots$
Therefore $\sim P$. ■

So the setup for contrapositive proof is very simple. The first line of the proof is the sentence *"Suppose Q is not true."* (Or something to that effect.) The last line is the sentence *"Therefore P is not true."* Between the first and last line we use logic and definitions to transform the statement $\sim Q$ to the statement $\sim P$.

To illustrate this new technique, and to contrast it with direct proof, we now prove a proposition in two ways: first with direct proof and then with contrapositive proof.

Proposition Suppose $x \in \mathbb{Z}$. If $7x+9$ is even, then x is odd.

Proof. (Direct) Suppose $7x+9$ is even.
Thus $7x+9 = 2a$ for some integer a.
Subtracting $6x+9$ from both sides, we get $x = 2a-6x-9$.
Thus $x = 2a-6x-9 = 2a-6x-10+1 = 2(a-3x-5)+1$.
Consequently $x = 2b+1$, where $b = a-3x-5 \in \mathbb{Z}$.
Therefore x is odd. ■

Here is a contrapositive proof of the same statement:

Proposition Suppose $x \in \mathbb{Z}$. If $7x+9$ is even, then x is odd.

Proof. (Contrapositive) Suppose x is not odd.
Thus x is even, so $x = 2a$ for some integer a.
Then $7x+9 = 7(2a)+9 = 14a+8+1 = 2(7a+4)+1$.
Therefore $7x+9 = 2b+1$, where b is the integer $7a+4$.
Consequently $7x+9$ is odd.
Therefore $7x+9$ is not even. ■

Though the proofs are of equal length, you may feel that the contrapositive proof flowed more smoothly. This is because it is easier to transform information about x into information about $7x+9$ than the other way around. For our next example, consider the following proposition concerning an integer x:

Proposition If x^2-6x+5 is even, then x is odd.

A direct proof would be problematic. We would begin by assuming that x^2-6x+5 is even, so $x^2-6x+5=2a$. Then we would need to transform this into $x=2b+1$ for $b \in \mathbb{Z}$. But it is not quite clear how that could be done, for it would involve isolating an x from the quadratic expression. However the proof becomes very simple if we use contrapositive proof.

Proposition Suppose $x \in \mathbb{Z}$. If x^2-6x+5 is even, then x is odd.

Proof. (Contrapositive) Suppose x is not odd.
Thus x is even, so $x=2a$ for some integer a.
So $x^2-6x+5=(2a)^2-6(2a)+5=4a^2-12a+5=4a^2-12a+4+1=2(2a^2-6a+2)+1$.
Therefore $x^2-6x+5=2b+1$, where b is the integer $2a^2-6a+2$.
Consequently x^2-6x+5 is odd.
Therefore x^2-6x+5 is not even. ■

In summary, since x being not odd ($\sim Q$) resulted in x^2-6x+5 being not even ($\sim P$), then x^2-6x+5 being even (P) means that x is odd (Q). Thus we have proved $P \Rightarrow Q$ by proving $\sim Q \Rightarrow \sim P$. Here is another example:

Proposition Suppose $x, y \in \mathbb{R}$. If $y^3+yx^2 \leq x^3+xy^2$, then $y \leq x$.

Proof. (Contrapositive) Suppose it is not true that $y \leq x$, so $y > x$.
Then $y-x>0$. Multiply both sides of $y-x>0$ by the positive value x^2+y^2.

$$
\begin{aligned}
(y-x)(x^2+y^2) &> 0(x^2+y^2)\\
yx^2+y^3-x^3-xy^2 &> 0\\
y^3+yx^2 &> x^3+xy^2
\end{aligned}
$$

Therefore $y^3+yx^2 > x^3+xy^2$, so it is not true that $y^3+yx^2 \leq x^3+xy^2$. ■

Proving *"If P, then Q,"* with the contrapositive approach necessarily involves the negated statements $\sim P$ and $\sim Q$. In working with these we may have to use the techniques for negating statements (e.g., DeMorgan's laws) discussed in Section 2.10. We consider such an example next.

Proposition Suppose $x, y \in \mathbb{Z}$. If $5 \nmid xy$, then $5 \nmid x$ and $5 \nmid y$.

Proof. (Contrapositive) Suppose it is not true that $5 \nmid x$ **and** $5 \nmid y$.
By DeMorgan's law, it is not true that $5 \nmid x$ **or** it is not true that $5 \nmid y$.
Therefore $5 \mid x$ or $5 \mid y$. We consider these possibilities separately.
Case 1. Suppose $5 \mid x$. Then $x = 5a$ for some $a \in \mathbb{Z}$.
From this we get $xy = 5(ay)$, and that means $5 \mid xy$.
Case 2. Suppose $5 \mid y$. Then $y = 5a$ for some $a \in \mathbb{Z}$.
From this we get $xy = 5(ax)$, and that means $5 \mid xy$.
The above cases show that $5 \mid xy$, so it is not true that $5 \nmid xy$. ■

5.2 Congruence of Integers

This is a good time to introduce a new definition. It is not necessarily related to contrapositive proof, but introducing it now ensures that we have a sufficient variety of exercises to practice all our proof techniques on. This new definition occurs in many branches of mathematics, and it will surely play a role in some of your later courses. But our primary reason for introducing it is that it will give us more practice in writing proofs.

Definition 5.1 Given integers a and b and an $n \in \mathbb{N}$, we say that a and b are **congruent modulo n** if $n \mid (a-b)$. We express this as $a \equiv b \pmod{n}$. If a and b are not congruent modulo n, we write this as $a \not\equiv b \pmod{n}$.

Example 5.1 Here are some examples:

1. $9 \equiv 1 \pmod{4}$ because $4 \mid (9-1)$.
2. $6 \equiv 10 \pmod{4}$ because $4 \mid (6-10)$.
3. $14 \not\equiv 8 \pmod{4}$ because $4 \nmid (14-8)$.
4. $20 \equiv 4 \pmod{8}$ because $8 \mid (20-4)$.
5. $17 \equiv -4 \pmod{3}$ because $3 \mid (17-(-4))$.

In practical terms, $a \equiv b \pmod{n}$ means that a and b have the same remainder when divided by n. For example, we saw above that $6 \equiv 10 \pmod{4}$ and indeed 6 and 10 both have remainder 2 when divided by 4. Also we saw $14 \not\equiv 8 \pmod{4}$, and sure enough 14 has remainder 2 when divided by 4, while 8 has remainder 0.

To see that this is true in general, note that if a and b both have the same remainder r when divided by n, then it follows that $a = kn + r$ and $b = \ell n + r$ for some $k, \ell \in \mathbb{Z}$. Then $a - b = (kn+r)-(\ell n+r) = n(k-\ell)$. But $a - b = n(k-\ell)$ means $n \mid (a-b)$, so $a \equiv b \pmod{n}$. Conversely, one of the exercises for this chapter asks you to show that if $a \equiv b \pmod{n}$, then a and b have the same remainder when divided by n.

We conclude this section with several proofs involving congruence of integers, but you will also test your skills with other proofs in the exercises.

Proposition Let $a, b \in \mathbb{Z}$ and $n \in \mathbb{N}$. If $a \equiv b \pmod{n}$, then $a^2 \equiv b^2 \pmod{n}$.

Proof. We will use direct proof. Suppose $a \equiv b \pmod{n}$.
By definition of congruence of integers, this means $n \mid (a-b)$.
Then by definition of divisibility, there is an integer c for which $a-b = nc$.
Now multiply both sides of this equation by $a+b$.

$$\begin{aligned} a-b &= nc \\ (a-b)(a+b) &= nc(a+b) \\ a^2-b^2 &= nc(a+b) \end{aligned}$$

Since $c(a+b) \in \mathbb{Z}$, the above equation tells us $n \mid (a^2-b^2)$.
According to Definition 5.1, this gives $a^2 \equiv b^2 \pmod{n}$. ■

Let's pause to consider this proposition's meaning. It says $a \equiv b \pmod{n}$ implies $a^2 \equiv b^2 \pmod{n}$. In other words, it says that if integers a and b have the same remainder when divided by n, then a^2 and b^2 also have the same remainder when divided by n. As an example of this, 6 and 10 have the same remainder (2) when divided by $n = 4$, and their squares 36 and 100 also have the same remainder (0) when divided by $n = 4$. The proposition promises this will happen for all a, b and n. In our examples we tend to concentrate more on how to prove propositions than on what the propositions mean. This is reasonable since our main goal is to learn how to prove statements. But it is helpful to sometimes also think about the meaning of what we prove.

Proposition Let $a, b, c \in \mathbb{Z}$ and $n \in \mathbb{N}$. If $a \equiv b \pmod{n}$, then $ac \equiv bc \pmod{n}$.

Proof. We employ direct proof. Suppose $a \equiv b \pmod{n}$. By Definition 5.1, it follows that $n \mid (a-b)$. Therefore, by definition of divisibility, there exists an integer k for which $a-b = nk$. Multiply both sides of this equation by c to get $ac-bc = nkc$. Thus $ac-bc = n(kc)$ where $kc \in \mathbb{Z}$, which means $n \mid (ac-bc)$. By Definition 5.1, we have $ac \equiv bc \pmod{n}$. ■

Contrapositive proof seems to be the best approach in the next example, since it will eliminate the symbols $\nmid$ and $\not\equiv$.

Proposition Suppose $a, b \in \mathbb{Z}$ and $n \in \mathbb{N}$. If $12a \not\equiv 12b \pmod{n}$, then $n \nmid 12$.

Proof. (Contrapositive) Suppose $n \mid 12$, so there is an integer c for which $12 = nc$. Now reason as follows.

$$\begin{aligned} 12 &= nc \\ 12(a-b) &= nc(a-b) \\ 12a - 12b &= n(ca - cb) \end{aligned}$$

Since $ca - cb \in \mathbb{Z}$, the equation $12a - 12b = n(ca - cb)$ implies $n \mid (12a - 12b)$. This in turn means $12a \equiv 12b \pmod{n}$. ■

5.3 Mathematical Writing

Now that we have begun writing proofs, it is a good time to contemplate the craft of writing. Unlike logic and mathematics, where there is a clear-cut distinction between what is right or wrong, the difference between good and bad writing is sometimes a matter of opinion. But there are some standard guidelines that will make your writing clearer. Some of these are listed below.

1. **Begin each sentence with a word, not a mathematical symbol.** The reason is that sentences begin with capital letters, but mathematical symbols are case sensitive. Because x and X can have entirely different meanings, putting such symbols at the beginning of a sentence can lead to ambiguity. Here are some examples of bad usage (marked with ×) and good usage (marked with ✓).

   A is a subset of B. ×
   The set A is a subset of B. ✓

   x is an integer, so $2x+5$ is an integer. ×
   Because x is an integer, $2x+5$ is an integer. ✓

   $x^2 - x + 2 = 0$ has two solutions. ×
   $X^2 - x + 2 = 0$ has two solutions. × (and silly too)
   The equation $x^2 - x + 2 = 0$ has two solutions. ✓

2. **End each sentence with a period,** even when the sentence ends with a mathematical symbol or expression.

   Euler proved that $\sum_{k=1}^{\infty} \frac{1}{k^s} = \prod_{p \in P} \frac{1}{1 - \frac{1}{p^s}}$ ×

   Euler proved that $\sum_{k=1}^{\infty} \frac{1}{k^s} = \prod_{p \in P} \frac{1}{1 - \frac{1}{p^s}}$. ✓

Mathematical statements (equations, etc.) are like English phrases that happen to contain special symbols, so use normal punctuation.

3. **Separate mathematical symbols and expressions with words.** Not doing this can cause confusion by making distinct expressions appear to merge into one. Compare the clarity of the following examples.

   Because $x^2 - 1 = 0$, $x = 1$ or $x = -1$. ×
   Because $x^2 - 1 = 0$, it follows that $x = 1$ or $x = -1$. ✓

   Unlike $A \cup B$, $A \cap B$ equals $\emptyset$. ×
   Unlike $A \cup B$, the set $A \cap B$ equals $\emptyset$. ✓

4. **Avoid misuse of symbols.** Symbols such as $=$, $\leq$, $\subseteq$, $\in$, etc., are not words. While it is appropriate to use them in mathematical expressions, they are out of place in other contexts.

   Since the two sets are $=$, one is a subset of the other. ×
   Since the two sets are equal, one is a subset of the other. ✓

   The empty set is a $\subseteq$ of every set. ×
   The empty set is a subset of every set. ✓

   Since a is odd and x odd $\Rightarrow x^2$ odd, a^2 is odd. ×
   Since a is odd and any odd number squared is odd, then a^2 is odd. ✓

5. **Avoid using unnecessary symbols.** Mathematics is confusing enough without them. Don't muddy the water even more.

   No set X has negative cardinality. ×
   No set has negative cardinality. ✓

6. **Use the first person plural.** In mathematical writing, it is common to use the words "we" and "us" rather than "I," "you" or "me." It is as if the reader and writer are having a conversation, with the writer guiding the reader through the details of the proof.

7. **Use the active voice.** This is just a suggestion, but the active voice makes your writing more lively.

   The value $x = 3$ is obtained through the division of both sides by 5. ×
   Dividing both sides by 5, we get the value $x = 3$. ✓

8. **Explain each new symbol.** In writing a proof, you must explain the meaning of every new symbol you introduce. Failure to do this can lead to ambiguity, misunderstanding and mistakes. For example, consider the following two possibilities for a sentence in a proof, where a and b have been introduced on a previous line.

Since $a \mid b$, it follows that $b = ac$. ×
Since $a \mid b$, it follows that $b = ac$ for some integer c. ✓

If you use the first form, then a reader who has been carefully following your proof may momentarily scan backwards looking for where the c entered into the picture, not realizing at first that it came from the definition of divides.

9. **Watch out for "it."** The pronoun "it" can cause confusion when it is unclear what it refers to. If there is any possibility of confusion, you should avoid the word "it." Here is an example:

   Since $X \subseteq Y$, and $0 < |X|$, we see that it is not empty. ×

   Is "it" X or Y? Either one would make sense, but which do we mean?

   Since $X \subseteq Y$, and $0 < |X|$, we see that Y is not empty. ✓

10. **Since, because, as, for, so.** In proofs, it is common to use these words as conjunctions joining two statements, and meaning that one statement is true and as a consequence the other true. The following statements all mean that P is true (or assumed to be true) and as a consequence Q is true also.

Q since P	Q because P	Q, as P	Q, for P	P, so Q
Since P, Q	Because P, Q	as P, Q		

    Notice that the meaning of these constructions is different from that of *"If P, then Q,"* for they are asserting not only that P implies Q, but **also** that P is true. Exercise care in using them. It must be the case that P and Q are both statements **and** that Q really does follow from P.

    $x \in \mathbb{N}$, so $\mathbb{Z}$ ×
    $x \in \mathbb{N}$, so $x \in \mathbb{Z}$ ✓

11. **Thus, hence, therefore consequently.** These adverbs precede a statement that follows logically from previous sentences or clauses. Be sure that a statement follows them.

    Therefore $2k + 1$. ×
    Therefore $a = 2k + 1$. ✓

12. **Clarity is the gold standard of mathematical writing.** If you believe breaking a rule makes your writing clearer, then break the rule.

Your mathematical writing will evolve with practice useage. One of the best ways to develop a good mathematical writing style is to read other people's proofs. Adopt what works and avoid what doesn't.

Exercises for Chapter 5

A. Use the method of contrapositive proof to prove the following statements. (In each case you should also think about how a direct proof would work. You will find in most cases that contrapositive is easier.)

1. Suppose $n \in \mathbb{Z}$. If n^2 is even, then n is even.

2. Suppose $n \in \mathbb{Z}$. If n^2 is odd, then n is odd.

3. Suppose $a,b \in \mathbb{Z}$. If $a^2(b^2-2b)$ is odd, then a and b are odd.

4. Suppose $a,b,c \in \mathbb{Z}$. If a does not divide bc, then a does not divide b.

5. Suppose $x \in \mathbb{R}$. If $x^2+5x<0$ then $x<0$.

6. Suppose $x \in \mathbb{R}$. If $x^3-x>0$ then $x>-1$.

7. Suppose $a,b \in \mathbb{Z}$. If both ab and $a+b$ are even, then both a and b are even.

8. Suppose $x \in \mathbb{R}$. If $x^5-4x^4+3x^3-x^2+3x-4 \geq 0$, then $x \geq 0$.

9. Suppose $n \in \mathbb{Z}$. If $3 \nmid n^2$, then $3 \nmid n$.

10. Suppose $x,y,z \in \mathbb{Z}$ and $x \neq 0$. If $x \nmid yz$, then $x \nmid y$ and $x \nmid z$.

11. Suppose $x,y \in \mathbb{Z}$. If $x^2(y+3)$ is even, then x is even or y is odd.

12. Suppose $a \in \mathbb{Z}$. If a^2 is not divisible by 4, then a is odd.

13. Suppose $x \in \mathbb{R}$. If $x^5+7x^3+5x \geq x^4+x^2+8$, then $x \geq 0$.

B. Prove the following statements using either direct or contrapositive proof. Sometimes one approach will be much easier than the other.

14. If $a,b \in \mathbb{Z}$ and a and b have the same parity, then $3a+7$ and $7b-4$ do not.

15. Suppose $x \in \mathbb{Z}$. If x^3-1 is even, then x is odd.

16. Suppose $x \in \mathbb{Z}$. If $x+y$ is even, then x and y have the same parity.

17. If n is odd, then $8 \mid (n^2-1)$.

18. For any $a,b \in \mathbb{Z}$, it follows that $(a+b)^3 \equiv a^3+b^3 \pmod 3$.

19. Let $a,b \in \mathbb{Z}$ and $n \in \mathbb{N}$. If $a \equiv b \pmod n$ and $a \equiv c \pmod n$, then $c \equiv b \pmod n$.

20. If $a \in \mathbb{Z}$ and $a \equiv 1 \pmod 5$, then $a^2 \equiv 1 \pmod 5$.

21. Let $a,b \in \mathbb{Z}$ and $n \in \mathbb{N}$. If $a \equiv b \pmod n$, then $a^3 \equiv b^3 \pmod n$

22. Let $a \in \mathbb{Z}$, $n \in \mathbb{N}$. If a has remainder r when divided by n, then $a \equiv r \pmod n$.

23. Let $a,b,c \in \mathbb{Z}$ and $n \in \mathbb{N}$. If $a \equiv b \pmod n$, then $ca \equiv cb \pmod n$.

24. If $a \equiv b \pmod n$ and $c \equiv d \pmod n$, then $ac \equiv bd \pmod n$.

25. If $n \in \mathbb{N}$ and 2^n-1 is prime, then n is prime.

26. If $n = 2^k-1$ for $k \in \mathbb{N}$, then every entry in Row n of Pascal's Triangle is odd.

27. If $a \equiv 0 \pmod 4$ or $a \equiv 1 \pmod 4$, then $\binom{a}{2}$ is even.

28. If $n \in \mathbb{Z}$, then $4 \nmid (n^2-3)$.

29. If integers a and b are not both zero, then $\gcd(a,b) = \gcd(a-b,b)$.

30. If $a \equiv b \pmod n$, then $\gcd(a,n) = \gcd(b,n)$.

31. Suppose the division algorithm applied to a and b yields $a = qb+r$. Then $\gcd(a,b) = \gcd(r,b)$.

CHAPTER 6

Proof by Contradiction

We now explore a third method of proof: **proof by contradiction**. This method is not limited to proving just conditional statements—it can be used to prove any kind of statement whatsoever. The basic idea is to assume that the statement we want to prove is *false*, and then show that this assumption leads to nonsense. We are then led to conclude that we were wrong to assume the statement was false, so the statement must be true. As an example, consider the following proposition and its proof.

Proposition If $a,b \in \mathbb{Z}$, then $a^2 - 4b \neq 2$.

Proof. Suppose this proposition is *false*.
This conditional statement being false means there exist numbers a and b for which $a,b \in \mathbb{Z}$ is true, but $a^2 - 4b \neq 2$ is false.
In other words, there exist integers $a,b \in \mathbb{Z}$ for which $\boxed{a^2 - 4b = 2.}$
From this equation we get $a^2 = 4b + 2 = 2(2b+1)$, so a^2 is even.
Because a^2 is even, it follows that a is even, so $a = 2c$ for some integer c.
Now plug $a = 2c$ back into the boxed equation to get $(2c)^2 - 4b = 2$,
so $4c^2 - 4b = 2$. Dividing by 2, we get $2c^2 - 2b = 1$.
Therefore $1 = 2(c^2 - b)$, and because $c^2 - b \in \mathbb{Z}$, it follows that 1 is even.
We know 1 is **not** even, so something went wrong.
But all the logic after the first line of the proof is correct, so it must be that the first line was incorrect. In other words, we were wrong to assume the proposition was false. Thus the proposition is true. ■

You may be a bit suspicious of this line of reasoning, but in the next section we will see that it is logically sound. For now, notice that at the end of the proof we deduced that 1 is even, which conflicts with our knowledge that 1 is odd. In essence, we have obtained the statement (1 is odd)$\wedge \sim$(1 is odd), which has the form $C \wedge \sim C$. Notice that no matter what statement C is, and whether or not it is true, the statement $C \wedge \sim C$ is false. A statement—like this one—that cannot be true is called a **contradiction**. Contradictions play a key role in our new technique.

6.1 Proving Statements with Contradiction

Let's now see why the proof on the previous page is logically valid. In that proof we needed to show that a statement $P : (a, b \in \mathbb{Z}) \Rightarrow (a^2 - 4b \neq 2)$ was true. The proof began with the assumption that P was false, that is that $\sim P$ was true, and from this we deduced $C \wedge \sim C$. In other words we proved that $\sim P$ being true forces $C \wedge \sim C$ to be true, and this means that we proved that the *conditional* statement $(\sim P) \Rightarrow (C \wedge \sim C)$ is true. To see that this is the same as proving P is true, look at the following truth table for $(\sim P) \Rightarrow (C \wedge \sim C)$. Notice that the columns for P and $(\sim P) \Rightarrow (C \wedge \sim C)$ are exactly the same, so P is logically equivalent to $(\sim P) \Rightarrow (C \wedge \sim C)$.

P	C	$\sim P$	$C \wedge \sim C$	$(\sim P) \Rightarrow (C \wedge \sim C)$
T	T	F	F	**T**
T	F	F	F	**T**
F	T	T	F	**F**
F	F	T	F	**F**

Therefore to prove a statement P, it suffices to instead prove the conditional statement $(\sim P) \Rightarrow (C \wedge \sim C)$. This can be done with direct proof: Assume $\sim P$ and deduce $C \wedge \sim C$. Here is the outline:

Outline for Proof by Contradiction

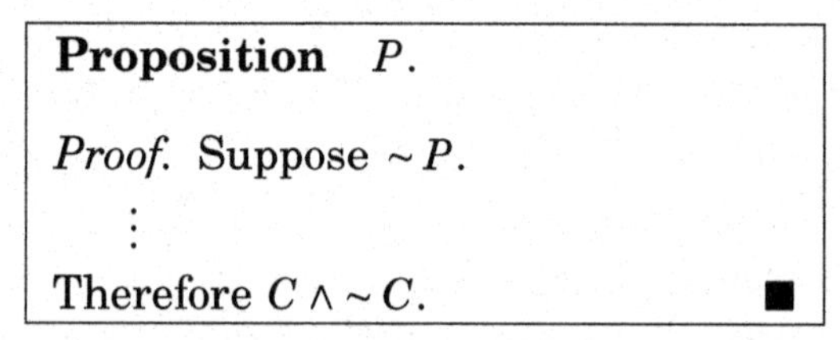

Proposition P.

Proof. Suppose $\sim P$.

$\vdots$

Therefore $C \wedge \sim C$. ■

One slightly unsettling feature of this method is that we may not know at the beginning of the proof what the statement C is going to be. In doing the scratch work for the proof, you assume that $\sim P$ is true, then deduce new statements until you have deduced some statement C *and* its negation $\sim C$.

If this method seems confusing, look at it this way. In the first line of the proof we suppose $\sim P$ is true, that is we assume P is *false*. But if P is really true then this contradicts our assumption that P is false. But we haven't yet *proved* P to be true, so the contradiction is not obvious. We use logic and reasoning to transform the non-obvious contradiction $\sim P$ to an obvious contradiction $C \wedge \sim C$.

The idea of proof by contradiction is quite ancient, and goes back at least as far as the Pythagoreans, who used it to prove that certain numbers are irrational. Our next example follows their logic to prove that $\sqrt{2}$ is irrational. Recall that a number is rational if it equals a fraction of two integers, and it is irrational if it cannot be expressed as a fraction of two integers. Here is the exact definition.

Definition 6.1 A real number x is **rational** if $x = \frac{a}{b}$ for some $a, b \in \mathbb{Z}$. Also, x is **irrational** if it is not rational, that is if $x \neq \frac{a}{b}$ for every $a, b \in \mathbb{Z}$.

We are now ready to use contradiction to prove that $\sqrt{2}$ is irrational. According to the outline, the first line of the proof should be "Suppose that it is not true that $\sqrt{2}$ is irrational." But it is helpful (though not mandatory) to tip our reader off to the fact that we are using proof by contradiction. One standard way of doing this is to make the first line "*Suppose for the sake of contradiction that it is not true that $\sqrt{2}$ is irrational.*"

Proposition The number $\sqrt{2}$ is irrational.

Proof. Suppose for the sake of contradiction that it is not true that $\sqrt{2}$ is irrational. Then $\sqrt{2}$ is rational, so there are integers a and b for which

$$\sqrt{2} = \frac{a}{b}. \tag{6.1}$$

Let this fraction be fully reduced; in particular, this means that a and b are not both even. (If they were both even, the fraction could be further reduced by factoring 2's from the numerator and denominator and canceling.) Squaring both sides of Equation 6.1 gives $2 = \frac{a^2}{b^2}$, and therefore

$$a^2 = 2b^2. \tag{6.2}$$

From this it follows that a^2 is even. But we proved earlier (Exercise 1 on page 110) that a^2 being even implies a is even. Thus, as we know that a and b are not both even, it follows that b **is odd.** Now, since a is even there is an integer c for which $a = 2c$. Plugging this value for a into Equation (6.2), we get $(2c)^2 = 2b^2$, so $4c^2 = 2b^2$, and hence $b^2 = 2c^2$. This means b^2 is even, so b is even also. But previously we deduced that b is odd. Thus we have the contradiction b is even **and** b is odd. ■

To appreciate the power of proof by contradiction, imagine trying to prove that $\sqrt{2}$ is irrational without it. Where would we begin? What would be our initial assumption? There are no clear answers to these questions.

Proof by contradiction gives us a starting point: Assume $\sqrt{2}$ is rational, and work from there.

In the above proof we got the contradiction (b is even) $\wedge \sim$(b is even) which has the form $C \wedge \sim C$. In general, your contradiction need not necessarily be of this form. Any statement that is clearly false is sufficient. For example $2 \neq 2$ would be a fine contradiction, as would be $4 \mid 2$, provided that you could deduce them.

Here is another ancient example, dating back at least as far as Euclid:

Proposition There are infinitely many prime numbers.

Proof. For the sake of contradiction, suppose there are only finitely many prime numbers. Then we can list all the prime numbers as $p_1, p_2, p_3, \ldots p_n$, where $p_1 = 2, p_2 = 3, p_3 = 5, p_4 = 7$ and so on. Thus p_n is the nth and largest prime number. Now consider the number $a = (p_1 p_2 p_3 \cdots p_n) + 1$, that is, a is the product of all prime numbers, plus 1. Now a, like any natural number greater than 1, has at least one prime divisor, and that means $p_k \mid a$ for at least one of our n prime numbers p_k. Thus there is an integer c for which $a = cp_k$, which is to say

$$(p_1 p_2 p_3 \cdots p_{k-1} p_k p_{k+1} \cdots p_n) + 1 = cp_k.$$

Dividing both sides of this by p_k gives us

$$(p_1 p_2 p_3 \cdots p_{k-1} p_{k+1} \cdots p_n) + \frac{1}{p_k} = c,$$

so

$$\frac{1}{p_k} = c - (p_1 p_2 p_3 \cdots p_{k-1} p_{k+1} \cdots p_n).$$

The expression on the right is an integer, while the expression on the left is not an integer. This is a contradiction. ■

Proof by contradiction often works well in proving statements of the form $\forall x, P(x)$. The reason is that the proof set-up involves assuming $\sim \forall x, P(x)$, which as we know from Section 2.10 is equivalent to $\exists x, \sim P(x)$. This gives us a specific x for which $\sim P(x)$ is true, and often that is enough to produce a contradiction. Here is an example:

Proposition For every real number $x \in [0, \pi/2]$, we have $\sin x + \cos x \geq 1$.

Proof. Suppose for the sake of contradiction that this is not true.
Then there exists an $x \in [0, \pi/2]$ for which $\sin x + \cos x < 1$.

Since $x \in [0, \pi/2]$, neither $\sin x$ nor $\cos x$ is negative, so $0 \le \sin x + \cos x < 1$. Thus $0^2 \le (\sin x + \cos x)^2 < 1^2$, which gives $0^2 \le \sin^2 x + 2\sin x \cos x + \cos^2 x < 1^2$. As $\sin^2 x + \cos^2 x = 1$, this becomes $0 \le 1 + 2\sin x \cos x < 1$, so $1 + 2\sin x \cos x < 1$. Subtracting 1 from both sides gives $2\sin x \cos x < 0$.
But this contradicts the fact that neither $\sin x$ nor $\cos x$ is negative. ■

6.2 Proving Conditional Statements by Contradiction

Since the previous two chapters dealt exclusively with proving conditional statements, we now formalize the procedure in which contradiction is used to prove a conditional statement. Suppose we want to prove a proposition of the following form.

Proposition If P, then Q.

Thus we need to prove that $P \Rightarrow Q$ is a true statement. Proof by contradiction begins with the assumption that $\sim (P \Rightarrow Q)$ is true, that is, that $P \Rightarrow Q$ is false. But we know that $P \Rightarrow Q$ being false means that it is possible that P can be true while Q is false. Thus the first step in the proof is to assume P and $\sim Q$. Here is an outline:

Outline for Proving a Conditional Statement with Contradiction

Proposition If P, then Q.

Proof. Suppose P and $\sim Q$.

$\vdots$

Therefore $C \wedge \sim C$. ■

To illustrate this new technique, we revisit a familiar result: If a^2 is even, then a is even. According to the outline, the first line of the proof should be "For the sake of contradiction, suppose a^2 is even and a is not even."

Proposition Suppose $a \in \mathbb{Z}$. If a^2 is even, then a is even.

Proof. For the sake of contradiction, suppose a^2 is even and a is not even.
Then a^2 is even, and a is odd.
Since a is odd, there is an integer c for which $a = 2c + 1$.
Then $a^2 = (2c+1)^2 = 4c^2 + 4c + 1 = 2(2c^2 + 2c) + 1$, so a^2 is odd.
Thus a^2 is even and a^2 is not even, a contradiction. ■

Here is another example.

Proposition If $a, b \in \mathbb{Z}$ and $a \geq 2$, then $a \nmid b$ or $a \nmid (b+1)$.

Proof. Suppose for the sake of contradiction there exist $a, b \in \mathbb{Z}$ with $a \geq 2$, and for which it is not true that $a \nmid b$ or $a \nmid (b+1)$.
By DeMorgan's law, we have $a \mid b$ and $a \mid (b+1)$.
The definition of divisibility says there are $c, d \in \mathbb{Z}$ with $b = ac$ and $b+1 = ad$.
Subtracting one equation from the other gives $ad - ac = 1$, so $a(d-c) = 1$.
Since a is positive, $d-c$ is also positive (otherwise $a(d-c)$ would be negative).
Then $d-c$ is a positive integer and $a(d-c) = 1$, so $a = 1/(d-c) < 2$.
Thus we have $a \geq 2$ and $a < 2$, a contradiction. ■

6.3 Combining Techniques

Often, especially in more complex proofs, several proof techniques are combined within a single proof. For example, in proving a conditional statement $P \Rightarrow Q$, we might begin with direct proof and thus assume P to be true with the aim of ultimately showing Q is true. But the truth of Q might hinge on the truth of some other statement R which—together with P—would imply Q. We would then need to prove R, and we would use whichever proof technique seems most appropriate. This can lead to "proofs inside of proofs." Consider the following example. The overall approach is direct, but inside the direct proof is a separate proof by contradiction.

Proposition Every non-zero rational number can be expressed as a product of two irrational numbers.

Proof. This proposition can be reworded as follows: If r is a non-zero rational number, then r is a product of two irrational numbers. In what follows, we prove this with direct proof.

Suppose r is a non-zero rational number. Then $r = \frac{a}{b}$ for integers a and b. Also, r can be written as a product of two numbers as follows:

$$r = \sqrt{2} \cdot \frac{r}{\sqrt{2}}.$$

We know $\sqrt{2}$ is irrational, so to complete the proof we must show $r/\sqrt{2}$ is also irrational.

To show this, assume for the sake of contradiction that $r/\sqrt{2}$ is rational. This means

$$\frac{r}{\sqrt{2}} = \frac{c}{d}$$

for integers c and d, so

$$\sqrt{2} = r\frac{d}{c}.$$

But we know $r = a/b$, which combines with the above equation to give

$$\sqrt{2} = r\frac{d}{c} = \frac{a}{b}\frac{d}{c} = \frac{ad}{bc}.$$

This means $\sqrt{2}$ is rational, which is a contradiction because we know it is irrational. Therefore $r/\sqrt{2}$ is irrational.

Consequently $r = \sqrt{2} \cdot r/\sqrt{2}$ is a product of two irrational numbers. ■

For another example of a proof-within-a-proof, try Exercise 5 at the end of this chapter (or see its solution). Exercise 5 asks you to prove that $\sqrt{3}$ is irrational. This turns out to be slightly trickier than proving that $\sqrt{2}$ is irrational.

6.4 Some Words of Advice

Despite the power of proof by contradiction, it's best to use it only when the direct and contrapositive approaches do not seem to work. The reason for this is that a proof by contradiction can often have hidden in it a simpler contrapositive proof, and if this is the case it's better to go with the simpler approach. Consider the following example.

Proposition Suppose $a \in \mathbb{Z}$. If $a^2 - 2a + 7$ is even, then a is odd.

Proof. To the contrary, suppose $a^2 - 2a + 7$ is even and a is not odd.
That is, suppose $a^2 - 2a + 7$ is even and a is even.
Since a is even, there is an integer c for which $a = 2c$.
Then $a^2 - 2a + 7 = (2c)^2 - 2(2c) + 7 = 2(2c^2 - 2c + 3) + 1$, so $a^2 - 2a + 7$ is odd.
Thus $a^2 - 2a + 7$ is both even and odd, a contradiction. ■

Though there is nothing really wrong with this proof, notice that part of it assumes a is not odd and deduces that $a^2 - 2a + 7$ is not even. That is the contrapositive approach! Thus it would be more efficient to proceed as follows, using contrapositive proof.

Proposition Suppose $a \in \mathbb{Z}$. If $a^2 - 2a + 7$ is even, then a is odd.

Proof. (Contrapositive) Suppose a is not odd.
Then a is even, so there is an integer c for which $a = 2c$.
Then $a^2 - 2a + 7 = (2c)^2 - 2(2c) + 7 = 2(2c^2 - 2c + 3) + 1$, so $a^2 - 2a + 7$ is odd.
Thus $a^2 - 2a + 7$ is not even. ■

Exercises for Chapter 6

A. Use the method of proof by contradiction to prove the following statements. (In each case, you should also think about how a direct or contrapositive proof would work. You will find in most cases that proof by contradiction is easier.)

1. Suppose $n \in \mathbb{Z}$. If n is odd, then n^2 is odd.
2. Suppose $n \in \mathbb{Z}$. If n^2 is odd, then n is odd.
3. Prove that $\sqrt[3]{2}$ is irrational.
4. Prove that $\sqrt{6}$ is irrational.
5. Prove that $\sqrt{3}$ is irrational.
6. If $a, b \in \mathbb{Z}$, then $a^2 - 4b - 2 \neq 0$.
7. If $a, b \in \mathbb{Z}$, then $a^2 - 4b - 3 \neq 0$.
8. Suppose $a, b, c \in \mathbb{Z}$. If $a^2 + b^2 = c^2$, then a or b is even.
9. Suppose $a, b \in \mathbb{R}$. If a is rational and ab is irrational, then b is irrational.
10. There exist no integers a and b for which $21a + 30b = 1$.
11. There exist no integers a and b for which $18a + 6b = 1$.
12. For every positive $x \in \mathbb{Q}$, there is a positive $y \in \mathbb{Q}$ for which $y < x$.
13. For every $x \in [\pi/2, \pi]$, $\sin x - \cos x \geq 1$.
14. If A and B are sets, then $A \cap (B - A) = \emptyset$.
15. If $b \in \mathbb{Z}$ and $b \nmid k$ for every $k \in \mathbb{N}$, then $b = 0$.
16. If a and b are positive real numbers, then $a + b \geq 2\sqrt{ab}$.
17. For every $n \in \mathbb{Z}$, $4 \nmid (n^2 + 2)$.
18. Suppose $a, b \in \mathbb{Z}$. If $4 \mid (a^2 + b^2)$, then a and b are not both odd.

B. Prove the following statements using any method from Chapters 4, 5 or 6.

19. The product of any five consecutive integers is divisible by 120. (For example, the product of 3,4,5,6 and 7 is 2520, and $2520 = 120 \cdot 21$.)
20. We say that a point $P = (x, y)$ in $\mathbb{R}^2$ is **rational** if both x and y are rational. More precisely, P is rational if $P = (x, y) \in \mathbb{Q}^2$. An equation $F(x, y) = 0$ is said to have a **rational point** if there exists $x_0, y_0 \in \mathbb{Q}$ such that $F(x_0, y_0) = 0$. For example, the curve $x^2 + y^2 - 1 = 0$ has rational point $(x_0, y_0) = (1, 0)$. Show that the curve $x^2 + y^2 - 3 = 0$ has no rational points.
21. Exercise 20 (above) involved showing that there are no rational points on the curve $x^2 + y^2 - 3 = 0$. Use this fact to show that $\sqrt{3}$ is irrational.
22. Explain why $x^2 + y^2 - 3 = 0$ not having any rational solutions (Exercise 20) implies $x^2 + y^2 - 3^k = 0$ has no rational solutions for k an odd, positive integer.
23. Use the above result to prove that $\sqrt{3^k}$ is irrational for all odd, positive k.
24. The number $\log_2 3$ is irrational.

Part III

More on Proof

CHAPTER 7

Proving Non-Conditional Statements

The last three chapters introduced three major proof techniques: direct, contrapositive and contradiction. These three techniques are used to prove statements of the form *"If P, then Q."* As we know, most theorems and propositions have this conditional form, or they can be reworded to have this form. Thus the three main techniques are quite important. But some theorems and propositions cannot be put into conditional form. For example, some theorems have form *"P if and only if Q."* Such theorems are biconditional statements, not conditional statements. In this chapter we examine ways to prove them. In addition to learning how to prove if-and-only-if theorems, we will also look at two other types of theorems.

7.1 If-and-Only-If Proof

Some propositions have the form

P if and only if Q.

We know from Section 2.4 that this statement asserts that **both** of the following conditional statements are true:

If P, then Q.
If Q, then P.

So to prove *"P if and only if Q,"* we must prove **two** conditional statements. Recall from Section 2.4 that $Q \Rightarrow P$ is called the *converse* of $P \Rightarrow Q$. Thus we need to prove both $P \Rightarrow Q$ and its converse. Since these are both conditional statements we may prove them with either direct, contrapositive or contradiction proof. Here is an outline:

Outline for If-and-Only-If Proof

Proposition P if and only if Q.

Proof.
[Prove $P \Rightarrow Q$ using direct, contrapositive or contradiction proof.]
[Prove $Q \Rightarrow P$ using direct, contrapositive or contradiction proof.] ■

Let's start with a very simple example. You already know that an integer n is odd if and only if n^2 is odd, but let's prove it anyway, just to illustrate the outline. In this example we prove (n is odd)$\Rightarrow$(n^2 is odd) using direct proof and (n^2 is odd)$\Rightarrow$(n is odd) using contrapositive proof.

Proposition The integer n is odd if and only if n^2 is odd.

Proof. First we show that n being odd implies that n^2 is odd. Suppose n is odd. Then, by definition of an odd number, $n = 2a+1$ for some integer a. Thus $n^2 = (2a+1)^2 = 4a^2+4a+1 = 2(2a^2+2a)+1$. This expresses n^2 as twice an integer, plus 1, so n^2 is odd.

Conversely, we need to prove that n^2 being odd implies that n is odd. We use contrapositive proof. Suppose n is not odd. Then n is even, so $n = 2a$ for some integer a (by definition of an even number). Thus $n^2 = (2a)^2 = 2(2a^2)$, so n^2 is even because it's twice an integer. Thus n^2 is not odd. We've now proved that if n is not odd, then n^2 is not odd, and this is a contrapositive proof that if n^2 is odd then n is odd. ■

In proving "P if and only if Q," you should begin a new paragraph when starting the proof of $Q \Rightarrow P$. Since this is the converse of $P \Rightarrow Q$, it's a good idea to begin the paragraph with the word "*Conversely*" (as we did above) to remind the reader that you've finished the first part of the proof and are moving on to the second. Likewise, it's a good idea to remind the reader of exactly what statement that paragraph is proving.

The next example uses direct proof in both parts of the proof.

Proposition Suppose a and b are integers. Then $a \equiv b \pmod{6}$ if and only if $a \equiv b \pmod{2}$ and $a \equiv b \pmod{3}$.

Proof. First we prove that if $a \equiv b \pmod{6}$, then $a \equiv b \pmod{2}$ and $a \equiv b \pmod{3}$. Suppose $a \equiv b \pmod{6}$. This means $6 \mid (a-b)$, so there is an integer n for which

$$a - b = 6n.$$

From this we get $a-b = 2(3n)$, which implies $2 \mid (a-b)$, so $a \equiv b \pmod{2}$. But we also get $a-b = 3(2n)$, which implies $3 \mid (a-b)$, so $a \equiv b \pmod{3}$. Therefore $a \equiv b \pmod{2}$ and $a \equiv b \pmod{3}$.

Conversely, suppose $a \equiv b \pmod{2}$ and $a \equiv b \pmod{3}$. Since $a \equiv b \pmod{2}$ we get $2 \mid (a-b)$, so there is an integer k for which $a-b = 2k$. Therefore $a-b$ is even. Also, from $a \equiv b \pmod{3}$ we get $3 \mid (a-b)$, so there is an integer ℓ for which

$$a - b = 3\ell.$$

But since we know $a-b$ is even, it follows that ℓ must be even also, for if it were odd then $a-b=3\ell$ would be odd (because $a-b$ would be the product of two odd integers). Hence $\ell = 2m$ for some integer m. Thus $a-b=3\ell=3\cdot 2m=6m$. This means $6 \mid (a-b)$, so $a \equiv b \pmod{6}$. ■

Since if-and-only-if proofs simply combine methods with which we are already familiar, we will not do any further examples in this section. However, it is of utmost importance that you practice your skill on some of this chapter's exercises.

7.2 Equivalent Statements

In other courses you will sometimes encounter a certain kind of theorem that is neither a conditional nor a biconditional statement. Instead, it asserts that a list of statements is "*equivalent*." You saw this (or will see it) in your linear algebra textbook, which featured the following theorem:

Theorem Suppose A is an $n \times n$ matrix. The following statements are equivalent:

- **(a)** The matrix A is invertible.
- **(b)** The equation $A\mathbf{x}=\mathbf{b}$ has a unique solution for every $\mathbf{b} \in \mathbb{R}^n$.
- **(c)** The equation $A\mathbf{x}=\mathbf{0}$ has only the trivial solution.
- **(d)** The reduced row echelon form of A is I_n.
- **(e)** $\det(A) \neq 0$.
- **(f)** The matrix A does not have 0 as an eigenvalue.

When a theorem asserts that a list of statements is "equivalent," it is asserting that either the statements are all true, or they are all false. Thus the above theorem tells us that whenever we are dealing with a particular $n \times n$ matrix A, then either the statements (a) through (f) are all true for A, or statements (a) through (f) are all false for A. For example, if we happen to know that $\det(A) \neq 0$, the theorem assures us that in addition to statement (e) being true, **all** the statements (a) through (f) are true. On the other hand, if it happens that $\det(A)=0$, the theorem tells us that all statements (a) through (f) are false. In this way, the theorem multiplies our knowledge of A by a factor of six. Obviously that can be very useful.

What method would we use to prove such a theorem? In a certain sense, the above theorem is like an if-and-only-if theorem. An if-and-only-if theorem of form $P \Leftrightarrow Q$ asserts that P and Q are either both true or both false, that is, that P and Q are equivalent. To prove $P \Leftrightarrow Q$ we prove $P \Rightarrow Q$ followed by $Q \Rightarrow P$, essentially making a "cycle" of implications from P to Q

and back to P. Similarly, one approach to proving the theorem about the $n \times n$ matrix would be to prove the conditional statement $(a) \Rightarrow (b)$, then $(b) \Rightarrow (c)$, then $(c) \Rightarrow (d)$, then $(d) \Rightarrow (e)$, then $(e) \Rightarrow (f)$ and finally $(f) \Rightarrow (a)$. This pattern is illustrated below.

$$\begin{array}{ccccc} (a) & \Longrightarrow & (b) & \Longrightarrow & (c) \\ \Uparrow & & & & \Downarrow \\ (f) & \Longleftarrow & (e) & \Longleftarrow & (d) \end{array}$$

Notice that if these six implications have been proved, then it really does follow that the statements (a) through (f) are either all true or all false. If one of them is true, then the circular chain of implications forces them all to be true. On the other hand, if one of them (say (c)) is false, the fact that $(b) \Rightarrow (c)$ is true forces (b) to be false. This combined with the truth of $(a) \Rightarrow (b)$ makes (a) false, and so on counterclockwise around the circle.

Thus to prove that n statements are equivalent, it suffices to prove n conditional statements showing each statement implies another, in circular pattern. But it is not necessary that the pattern be circular. The following schemes would also do the job:

$$\begin{array}{ccccc} (a) & \Longrightarrow & (b) & \Longleftrightarrow & (c) \\ \Uparrow & & \Downarrow & & \\ (f) & \Longleftarrow & (e) & \Longleftrightarrow & (d) \end{array}$$

$$\begin{array}{ccccc} (a) & \Longleftrightarrow & (b) & \Longleftrightarrow & (c) \\ & & \Updownarrow & & \\ (f) & \Longleftrightarrow & (e) & \Longleftrightarrow & (d) \end{array}$$

But a circular pattern yields the fewest conditional statements that must be proved. Whatever the pattern, each conditional statement can be proved with either direct, contrapositive or contradiction proof.

Though we shall not do any of these proofs in this text, you are sure to encounter them in subsequent courses.

7.3 Existence Proofs; Existence and Uniqueness Proofs

Up until this point, we have dealt with proving conditional statements or with statements that can be expressed with two or more conditional statements. Generally, these conditional statements have form $P(x) \Rightarrow Q(x)$. (Possibly with more than one variable.) We saw in Section 2.8 that this can be interpreted as a universally quantified statement $\forall x, P(x) \Rightarrow Q(x)$.

Thus, conditional statements are universally quantified statements, so in proving a conditional statement—whether we use direct, contrapositive or contradiction proof—we are really proving a universally quantified statement.

But how would we prove an *existentially* quantified statement? What technique would we employ to prove a theorem of the following form?

$$\exists x, R(x)$$

This statement asserts that there exists some specific object x for which $R(x)$ is true. To prove $\exists x, R(x)$ is true, all we would have to do is find and display an *example* of a specific x that makes $R(x)$ true.

Though most theorems and propositions are conditional (or if-and-only-if) statements, a few have the form $\exists x, R(x)$. Such statements are called **existence statements**, and theorems that have this form are called **existence theorems**. To prove an existence theorem, all you have to do is provide a particular example that shows it is true. This is often quite simple. (But not always!) Here are some examples:

Proposition There exists an even prime number.

Proof. Observe that 2 is an even prime number. ■

Admittedly, this last proposition was a bit of an oversimplification. The next one is slightly more challenging.

Proposition There exists an integer that can be expressed as the sum of two perfect cubes in two different ways.

Proof. Consider the number 1729. Note that $1^3 + 12^3 = 1729$ and $9^3 + 10^3 = 1729$. Thus the number 1729 can be expressed as the sum of two perfect cubes in two different ways. ■

Sometimes in the proof of an existence statement, a little verification is needed to show that the example really does work. For example, the above proof would be incomplete if we just asserted that 1729 can be written as a sum of two cubes in two ways without showing *how* this is possible.

WARNING: Although an example suffices to prove an existence statement, a single example does not prove a conditional statement.

Often an existence statement will be embedded inside of a conditional statement. Consider the following. (Recall the definition of gcd on page 90.)

If $a,b \in \mathbb{N}$, then there exist integers k and ℓ for which $\gcd(a,b) = ak+b\ell$.

This is a conditional statement that has the form

$$a,b \in \mathbb{N} \implies \exists\, k,\ell \in \mathbb{Z},\ \gcd(a,b) = ak+b\ell.$$

To prove it with direct proof, we would first assume that $a,b \in \mathbb{N}$, then prove the existence statement $\exists\, k,\ell \in \mathbb{Z}$, $\gcd(a,b) = ak+b\ell$. That is, we would produce two integers k and ℓ (which depend on a and b) for which $\gcd(a,b) = ak+b\ell$. Let's carry out this plan. (We will use this fundamental proposition several times later, so it is given a number.)

Proposition 7.1 If $a,b \in \mathbb{N}$, then there exist integers k and ℓ for which $\gcd(a,b) = ak+b\ell$.

Proof. (Direct) Suppose $a,b \in \mathbb{N}$. Consider the set $A = \{ax+by : x,y \in \mathbb{Z}\}$. This set contains both positive and negative integers, as well as 0. (Reason: Let $y = 0$ and let x range over all integers. Then $ax+by = ax$ ranges over all multiples of a, both positive, negative and zero.) Let d be the smallest positive element of A. Then, because d is in A, it must have the form $d = ak+b\ell$ for some specific $k,\ell \in \mathbb{Z}$.

To finish, we will show $d = \gcd(a,b)$. We will first argue that d is a common divisor of a and b, and then that it is the *greatest* common divisor.

To see that $d \mid a$, use the division algorithm (page 29) to write $a = qd+r$ for integers q and r with $0 \le r < d$. The equation $a = qd+r$ yields

$$\begin{aligned} r &= a - qd \\ &= a - q(ak+b\ell) \\ &= a(1-qk) + b(-q\ell). \end{aligned}$$

Therefore r has form $r = ax+by$, so it belongs to A. But $0 \le r < d$ and d is the smallest positive number in A, so r can't be positive; hence $r = 0$. Updating our equation $a = qd+r$, we get $a = qd$, so $d \mid a$. Repeating this argument with $b = qd+r$ shows $d \mid b$. Thus d is indeed a common divisor of a and b. It remains to show that it is the *greatest* common divisor.

As $\gcd(a,b)$ divides a and b, we have $a = \gcd(a,b)\cdot m$ and $b = \gcd(a,b)\cdot n$ for some $m,n \in \mathbb{Z}$. So $d = ak+b\ell = \gcd(a,b)\cdot mk + \gcd(a,b)\cdot n\ell = \gcd(a,b)(mk+n\ell)$, and thus d is a multiple of $\gcd(a,b)$. Therefore $d \ge \gcd(a,b)$. But d can't be a larger common divisor of a and b than $\gcd(a,b)$, so $d = \gcd(a,b)$. ■

We conclude this section with a discussion of so-called *uniqueness proofs*. Some existence statements have form "*There is a* unique x *for which* $P(x)$." Such a statement asserts that there is *exactly one* example x for which $P(x)$ is true. To prove it, you must produce an example $x = d$ for which $P(d)$ is true, **and** you must show that d is the only such example. The next proposition illustrates this. In essence, it asserts that the set $\{ax+by : x, y \in \mathbb{Z}\}$ consists precisely of all the multiples of $\gcd(a,b)$.

Proposition Suppose $a,b \in \mathbb{N}$. Then there exists a unique $d \in \mathbb{N}$ for which: An integer m is a multiple of d if and only if $m = ax + by$ for some $x, y \in \mathbb{Z}$.

Proof. Suppose $a,b \in \mathbb{N}$. Let $d = \gcd(a,b)$. We now show that an integer m is a multiple of d if and only if $m = ax+by$ for some $x, y \in \mathbb{Z}$. Let $m = dn$ be a multiple of d. By Proposition 7.1 (on the previous page), there are integers k and ℓ for which $d = ak + b\ell$. Then $m = dn = (ak + b\ell)n = a(kn) + b(\ell n)$, so $m = ax + by$ for integers $x = kn$ and $y = \ell n$.

Conversely, suppose $m = ax + by$ for some $x, y \in \mathbb{Z}$. Since $d = \gcd(a,b)$ is a divisor of both a and b, we have $a = dc$ and $b = de$ for some $c, e \in \mathbb{Z}$. Then $m = ax + by = dcx + dey = d(cx + ey)$, and this is a multiple of d.

We have now shown that there is a natural number d with the property that m is a multiple of d if and only if $m = ax + by$ for some $x, y \in \mathbb{Z}$. It remains to show that d is the *unique* such natural number. To do this, suppose d' is *any* natural number with the property that d has:

$$m \text{ is a multiple of } d' \iff m = ax + by \text{ for some } x, y \in \mathbb{Z}. \tag{7.1}$$

We next argue that $d' = d$; that is, d is the *unique* natural number with the stated property. Because of (7.1), $m = a \cdot 1 + b \cdot 0 = a$ is a multiple of d'. Likewise $m = a \cdot 0 + b \cdot 1 = b$ is a multiple of d'. Hence a and b are both multiples of d', so d' is a common divisor of a and b, and therefore

$$d' \leq \gcd(a,b) = d.$$

But also, by (7.1), the multiple $m = d' \cdot 1 = d'$ of d' can be expressed as $d' = ax + by$ for some $x, y \in \mathbb{Z}$. As noted in the second paragraph of the proof, $a = dc$ and $b = de$ for some $c, e \in \mathbb{Z}$. Thus $d' = ax + by = dcx + dey = d(cx + ey)$, so d' is a multiple d. As d' and d are both positive, it follows that

$$d \leq d'.$$

We've now shown that $d' \leq d$ and $d \leq d'$, so $d = d'$. The proof is complete. ■

7.4 Constructive Versus Non-Constructive Proofs

Existence proofs fall into two categories: constructive and non-constructive. Constructive proofs display an explicit example that proves the theorem; non-constructive proofs prove an example exists without actually giving it. We illustrate the difference with two proofs of the same fact: There exist *irrational* numbers x and y (possibly equal) for which x^y is *rational*.

Proposition There exist irrational numbers x, y for which x^y is rational.

Proof. Let $x = \sqrt{2}^{\sqrt{2}}$ and $y = \sqrt{2}$. We know y is irrational, but it is not clear whether x is rational or irrational. On one hand, if x is irrational, then we have an irrational number to an irrational power that is rational:

$$x^y = \left(\sqrt{2}^{\sqrt{2}}\right)^{\sqrt{2}} = \sqrt{2}^{\sqrt{2}\sqrt{2}} = \sqrt{2}^2 = 2.$$

On the other hand, if x is rational, then $y^y = \sqrt{2}^{\sqrt{2}} = x$ is rational. Either way, we have a irrational number to an irrational power that is rational. ■

The above is a classic example of a **non-constructive** proof. It shows that there exist irrational numbers x and y for which x^y is rational without actually producing (or constructing) an example. It convinces us that one of $\left(\sqrt{2}^{\sqrt{2}}\right)^{\sqrt{2}}$ or $\sqrt{2}^{\sqrt{2}}$ is an irrational number to an irrational power that is rational, but it does not say which one is the correct example. It thus proves that an example exists without explicitly stating one.

Next comes a **constructive proof** of this statement, one that produces (or constructs) two explicit irrational numbers x, y for which x^y is rational.

Proposition There exist irrational numbers x, y for which x^y is rational.

Proof. Let $x = \sqrt{2}$ and $y = \log_2 9$. Then

$$x^y = \sqrt{2}^{\log_2 9} = \sqrt{2}^{\log_2 3^2} = \sqrt{2}^{2\log_2 3} = \left(\sqrt{2}^2\right)^{\log_2 3} = 2^{\log_2 3} = 3.$$

As 3 is rational, we have shown that $x^y = 3$ is rational.

We know that $x = \sqrt{2}$ is irrational. The proof will be complete if we can show that $y = \log_2 9$ is irrational. Suppose for the sake of contradiction that $\log_2 9$ is rational, so there are integers a and b for which $\frac{a}{b} = \log_2 9$. This means $2^{a/b} = 9$, so $\left(2^{a/b}\right)^b = 9^b$, which reduces to $2^a = 9^b$. But 2^a is even, while 9^b is odd (because it is the product of the odd number 9 with itself b times). This is a contradiction; the proof is complete. ■

This existence proof has inside of it a separate proof (by contradiction) that $\log_2 9$ is irrational. Such combinations of proof techniques are, of course, typical.

Be alert to constructive and non-constructive proofs as you read proofs in other books and articles, as well as to the possibility of crafting such proofs of your own.

Exercises for Chapter 7

Prove the following statements. These exercises are cumulative, covering all techniques addressed in Chapters 4–7.

1. Suppose $x \in \mathbb{Z}$. Then x is even if and only if $3x+5$ is odd.
2. Suppose $x \in \mathbb{Z}$. Then x is odd if and only if $3x+6$ is odd.
3. Given an integer a, then a^3+a^2+a is even if and only if a is even.
4. Given an integer a, then a^2+4a+5 is odd if and only if a is even.
5. An integer a is odd if and only if a^3 is odd.
6. Suppose $x, y \in \mathbb{R}$. Then $x^3+x^2y = y^2+xy$ if and only if $y = x^2$ or $y = -x$.
7. Suppose $x, y \in \mathbb{R}$. Then $(x+y)^2 = x^2+y^2$ if and only if $x = 0$ or $y = 0$.
8. Suppose $a, b \in \mathbb{Z}$. Prove that $a \equiv b \pmod{10}$ if and only if $a \equiv b \pmod 2$ and $a \equiv b \pmod 5$.
9. Suppose $a \in \mathbb{Z}$. Prove that $14 \mid a$ if and only if $7 \mid a$ and $2 \mid a$.
10. If $a \in \mathbb{Z}$, then $a^3 \equiv a \pmod 3$.
11. Suppose $a, b \in \mathbb{Z}$. Prove that $(a-3)b^2$ is even if and only if a is odd or b is even.
12. There exist a positive real number x for which $x^2 < \sqrt{x}$.
13. Suppose $a, b \in \mathbb{Z}$. If $a+b$ is odd, then a^2+b^2 is odd.
14. Suppose $a \in \mathbb{Z}$. Then $a^2 \mid a$ if and only if $a \in \{-1, 0, 1\}$.
15. Suppose $a, b \in \mathbb{Z}$. Prove that $a+b$ is even if and only if a and b have the same parity.
16. Suppose $a, b \in \mathbb{Z}$. If ab is odd, then a^2+b^2 is even.
17. There is a prime number between 90 and 100.
18. There is a set X for which $\mathbb{N} \in X$ and $\mathbb{N} \subseteq X$.
19. If $n \in \mathbb{N}$, then $2^0+2^1+2^2+2^3+2^4+\cdots+2^n = 2^{n+1}-1$.
20. There exists an $n \in \mathbb{N}$ for which $11 \mid (2^n-1)$.
21. Every real solution of $x^3+x+3=0$ is irrational.
22. If $n \in \mathbb{Z}$, then $4 \mid n^2$ or $4 \mid (n^2-1)$.
23. Suppose a, b and c are integers. If $a \mid b$ and $a \mid (b^2-c)$, then $a \mid c$.
24. If $a \in \mathbb{Z}$, then $4 \nmid (a^2-3)$.

25. If $p > 1$ is an integer and $n \nmid p$ for each integer n for which $2 \le n \le \sqrt{p}$, then p is prime.

26. The product of any n consecutive positive integers is divisible by $n!$.

27. Suppose $a, b \in \mathbb{Z}$. If $a^2 + b^2$ is a perfect square, then a and b are not both odd.

28. Prove the division algorithm: If $a, b \in \mathbb{N}$, there exist *unique* integers q, r for which $a = bq + r$, and $0 \le r < b$. (A proof of existence is given in Section 1.9, but uniqueness needs to be established too.)

29. If $a \mid bc$ and $\gcd(a, b) = 1$, then $a \mid c$.
(Suggestion: Use the proposition on page 126.)

30. Suppose $a, b, p \in \mathbb{Z}$ and p is prime. Prove that if $p \mid ab$ then $p \mid a$ or $p \mid b$.
(Suggestion: Use the proposition on page 126.)

31. If $n \in \mathbb{Z}$, then $\gcd(n, n+1) = 1$.

32. If $n \in \mathbb{Z}$, then $\gcd(n, n+2) \in \{1, 2\}$.

33. If $n \in \mathbb{Z}$, then $\gcd(2n+1, 4n^2+1) = 1$.

34. If $\gcd(a, c) = \gcd(b, c) = 1$, then $\gcd(ab, c) = 1$.
(Suggestion: Use the proposition on page 126.)

35. Suppose $a, b \in \mathbb{N}$. Then $a = \gcd(a, b)$ if and only if $a \mid b$.

36. Suppose $a, b \in \mathbb{N}$. Then $a = \operatorname{lcm}(a, b)$ if and only if $b \mid a$.

CHAPTER 8

Proofs Involving Sets

Students in their first advanced mathematics classes are often surprised by the extensive role that sets play and by the fact that most of the proofs they encounter are proofs about sets. Perhaps you've already seen such proofs in your linear algebra course, where a **vector space** was defined to be a *set* of objects (called vectors) that obey certain properties. Your text proved many things about vector spaces, such as the fact that the intersection of two vector spaces is also a vector space, and the proofs used ideas from set theory. As you go deeper into mathematics, you will encounter more and more ideas, theorems and proofs that involve sets. The purpose of this chapter is to give you a foundation that will prepare you for this new outlook.

We will discuss how to show that an object is an element of a set, how to prove one set is a subset of another and how to prove two sets are equal. As you read this chapter, you may need to occasionally refer back to Chapter 1 to refresh your memory. For your convenience, the main definitions from Chapter 1 are summarized below. If A and B are sets, then:

$$\begin{aligned} A \times B &= \{(x,y) : x \in A,\ y \in B\}, \\ A \cup B &= \{x : (x \in A) \vee (x \in B)\}, \\ A \cap B &= \{x : (x \in A) \wedge (x \in B)\}, \\ A - B &= \{x : (x \in A) \wedge (x \notin B)\}, \\ \overline{A} &= U - A. \end{aligned}$$

Recall that $A \subseteq B$ means that every element of A is also an element of B.

8.1 How to Prove $a \in A$

We will begin with a review of set-builder notation, and then review how to show that a given object a is an element of some set A.

Generally, a set A will be expressed in set-builder notation $A = \{x : P(x)\}$, where $P(x)$ is some statement (or open sentence) about x. The set A is understood to have as elements all those things x for which $P(x)$ is true. For example,

$$\{x : x \text{ is an odd integer}\} = \{\ldots, -5, -3, -1, 1, 3, 5, \ldots\}.$$

A common variation of this notation is to express a set as $A = \{x \in S : P(x)\}$. Here it is understood that A consists of all elements x of the (predetermined) set S for which $P(x)$ is true. Keep in mind that, depending on context, x could be any kind of object (integer, ordered pair, set, function, etc.). There is also nothing special about the particular variable x; any reasonable symbol x, y, k, etc., would do. Some examples follow.

$$\begin{aligned}
\{n \in \mathbb{Z} : n \text{ is odd}\} &= \{\ldots, -5, -3, -1, 1, 3, 5, \ldots\} \\
\{x \in \mathbb{N} : 6 \mid x\} &= \{6, 12, 18, 24, 30, \ldots\} \\
\{(a,b) \in \mathbb{Z} \times \mathbb{Z} : b = a + 5\} &= \{\ldots, (-2,3), (-1,4), (0,5), (1,6), \ldots\} \\
\{X \in \mathscr{P}(\mathbb{Z}) : |X| = 1\} &= \{\ldots, \{-1\}, \{0\}, \{1\}, \{2\}, \{3\}, \{4\}, \ldots\}
\end{aligned}$$

Now it should be clear how to prove that an object a belongs to a set $\{x : P(x)\}$. Since $\{x : P(x)\}$ consists of all things x for which $P(x)$ is true, to show that $a \in \{x : P(x)\}$ we just need to show that $P(a)$ is true. Likewise, to show $a \in \{x \in S : P(x)\}$, we need to confirm that $a \in S$ *and* that $P(a)$ is true. These ideas are summarized below. However, you should **not** memorize these methods, you should **understand** them. With contemplation and practice, using them becomes natural and intuitive.

How to show $\mathbf{a} \in \{\mathbf{x} : \mathbf{P(x)}\}$	**How to show $\mathbf{a} \in \{\mathbf{x} \in \mathbf{S} : \mathbf{P(x)}\}$**
Show that $P(a)$ is true.	1. Verify that $a \in S$. 2. Show that $P(a)$ is true.

Example 8.1 Let's investigate elements of $A = \{x : x \in \mathbb{N} \text{ and } 7 \mid x\}$. This set has form $A = \{x : P(x)\}$ where $P(x)$ is the open sentence $(x \in \mathbb{N}) \wedge (7 \mid x)$. Thus $21 \in A$ because $P(21)$ is true. Similarly, $7, 14, 28, 35$, etc., are all elements of A. But $8 \notin A$ (for example) because $P(8)$ is false. Likewise $-14 \notin A$ because $P(-14)$ is false.

Example 8.2 Consider the set $A = \{X \in \mathscr{P}(\mathbb{N}) : |X| = 3\}$. We know that $\{4, 13, 45\} \in A$ because $\{4, 13, 45\} \in \mathscr{P}(\mathbb{N})$ and $\left|\{4, 13, 45\}\right| = 3$. Also $\{1, 2, 3\} \in A$, $\{10, 854, 3\} \in A$, etc. However $\{1, 2, 3, 4\} \notin A$ because $\left|\{1, 2, 3, 4\}\right| \neq 3$. Further, $\{-1, 2, 3\} \notin A$ because $\{-1, 2, 3\} \notin \mathscr{P}(\mathbb{N})$.

Example 8.3 Consider the set $B = \{(x,y) \in \mathbb{Z} \times \mathbb{Z} : x \equiv y \pmod{5}\}$. Notice $(8,23) \in B$ because $(8,23) \in \mathbb{Z} \times \mathbb{Z}$ and $8 \equiv 23 \pmod{5}$. Likewise, $(100,75) \in B$, $(102,77) \in B$, etc., but $(6,10) \notin B$.

Now suppose $n \in \mathbb{Z}$ and consider the ordered pair $(4n+3, 9n-2)$. Does this ordered pair belong to B? To answer this, we first observe that $(4n+3, 9n-2) \in \mathbb{Z} \times \mathbb{Z}$. Next, we observe that $(4n+3)-(9n-2) = -5n+5 = 5(1-n)$, so $5 \mid ((4n+3)-(9n-2))$, which means $(4n+3) \equiv (9n-2) \pmod{5}$. Therefore we have established that $(4n+3, 9n-2)$ meets the requirements for belonging to B, so $(4n+3, 9n-2) \in B$ for every $n \in \mathbb{Z}$.

Example 8.4 This illustrates another common way of defining a set. Consider the set $C = \{3x^3 + 2 : x \in \mathbb{Z}\}$. Elements of this set consist of all the values $3x^3 + 2$ where x is an integer. Thus $-22 \in C$ because $-22 = 3(-2)^3 + 2$. You can confirm $-1 \in C$ and $5 \in C$, etc. Also $0 \notin C$ and $\frac{1}{2} \notin C$, etc.

8.2 How to Prove A ⊆ B

In this course (and more importantly, beyond it) you will encounter many circumstances where it is necessary to prove that one set is a subset of another. This section explains how to do this. The methods we discuss should improve your skills in both writing your own proofs and in comprehending the proofs that you read.

Recall (Definition 1.3) that if A and B are sets, then $A \subseteq B$ means that every element of A is also an element of B. In other words, it means *if* $a \in A$, *then* $a \in B$. Therefore to prove that $A \subseteq B$, we just need to prove that the conditional statement

"*If* $a \in A$, *then* $a \in B$"

is true. This can be proved directly, by assuming $a \in A$ and deducing $a \in B$. The contrapositive approach is another option: Assume $a \notin B$ and deduce $a \notin A$. Each of these two approaches is outlined below.

How to Prove A ⊆ B
(Direct approach)

Proof. Suppose $a \in A$.
⋮
Therefore $a \in B$.
Thus $a \in A$ implies $a \in B$,
so it follows that $A \subseteq B$. ■

How to Prove A ⊆ B
(Contrapositive approach)

Proof. Suppose $a \notin B$.
⋮
Therefore $a \notin A$.
Thus $a \notin B$ implies $a \notin A$,
so it follows that $A \subseteq B$. ■

In practice, the direct approach usually results in the most straightforward and easy proof, though occasionally the contrapositive is the most expedient. (You can even prove $A \subseteq B$ by contradiction: Assume $(a \in A) \wedge (a \notin B)$, and deduce a contradiction.) The remainder of this section consists of examples with occasional commentary. Unless stated otherwise, we will use the direct approach in all proofs; pay special attention to how the above outline for the direct approach is used.

Example 8.5 Prove that $\{x \in \mathbb{Z} : 18 \mid x\} \subseteq \{x \in \mathbb{Z} : 6 \mid x\}$.

Proof. Suppose $a \in \{x \in \mathbb{Z} : 18 \mid x\}$.
This means that $a \in \mathbb{Z}$ and $18 \mid a$.
By definition of divisibility, there is an integer c for which $a = 18c$.
Consequently $a = 6(3c)$, and from this we deduce that $6 \mid a$.
Therefore a is one of the integers that 6 divides, so $a \in \{x \in \mathbb{Z} : 6 \mid x\}$.
We've shown $a \in \{x \in \mathbb{Z} : 18 \mid x\}$ implies $a \in \{n \in \mathbb{Z} : 6 \mid x\}$, so it follows that $\{x \in \mathbb{Z} : 18 \mid x\} \subseteq \{x \in \mathbb{Z} : 6 \mid x\}$. ■

Example 8.6 Prove that $\{x \in \mathbb{Z} : 2 \mid x\} \cap \{x \in \mathbb{Z} : 9 \mid x\} \subseteq \{x \in \mathbb{Z} : 6 \mid x\}$.

Proof. Suppose $a \in \{x \in \mathbb{Z} : 2 \mid x\} \cap \{x \in \mathbb{Z} : 9 \mid x\}$.
By definition of intersection, this means $a \in \{x \in \mathbb{Z} : 2 \mid x\}$ and $a \in \{x \in \mathbb{Z} : 9 \mid x\}$.
Since $a \in \{x \in \mathbb{Z} : 2 \mid x\}$ we know $2 \mid a$, so $a = 2c$ for some $c \in \mathbb{Z}$. Thus a is even.
Since $a \in \{x \in \mathbb{Z} : 9 \mid x\}$ we know $9 \mid a$, so $a = 9d$ for some $d \in \mathbb{Z}$.
As a is even, $a = 9d$ implies d is even. (Otherwise $a = 9d$ would be odd.)
Then $d = 2e$ for some integer e, and we have $a = 9d = 9(2e) = 6(3e)$.
From $a = 6(3e)$, we conclude $6 \mid a$, and this means $a \in \{x \in \mathbb{Z} : 6 \mid x\}$.
We have shown that $a \in \{x \in \mathbb{Z} : 2 \mid x\} \cap \{x \in \mathbb{Z} : 9 \mid x\}$ implies $a \in \{x \in \mathbb{Z} : 6 \mid x\}$, so it follows that $\{x \in \mathbb{Z} : 2 \mid x\} \cap \{x \in \mathbb{Z} : 9 \mid x\} \subseteq \{x \in \mathbb{Z} : 6 \mid x\}$. ■

Example 8.7 Show $\{(x,y) \in \mathbb{Z} \times \mathbb{Z} : x \equiv y \pmod 6\} \subseteq \{(x,y) \in \mathbb{Z} \times \mathbb{Z} : x \equiv y \pmod 3\}$.

Proof. Suppose $(a,b) \in \{(x,y) \in \mathbb{Z} \times \mathbb{Z} : x \equiv y \pmod 6\}$.
This means $(a,b) \in \mathbb{Z} \times \mathbb{Z}$ and $a \equiv b \pmod 6$.
Consequently $6 \mid (a-b)$, so $a - b = 6c$ for some integer c.
It follows that $a - b = 3(2c)$, and this means $3 \mid (a-b)$, so $a \equiv b \pmod 3$.
Thus $(a,b) \in \{(x,y) \in \mathbb{Z} \times \mathbb{Z} : x \equiv y \pmod 3\}$.
We've now seen that $(a,b) \in \{(x,y) \in \mathbb{Z} \times \mathbb{Z} : x \equiv y \pmod 6\}$ implies $(a,b) \in \{(x,y) \in \mathbb{Z} \times \mathbb{Z} : x \equiv y \pmod 3\}$, so it follows that $\{(x,y) \in \mathbb{Z} \times \mathbb{Z} : x \equiv y \pmod 6\} \subseteq \{(x,y) \in \mathbb{Z} \times \mathbb{Z} : x \equiv y \pmod 3\}$. ■

Some statements involving subsets are transparent enough that we often accept (and use) them without proof. For example, if A and B are any sets, then it's very easy to confirm $A \cap B \subseteq A$. (Reason: Suppose $x \in A \cap B$. Then $x \in A$ and $x \in B$ by definition of intersection, so in particular $x \in A$. Thus $x \in A \cap B$ implies $x \in A$, so $A \cap B \subseteq A$.) Other statements of this nature include $A \subseteq A \cup B$ and $A - B \subseteq A$, as well as conditional statements such as $\big((A \subseteq B) \wedge (B \subseteq C)\big) \Rightarrow (A \subseteq C)$ and $(X \subseteq A) \Rightarrow (X \subseteq A \cup B)$. Our point of view in this text is that we do not need to prove such obvious statements unless we are explicitly asked to do so in an exercise. (Still, you should do some quick mental proofs to convince yourself that the above statements are true. If you don't see that $A \cap B \subseteq A$ is true but that $A \subseteq A \cap B$ is not necessarily true, then you need to spend more time on this topic.)

The next example will show that if A and B are sets, then $\mathscr{P}(A) \cup \mathscr{P}(B) \subseteq \mathscr{P}(A \cup B)$. Before beginning our proof, let's look at an example to see if this statement really makes sense. Suppose $A = \{1,2\}$ and $B = \{2,3\}$. Then

$$\begin{aligned} \mathscr{P}(A) \cup \mathscr{P}(B) &= \{\emptyset,\{1\},\{2\},\{1,2\}\} \cup \{\emptyset,\{2\},\{3\},\{2,3\}\} \\ &= \{\emptyset,\{1\},\{2\},\{3\},\{1,2\},\{2,3\}\}. \end{aligned}$$

Also $\mathscr{P}(A \cup B) = \mathscr{P}(\{1,2,3\}) = \{\emptyset,\{1\},\{2\},\{3\},\{1,2\},\{2,3\},\{1,3\},\{1,2,3\}\}$. Thus, even though $\mathscr{P}(A) \cup \mathscr{P}(B) \neq \mathscr{P}(A \cup B)$, it is true that $\mathscr{P}(A) \cup \mathscr{P}(B) \subseteq \mathscr{P}(A \cup B)$ for this particular A and B. Now let's prove $\mathscr{P}(A) \cup \mathscr{P}(B) \subseteq \mathscr{P}(A \cup B)$ no matter what sets A and B are.

Example 8.8 Prove that if A and B are sets, then $\mathscr{P}(A) \cup \mathscr{P}(B) \subseteq \mathscr{P}(A \cup B)$.

Proof. Suppose $X \in \mathscr{P}(A) \cup \mathscr{P}(B)$.
By definition of union, this means $X \in \mathscr{P}(A)$ or $X \in \mathscr{P}(B)$.
Therefore $X \subseteq A$ or $X \subseteq B$ (by definition of power sets). We consider cases.
Case 1. Suppose $X \subseteq A$. Then $X \subseteq A \cup B$, and this means $X \in \mathscr{P}(A \cup B)$.
Case 2. Suppose $X \subseteq B$. Then $X \subseteq A \cup B$, and this means $X \in \mathscr{P}(A \cup B)$.
(We do not need to consider the case where $X \subseteq A$ *and* $X \subseteq B$ because that is taken care of by either of cases 1 or 2.) The above cases show that $X \in \mathscr{P}(A \cup B)$.

Thus we've shown that $X \in \mathscr{P}(A) \cup \mathscr{P}(B)$ implies $X \in \mathscr{P}(A \cup B)$, and this completes the proof that $\mathscr{P}(A) \cup \mathscr{P}(B) \subseteq \mathscr{P}(A \cup B)$. ■

In our next example, we prove a conditional statement. Direct proof is used, and in the process we use our new technique for showing $A \subseteq B$.

Example 8.9 Suppose A and B are sets. If $\mathscr{P}(A) \subseteq \mathscr{P}(B)$, then $A \subseteq B$.

Proof. We use direct proof. Assume $\mathscr{P}(A) \subseteq \mathscr{P}(B)$.
Based on this assumption, we must now show that $A \subseteq B$.
To show $A \subseteq B$, suppose that $a \in A$.
Then the one-element set $\{a\}$ is a subset of A, so $\{a\} \in \mathscr{P}(A)$.
But then, since $\mathscr{P}(A) \subseteq \mathscr{P}(B)$, it follows that $\{a\} \in \mathscr{P}(B)$.
This means that $\{a\} \subseteq B$, hence $a \in B$.
We've shown that $a \in A$ implies $a \in B$, so therefore $A \subseteq B$. ■

8.3 How to Prove A = B

In proofs it is often necessary to show that two sets are equal. There is a standard way of doing this. Suppose we want to show $A = B$. If we show $A \subseteq B$, then every element of A is also in B, but there is still a possibility that B could have some elements that are not in A, so we can't conclude $A = B$. But if *in addition* we also show $B \subseteq A$, then B can't contain anything that is not in A, so $A = B$. This is the standard procedure for proving $A = B$: Prove both $A \subseteq B$ and $B \subseteq A$.

How to Prove A = B

Proof.
[Prove that $A \subseteq B$.]
[Prove that $B \subseteq A$.]

Therefore, since $A \subseteq B$ and $B \subseteq A$,
it follows that $A = B$. ■

Example 8.10 Prove that $\{n \in \mathbb{Z} : 35 \mid n\} = \{n \in \mathbb{Z} : 5 \mid n\} \cap \{n \in \mathbb{Z} : 7 \mid n\}$.

Proof. First we show $\{n \in \mathbb{Z} : 35 \mid n\} \subseteq \{n \in \mathbb{Z} : 5 \mid n\} \cap \{n \in \mathbb{Z} : 7 \mid n\}$. Suppose $a \in \{n \in \mathbb{Z} : 35 \mid n\}$. This means $35 \mid a$, so $a = 35c$ for some $c \in \mathbb{Z}$. Thus $a = 5(7c)$ and $a = 7(5c)$. From $a = 5(7c)$ it follows that $5 \mid a$, so $a \in \{n \in \mathbb{Z} : 5 \mid n\}$. From $a = 7(5c)$ it follows that $7 \mid a$, which means $a \in \{n \in \mathbb{Z} : 7 \mid n\}$. As a belongs to both $\{n \in \mathbb{Z} : 5 \mid n\}$ and $\{n \in \mathbb{Z} : 7 \mid n\}$, we get $a \in \{n \in \mathbb{Z} : 5 \mid n\} \cap \{n \in \mathbb{Z} : 7 \mid n\}$. Thus we've shown that $\{n \in \mathbb{Z} : 35 \mid n\} \subseteq \{n \in \mathbb{Z} : 5 \mid n\} \cap \{n \in \mathbb{Z} : 7 \mid n\}$.

Next we show $\{n \in \mathbb{Z} : 5 \mid n\} \cap \{n \in \mathbb{Z} : 7 \mid n\} \subseteq \{n \in \mathbb{Z} : 35 \mid n\}$. Suppose that $a \in \{n \in \mathbb{Z} : 5 \mid n\} \cap \{n \in \mathbb{Z} : 7 \mid n\}$. By definition of intersection, this means that $a \in \{n \in \mathbb{Z} : 5 \mid n\}$ and $a \in \{n \in \mathbb{Z} : 7 \mid n\}$. Therefore it follows that $5 \mid a$ and $7 \mid a$. By definition of divisibility, there are integers c and d with $a = 5c$ and $a = 7d$. Then a has both 5 and 7 as prime factors, so the prime factorization of a

must include factors of 5 and 7. Hence $5 \cdot 7 = 35$ divides a, so $a \in \{n \in \mathbb{Z} : 35 \mid n\}$. We've now shown that $\{n \in \mathbb{Z} : 5 \mid n\} \cap \{n \in \mathbb{Z} : 7 \mid n\} \subseteq \{n \in \mathbb{Z} : 35 \mid n\}$.

At this point we've shown that $\{n \in \mathbb{Z} : 35 \mid n\} \subseteq \{n \in \mathbb{Z} : 5 \mid n\} \cap \{n \in \mathbb{Z} : 7 \mid n\}$ and $\{n \in \mathbb{Z} : 5 \mid n\} \cap \{n \in \mathbb{Z} : 7 \mid n\} \subseteq \{n \in \mathbb{Z} : 35 \mid n\}$, so we've proved $\{n \in \mathbb{Z} : 35 \mid n\} = \{n \in \mathbb{Z} : 5 \mid n\} \cap \{n \in \mathbb{Z} : 7 \mid n\}$. ■

You know from algebra that if $c \neq 0$ and $ac = bc$, then $a = b$. The next example shows that an analogous statement holds for sets A, B and C. The example asks us to prove a conditional statement. We will prove it with direct proof. In carrying out the process of direct proof, we will have to use the new techniques from this section.

Example 8.11 Suppose A, B, and C are sets, and $C \neq \emptyset$. Prove that if $A \times C = B \times C$, then $A = B$.

Proof. Suppose $A \times C = B \times C$. We must now show $A = B$.

First we will show $A \subseteq B$. Suppose $a \in A$. Since $C \neq \emptyset$, there exists an element $c \in C$. Thus, since $a \in A$ and $c \in C$, we have $(a, c) \in A \times C$, by definition of the Cartesian product. But then, since $A \times C = B \times C$, it follows that $(a, c) \in B \times C$. Again by definition of the Cartesian product, it follows that $a \in B$. We have shown $a \in A$ implies $a \in B$, so $A \subseteq B$.

Next we show $B \subseteq A$. We use the same argument as above, with the roles of A and B reversed. Suppose $a \in B$. Since $C \neq \emptyset$, there exists an element $c \in C$. Thus, since $a \in B$ and $c \in C$, we have $(a, c) \in B \times C$. But then, since $B \times C = A \times C$, we have $(a, c) \in A \times C$. It follows that $a \in A$. We have shown $a \in B$ implies $a \in A$, so $B \subseteq A$.

The previous two paragraphs have shown $A \subseteq B$ and $B \subseteq A$, so $A = B$. In summary, we have shown that if $A \times C = B \times C$, then $A = B$. This completes the proof. ■

Now we'll look at another way that set operations are similar to operations on numbers. From algebra you are familiar with the distributive property $a \cdot (b + c) = a \cdot b + a \cdot c$. Replace the numbers a, b, c with sets A, B, C, and replace $\cdot$ with $\times$ and $+$ with $\cap$. We get $A \times (B \cap C) = (A \times B) \cap (A \times C)$. This statement turns out to be true, as we now prove.

Example 8.12 Given sets A, B, and C, prove $A \times (B \cap C) = (A \times B) \cap (A \times C)$.

Proof. First we will show that $A \times (B \cap C) \subseteq (A \times B) \cap (A \times C)$.
Suppose $(a, b) \in A \times (B \cap C)$.
By definition of the Cartesian product, this means $a \in A$ and $b \in B \cap C$.
By definition of intersection, it follows that $b \in B$ and $b \in C$.

Thus, since $a \in A$ and $b \in B$, it follows that $(a,b) \in A \times B$ (by definition of $\times$).
Also, since $a \in A$ and $b \in C$, it follows that $(a,b) \in A \times C$ (by definition of $\times$).
Now we have $(a,b) \in A \times B$ and $(a,b) \in A \times C$, so $(a,b) \in (A \times B) \cap (A \times C)$.
We've shown that $(a,b) \in A \times (B \cap C)$ implies $(a,b) \in (A \times B) \cap (A \times C)$ so we have $A \times (B \cap C) \subseteq (A \times B) \cap (A \times C)$.

Next we will show that $(A \times B) \cap (A \times C) \subseteq A \times (B \cap C)$.
Suppose $(a,b) \in (A \times B) \cap (A \times C)$.
By definition of intersection, this means $(a,b) \in A \times B$ and $(a,b) \in A \times C$.
By definition of the Cartesian product, $(a,b) \in A \times B$ means $a \in A$ and $b \in B$.
By definition of the Cartesian product, $(a,b) \in A \times C$ means $a \in A$ and $b \in C$.
We now have $b \in B$ and $b \in C$, so $b \in B \cap C$, by definition of intersection.
Thus we've deduced that $a \in A$ and $b \in B \cap C$, so $(a,b) \in A \times (B \cap C)$.
In summary, we've shown that $(a,b) \in (A \times B) \cap (A \times C)$ implies $(a,b) \in A \times (B \cap C)$ so we have $(A \times B) \cap (A \times C) \subseteq A \times (B \cap C)$.

The previous two paragraphs show that $A \times (B \cap C) \subseteq (A \times B) \cap (A \times C)$ and $(A \times B) \cap (A \times C) \subseteq A \times (B \cap C)$, so it follows that $(A \times B) \cap (A \times C) = A \times (B \cap C)$. ■

Occasionally you can prove two sets are equal by working out a series of equalities leading from one set to the other. This is analogous to showing two algebraic expressions are equal by manipulating one until you obtain the other. We illustrate this in the following example, which gives an alternate solution to the previous example. You are cautioned that this approach is sometimes difficult to apply, but when it works it can shorten a proof dramatically.

Before beginning the example, a note is in order. Notice that any statement P is logically equivalent to $P \wedge P$. (Write out a truth table if you are in doubt.) At one point in the following example we will replace the expression $x \in A$ with the logically equivalent statement $(x \in A) \wedge (x \in A)$.

Example 8.13 Given sets A, B, and C, prove $A \times (B \cap C) = (A \times B) \cap (A \times C)$.

Proof. Just observe the following sequence of equalities.

$$\begin{aligned}
A \times (B \cap C) &= \{(x,y) : (x \in A) \wedge (y \in B \cap C)\} && \text{(def. of } \times) \\
&= \{(x,y) : (x \in A) \wedge (y \in B) \wedge (y \in C)\} && \text{(def. of } \cap) \\
&= \{(x,y) : (x \in A) \wedge (x \in A) \wedge (y \in B) \wedge (y \in C)\} && (P = P \wedge P) \\
&= \{(x,y) : ((x \in A) \wedge (y \in B)) \wedge ((x \in A) \wedge (y \in C))\} && \text{(rearrange)} \\
&= \{(x,y) : (x \in A) \wedge (y \in B)\} \cap \{(x,y) : (x \in A) \wedge (y \in C)\} && \text{(def. of } \cap) \\
&= (A \times B) \cap (A \times C) && \text{(def. of } \times)
\end{aligned}$$

The proof is complete. ■

The equation $A \times (B \cap C) = (A \times B) \cap (A \times C)$ just obtained is a fundamental law that you may actually use fairly often as you continue with mathematics. Some similar equations are listed below. Each of these can be proved with this section's techniques, and the exercises will ask that you do so.

$$\left.\begin{array}{r} \overline{A \cap B} = \overline{A} \cup \overline{B} \\ \overline{A \cup B} = \overline{A} \cap \overline{B} \end{array}\right\} \text{ DeMorgan's laws for sets}$$

$$\left.\begin{array}{r} A \cap (B \cup C) = (A \cap B) \cup (A \cap C) \\ A \cup (B \cap C) = (A \cup B) \cap (A \cup C) \end{array}\right\} \text{ Distributive laws for sets}$$

$$\left.\begin{array}{r} A \times (B \cup C) = (A \times B) \cup (A \times C) \\ A \times (B \cap C) = (A \times B) \cap (A \times C) \end{array}\right\} \text{ Distributive laws for sets}$$

It is very good practice to prove these equations. Depending on your learning style, it is probably not necessary to commit them to memory. But don't forget them entirely. They may well be useful later in your mathematical education. If so, you can look them up or re-derive them on the spot. If you go on to study mathematics deeply, you will at some point realize that you've internalized them without even being cognizant of it.

8.4 Examples: Perfect Numbers

Sometimes it takes a good bit of work and creativity to show that one set is a subset of another or that they are equal. We illustrate this now with examples from number theory involving what are called perfect numbers. Even though this topic is quite old, dating back more than 2000 years, it leads to some questions that are unanswered even today.

The problem involves adding up the positive divisors of a natural number. To begin the discussion, consider the number 12. If we add up the positive divisors of 12 that are less than 12, we obtain $1+2+3+4+6=16$, which is greater than 12. Doing the same thing for 15, we get $1+3+5=9$ which is less than 15. For the most part, given a natural number p, the sum of its positive divisors less than itself will either be greater than p or less than p. But occasionally the divisors add up to exactly p. If this happens, then p is said to be a *perfect number*.

Definition 8.1 A number $p \in \mathbb{N}$ is **perfect** if it equals the sum of its positive divisors less than itself. Some examples follow.

- The number 6 is perfect since $6 = 1+2+3$.
- The number 28 is perfect since $28 = 1+2+4+7+14$.
- The number 496 is perfect since $496 = 1+2+4+8+16+31+62+124+248$.

Though it would take a while to find it by trial-and-error, the next perfect number after 496 is 8128. You can check that 8128 is perfect. Its divisors are 1, 2, 4, 8, 16, 32, 64, 127, 254, 508, 1016, 2032, 4064 and indeed

$$8128 = 1+2+4+8+16+32+64+127+254+508+1016+2032+4064.$$

Are there other perfect numbers? How can they be found? Do they obey any patterns? These questions fascinated the ancient Greek mathematicians. In what follows we will develop an idea—recorded by Euclid—that partially answers these questions. Although Euclid did not use sets,[1] we will nonetheless phrase his idea using the language of sets.

Since our goal is to understand what numbers are perfect, let's define the following set:

$$P = \{p \in \mathbb{N} : p \text{ is perfect}\}.$$

Therefore $P = \{6, 28, 496, 8128, \ldots\}$, but it is unclear what numbers are in P other than the ones listed. Our goal is to gain a better understanding of just which numbers the set P includes. To do this, we will examine the following set A. It looks more complicated than P, but it will be very helpful for understanding P, as we will soon see.

$$A = \{2^{n-1}(2^n - 1) : n \in \mathbb{N}, \text{ and } 2^n - 1 \text{ is prime}\}$$

In words, A consists of every natural number of form $2^{n-1}(2^n-1)$, where 2^n-1 is prime. To get a feel for what numbers belong to A, look at the following table. For each natural number n, it tallies the corresponding numbers 2^{n-1} and 2^n-1. If 2^n-1 happens to be prime, then the product $2^{n-1}(2^n-1)$ is given; otherwise that entry is labeled with an $*$.

n	2^{n-1}	2^n-1	$2^{n-1}(2^n-1)$
1	1	1	*
2	2	3	6
3	4	7	28
4	8	15	*
5	16	31	496
6	32	63	*
7	64	127	8128
8	128	255	*
9	256	511	*
10	512	1023	*
11	1024	2047	*
12	2048	4095	*
13	4096	8191	33,550,336

[1]Set theory was invented over 2000 years after Euclid died.

Notice that the first four entries of A are the perfect numbers 6, 28, 496 and 8128. At this point you may want to jump to the conclusion that $A = P$. But it is a shocking fact that in over 2000 years no one has ever been able to determine whether or not $A = P$. But it is known that $A \subseteq P$, and we will now prove it. In other words, we are going to show that every element of A is perfect. (But by itself, that leaves open the possibility that there may be some perfect numbers in P that are not in A.)

The main ingredient for the proof will be the formula for the sum of a geometric series with common ratio r. You probably saw this most recently in Calculus II. The formula is

$$\sum_{k=0}^{n} r^k = \frac{r^{n+1}-1}{r-1}.$$

We will need this for the case $r = 2$, which is

$$\sum_{k=0}^{n} 2^k = 2^{n+1} - 1. \tag{8.1}$$

(See the solution for Exercise 19 in Section 7.4 for a proof of this formula.) Now we are ready to prove our result. Let's draw attention to its significance by calling it a theorem rather than a proposition.

Theorem 8.1 If $A = \{2^{n-1}(2^n - 1) : n \in \mathbb{N}, \text{ and } 2^n - 1 \text{ is prime}\}$ and $P = \{p \in \mathbb{N} : p \text{ is perfect}\}$, then $A \subseteq P$.

Proof. Assume A and P are as stated. To show $A \subseteq P$, we must show that $p \in A$ implies $p \in P$. Thus suppose $p \in A$. By definition of A, this means

$$p = 2^{n-1}(2^n - 1) \tag{8.2}$$

for some $n \in \mathbb{N}$ for which $2^n - 1$ is prime. We want to show that $p \in P$, that is, we want to show p is perfect. Thus, we need to show that the sum of the positive divisors of p that are less than p add up to p. Notice that since $2^n - 1$ is prime, any divisor of $p = 2^{n-1}(2^n - 1)$ must have the form 2^k or $2^k(2^n - 1)$ for $0 \le k \le n-1$. Thus the positive divisors of p are as follows:

$$\begin{array}{llllll} 2^0, & 2^1, & 2^2, & \ldots & 2^{n-2}, & 2^{n-1}, \\ 2^0(2^n-1), & 2^1(2^n-1), & 2^2(2^n-1), & \ldots & 2^{n-2}(2^n-1), & 2^{n-1}(2^n-1). \end{array}$$

Notice that this list starts with $2^0 = 1$ and ends with $2^{n-1}(2^n - 1) = p$.

If we add up all these divisors except for the last one (which equals p) we get the following:

$$
\begin{aligned}
\sum_{k=0}^{n-1} 2^k + \sum_{k=0}^{n-2} 2^k(2^n - 1) &= \sum_{k=0}^{n-1} 2^k + (2^n - 1)\sum_{k=0}^{n-2} 2^k \\
&= (2^n - 1) + (2^n - 1)(2^{n-1} - 1) \qquad \text{(by Equation (8.1))} \\
&= [1 + (2^{n-1} - 1)](2^n - 1) \\
&= 2^{n-1}(2^n - 1) \\
&= p \qquad \text{(by Equation (8.2))}.
\end{aligned}
$$

This shows that the positive divisors of p that are less than p add up to p. Therefore p is perfect, by definition of a perfect number. Thus $p \in P$, by definition of P.

We have shown that $p \in A$ implies $p \in P$, which means $A \subseteq P$. ■

Combined with the chart on the previous page, this theorem gives us a new perfect number! The element $p = 2^{13-1}(2^{13} - 1) = 33,550,336$ in A is perfect.

Observe also that every element of A is a multiple of a power of 2, and therefore even. But this does not necessarily mean every perfect number is even, because we've only shown $A \subseteq P$, not $A = P$. For all we know there may be odd perfect numbers in $P - A$ that are not in A.

Are there any odd perfect numbers? No one knows.

In over 2000 years, no one has ever found an odd perfect number, nor has anyone been able to prove that there are none. But it *is* known that the set A does contain every *even* perfect number. This fact was first proved by Euler, and we duplicate his reasoning in the next theorem, which proves that $A = E$, where E is the set of all *even* perfect numbers. It is a good example of how to prove two sets are equal.

For convenience, we are going to use a slightly different definition of a perfect number. A number $p \in \mathbb{N}$ is **perfect** if its positive divisors add up to $2p$. For example, the number 6 is perfect since the sum of its divisors is $1+2+3+6 = 2\cdot 6$. This definition is simpler than the first one because we do not have to stipulate that we are adding up the divisors that are *less than p*. Instead we add in the last divisor p, and that has the effect of adding an additional p, thereby doubling the answer.

Theorem 8.2 If $A = \{2^{n-1}(2^n - 1) : n \in \mathbb{N}$, and $2^n - 1$ is prime$\}$ and $E = \{p \in \mathbb{N} : p$ is perfect and even$\}$, then $A = E$.

Proof. To show that $A = E$, we need to show $A \subseteq E$ and $E \subseteq A$.

First we will show that $A \subseteq E$. Suppose $p \in A$. This means p is even, because the definition of A shows that every element of A is a multiple of a power of 2. Also, p is a perfect number because Theorem 8.1 states that every element of A is also an element of P, hence perfect. Thus p is an even perfect number, so $p \in E$. Therefore $A \subseteq E$.

Next we show that $E \subseteq A$. Suppose $p \in E$. This means p is an even perfect number. Write the prime factorization of p as $p = 2^k 3^{n_1} 5^{n_2} 7^{n_2} \ldots$, where some of the powers n_1, n_2, $n_3 \ldots$ may be zero. But, as p is even, the power k must be greater than zero. It follows $p = 2^k q$ for some positive integer k and an odd integer q. Now, our aim is to show that $p \in A$, which means we must show p has form $p = 2^{n-1}(2^n - 1)$. To get our current $p = 2^k q$ closer to this form, let $n = k + 1$, so we now have

$$p = 2^{n-1} q. \tag{8.3}$$

List the positive divisors of q as $d_1, d_2, d_3, \ldots, d_m$. (Where $d_1 = 1$ and $d_m = q$.) Then the divisors of p are:

$$\begin{array}{ccccc}
2^0 d_1 & 2^0 d_2 & 2^0 d_3 & \cdots & 2^0 d_m \\
2^1 d_1 & 2^1 d_2 & 2^1 d_3 & \cdots & 2^1 d_m \\
2^2 d_1 & 2^2 d_2 & 2^2 d_3 & \cdots & 2^2 d_m \\
2^3 d_1 & 2^3 d_2 & 2^3 d_3 & \cdots & 2^3 d_m \\
\vdots & \vdots & \vdots & & \vdots \\
2^{n-1} d_1 & 2^{n-1} d_2 & 2^{n-1} d_3 & \cdots & 2^{n-1} d_m
\end{array}$$

Since p is perfect, these divisors add up to $2p$. By Equation (8.3), their sum is $2p = 2(2^{n-1} q) = 2^n q$. Adding the divisors column-by-column, we get

$$\sum_{k=0}^{n-1} 2^k d_1 + \sum_{k=0}^{n-1} 2^k d_2 + \sum_{k=0}^{n-1} 2^k d_3 + \cdots + \sum_{k=0}^{n-1} 2^k d_m \quad = \quad 2^n q.$$

Applying Equation (8.1), this becomes

$$\begin{aligned}
(2^n - 1)d_1 + (2^n - 1)d_2 + (2^n - 1)d_3 + \cdots + (2^n - 1)d_m &= 2^n q \\
(2^n - 1)(d_1 + d_2 + d_3 + \cdots + d_m) &= 2^n q \\
d_1 + d_2 + d_3 + \cdots + d_m &= \frac{2^n q}{2^n - 1},
\end{aligned}$$

so that

$$d_1+d_2+d_3+\cdots+d_m=\frac{(2^n-1+1)q}{2^n-1}=\frac{(2^n-1)q+q}{2^n-1}=q+\frac{q}{2^n-1}.$$

From this we see that $\frac{q}{2^n-1}$ is an integer. It follows that both q and $\frac{q}{2^n-1}$ are positive divisors of q. Since their sum equals the sum of *all* positive divisors of q, it follows that q has only two positive divisors, q and $\frac{q}{2^n-1}$. Since one of its divisors must be 1, it must be that $\frac{q}{2^n-1}=1$, which means $q=2^n-1$. Now a number with just two positive divisors is prime, so $q=2^n-1$ is prime. Plugging this into Equation (8.3) gives $p=2^{n-1}(2^n-1)$, where 2^n-1 is prime. This means $p\in A$, by definition of A. We have now shown that $p\in E$ implies $p\in A$, so $E\subseteq A$.

Since $A\subseteq E$ and $E\subseteq A$, it follows that $A=E$. ■

Do not be alarmed if you feel that you wouldn't have thought of this proof. It took the genius of Euler to discover this approach.

We'll conclude this chapter with some facts about perfect numbers.

- The sixth perfect number is $p=2^{17-1}(2^{17}-1)=8589869056$.
- The seventh perfect number is $p=2^{19-1}(2^{19}-1)=137438691328$.
- The eighth perfect number is $p=2^{31-1}(2^{31}-1)=2305843008139952128$.
- The twentieth perfect number is $p=2^{4423-1}(2^{4423}-1)$. It has 2663 digits.
- The twenty-third perfect number is $p=2^{11213-1}(2^{11213}-1)$. It has 6957 digits.

As mentioned earlier, no one knows whether or not there are any odd perfect numbers. It is not even known whether there are finitely many or infinitely many perfect numbers. It **is** known that the last digit of every even perfect number is either a 6 or an 8. Perhaps this is something you'd enjoy proving.

We've seen that perfect numbers are closely related to prime numbers that have the form 2^n-1. Such prime numbers are called **Mersenne primes**, after the French scholar Marin Mersenne (1588–1648), who popularized them. The first several Mersenne primes are $2^2-1=3$, $2^3-1=7$, $2^5-1=31$, $2^7-1=127$ and $2^{13}-1=8191$. To date, only 48 Mersenne primes are known, the largest of which is $2^{57,885,161}-1$. There is a substantial cash prize for anyone who finds a 49th. (See http://www.mersenne.org/prime.htm.) You will probably have better luck with the exercises.

Exercises for Chapter 8

Use the methods introduced in this chapter to prove the following statements.

1. Prove that $\{12n : n \in \mathbb{Z}\} \subseteq \{2n : n \in \mathbb{Z}\} \cap \{3n : n \in \mathbb{Z}\}$.
2. Prove that $\{6n : n \in \mathbb{Z}\} = \{2n : n \in \mathbb{Z}\} \cap \{3n : n \in \mathbb{Z}\}$.
3. If $k \in \mathbb{Z}$, then $\{n \in \mathbb{Z} : n \mid k\} \subseteq \{n \in \mathbb{Z} : n \mid k^2\}$.
4. If $m,n \in \mathbb{Z}$, then $\{x \in \mathbb{Z} : mn \mid x\} \subseteq \{x \in \mathbb{Z} : m \mid x\} \cap \{x \in \mathbb{Z} : n \mid x\}$.
5. If p and q are positive integers, then $\{pn : n \in \mathbb{N}\} \cap \{qn : n \in \mathbb{N}\} \neq \emptyset$.
6. Suppose A,B and C are sets. Prove that if $A \subseteq B$, then $A - C \subseteq B - C$.
7. Suppose A,B and C are sets. If $B \subseteq C$, then $A \times B \subseteq A \times C$.
8. If A,B and C are sets, then $A \cup (B \cap C) = (A \cup B) \cap (A \cup C)$.
9. If A,B and C are sets, then $A \cap (B \cup C) = (A \cap B) \cup (A \cap C)$.
10. If A and B are sets in a universal set U, then $\overline{A \cap B} = \overline{A} \cup \overline{B}$.
11. If A and B are sets in a universal set U, then $\overline{A \cup B} = \overline{A} \cap \overline{B}$.
12. If A,B and C are sets, then $A - (B \cap C) = (A - B) \cup (A - C)$.
13. If A,B and C are sets, then $A - (B \cup C) = (A - B) \cap (A - C)$.
14. If A,B and C are sets, then $(A \cup B) - C = (A - C) \cup (B - C)$.
15. If A,B and C are sets, then $(A \cap B) - C = (A - C) \cap (B - C)$.
16. If A,B and C are sets, then $A \times (B \cup C) = (A \times B) \cup (A \times C)$.
17. If A,B and C are sets, then $A \times (B \cap C) = (A \times B) \cap (A \times C)$.
18. If A,B and C are sets, then $A \times (B - C) = (A \times B) - (A \times C)$.
19. Prove that $\{9^n : n \in \mathbb{Z}\} \subseteq \{3^n : n \in \mathbb{Z}\}$, but $\{9^n : n \in \mathbb{Z}\} \neq \{3^n : n \in \mathbb{Z}\}$
20. Prove that $\{9^n : n \in \mathbb{Q}\} = \{3^n : n \in \mathbb{Q}\}$.
21. Suppose A and B are sets. Prove $A \subseteq B$ if and only if $A - B = \emptyset$.
22. Let A and B be sets. Prove that $A \subseteq B$ if and only if $A \cap B = A$.
23. For each $a \in \mathbb{R}$, let $A_a = \{(x, a(x^2-1)) \in \mathbb{R}^2 : x \in \mathbb{R}\}$. Prove that $\bigcap_{a \in \mathbb{R}} A_a = \{(-1,0),(1,0)\}$.
24. Prove that $\bigcap_{x \in \mathbb{R}} [3 - x^2, 5 + x^2] = [3,5]$.
25. Suppose A,B,C and D are sets. Prove that $(A \times B) \cup (C \times D) \subseteq (A \cup C) \times (B \cup D)$.
26. Prove $\{4k+5 : k \in \mathbb{Z}\} = \{4k+1 : k \in \mathbb{Z}\}$.
27. Prove $\{12a + 4b : a,b \in \mathbb{Z}\} = \{4c : c \in \mathbb{Z}\}$.
28. Prove $\{12a + 25b : a,b \in \mathbb{Z}\} = \mathbb{Z}$.
29. Suppose $A \neq \emptyset$. Prove that $A \times B \subseteq A \times C$, if and only if $B \subseteq C$.
30. Prove that $(\mathbb{Z} \times \mathbb{N}) \cap (\mathbb{N} \times \mathbb{Z}) = \mathbb{N} \times \mathbb{N}$.
31. Suppose $B \neq \emptyset$ and $A \times B \subseteq B \times C$. Prove $A \subseteq C$.

CHAPTER 9

Disproof

Ever since Chapter 4 we have dealt with one major theme: Given a statement, prove that is it true. In every example and exercise we were handed a true statement and charged with the task of proving it. Have you ever wondered what would happen if you were given a *false* statement to prove? The answer is that no (correct) proof would be possible, for if it were, the statement would be true, not false.

But how would you convince someone that a statement is false? The mere fact that you could not produce a proof does not automatically mean the statement is false, for you know (perhaps all too well) that proofs can be difficult to construct. It turns out that there is a very simple and utterly convincing procedure that proves a statement is false. The process of carrying out this procedure is called **disproof**. Thus, this chapter is concerned with **disproving** statements.

Before describing the new method, we will set the stage with some relevant background information. First, we point out that mathematical statements can be divided into three categories, described below.

One category consists of all those statements that have been proved to be true. For the most part we regard these statements as significant enough to be designated with special names such as "theorem," "proposition," "lemma" and "corollary." Some examples of statements in this category are listed in the left-hand box in the diagram on the following page. There are also some wholly uninteresting statements (such as $2 = 2$) in this category, and although we acknowledge their existence we certainly do not dignify them with terms such as "theorem" or "proposition."

At the other extreme is a category consisting of statements that are known to be false. Examples are listed in the box on the right. Since mathematicians are not very interested in them, these types of statements do not get any special names, other than the blanket term "false statement."

But there is a third (and quite interesting) category between these two extremes. It consists of statements whose truth or falsity has not been determined. Examples include things like *"Every perfect number is even,"* or *"Every even integer greater than 2 is the sum of two primes."* (The latter statement is called the *Goldbach conjecture*. See Section 2.1.) Mathematicians have a special name for the statements in this category that they suspect (but haven't yet proved) are true. Such statements are called **conjectures**.

THREE TYPES OF STATEMENTS:

Known to be true (Theorems & propositions)	**Truth unknown** (Conjectures)	**Known to be false**
Examples: • Pythagorean theorem • Fermat's last theorem (Section 2.1) • The square of an odd number is odd. • The series $\sum_{k=1}^{\infty} \frac{1}{k}$ diverges.	Examples: • All perfect numbers are even. • Any even number greater than 2 is the sum of two primes. (Goldbach's conjecture, Section 2.1) • There are infinitely many prime numbers of form $2^n - 1$, with $n \in \mathbb{N}$.	Examples: • All prime numbers are odd. • Some quadratic equations have three solutions. • $0 = 1$ • There exist natural numbers a, b and c for which $a^3 + b^3 = c^3$.

Mathematicians spend much of their time and energy attempting to prove or disprove conjectures. (They also expend considerable mental energy in creating new conjectures based on collected evidence or intuition.) When a conjecture is proved (or disproved) the proof or disproof will typically appear in a published paper, provided the conjecture is of sufficient interest. If it is proved, the conjecture attains the status of a theorem or proposition. If it is disproved, then no one is really very interested in it anymore—mathematicians do not care much for false statements.

Most conjectures that mathematicians are interested in are quite difficult to prove or disprove. We are not at that level yet. In this text, the "conjectures" that you will encounter are the kinds of statements that an experienced mathematician would immediately spot as true or false, but you may have to do some work before figuring out a proof or disproof. But in keeping with the cloud of uncertainty that surrounds conjectures at the advanced levels of mathematics, most exercises in this chapter (and many beyond it) will ask you to prove or disprove statements without giving any hint as to whether they are true or false. Your job will be to decide whether or not they are true and to either prove or disprove them. The examples in this chapter will illustrate the processes one typically goes through in deciding whether a statement is true or false, and then verifying that it's true or false.

You know the three major methods of proving a statement: direct proof, contrapositive proof and proof by contradiction. Now we are ready to understand the method of disproving a statement. Suppose you want to disprove a statement P. In other words you want to prove that P is *false*. The way to do this is to prove that $\sim P$ is *true*, for if $\sim P$ is true, it follows immediately that P has to be false.

How to disprove P: Prove $\sim P$.

Our approach is incredibly simple. To disprove P, prove $\sim P$. In theory, this proof can be carried out by direct, contrapositive or contradiction approaches. However, in practice things can be even easier than that if we are disproving a universally quantified statement or a conditional statement. That is our next topic.

9.1 Disproving Universal Statements: Counterexamples

A conjecture may be described as a statement that we hope is a theorem. As we know, many theorems (hence many conjectures) are universally quantified statements. Thus it seems reasonable to begin our discussion by investigating how to disprove a universally quantified statement such as

$$\forall x \in S, P(x).$$

To disprove this statement, we must prove its negation. Its negation is

$$\sim (\forall x \in S, P(x)) \quad = \quad \exists x \in S, \sim P(x).$$

The negation is an existence statement. To prove the negation is true, we just need to produce an *example* of an $x \in S$ that makes $\sim P(x)$ true, that is, an x that makes $P(x)$ false. This leads to the following outline for disproving a universally quantified statement.

How to disprove $\forall x \in S, P(x)$.
Produce an example of an $x \in S$ that makes $P(x)$ false.

Things are even simpler if we want to disprove a conditional statement $P(x) \Rightarrow Q(x)$. This statement asserts that for every x that makes $P(x)$ true, $Q(x)$ will also be true. The statement can only be false if there is an x that makes $P(x)$ true and $Q(x)$ false. This leads to our next outline for disproof.

> **How to disprove** $P(x) \Rightarrow Q(x)$.
> Produce an example of an x that makes $P(x)$ true and $Q(x)$ false.

In both of the above outlines, the statement is disproved simply by exhibiting an example that shows the statement is not always true. (Think of it as an example that proves the statement is a promise that can be broken.) There is a special name for an example that disproves a statement: It is called a **counterexample**.

Example 9.1 As our first example, we will work through the process of deciding whether or not the following conjecture is true.

Conjecture: For every $n \in \mathbb{Z}$, the integer $f(n) = n^2 - n + 11$ is prime.

In resolving the truth or falsity of a conjecture, it's a good idea to gather as much information about the conjecture as possible. In this case let's start by making a table that tallies the values of $f(n)$ for some integers n.

n	−3	−2	−1	0	1	2	3	4	5	6	7	8	9	10
$f(n)$	23	17	13	11	11	13	17	23	31	41	53	67	83	101

In every case, $f(n)$ is prime, so you may begin to suspect that the conjecture is true. Before attempting a proof, let's try one more n. Unfortunately, $f(11) = 11^2 - 11 + 11 = 11^2$ is not prime. The conjecture is false because $n = 11$ is a counterexample. We summarize our disproof as follows:

Disproof. The statement "*For every* $n \in \mathbb{Z}$, *the integer* $f(n) = n^2 - n + 11$ *is prime,*" is **false**. For a counterexample, note that for $n = 11$, the integer $f(11) = 121 = 11 \cdot 11$ is not prime. ■

In disproving a statement with a counterexample, it is important to explain exactly how the counterexample makes the statement false. Our work would not have been complete if we had just said "*for a counterexample, consider* $n = 11$," and left it at that. We need to show that the answer $f(11)$ is not prime. Showing the factorization $f(11) = 11 \cdot 11$ suffices for this.

Example 9.2 Either prove or disprove the following conjecture.

Conjecture If A, B and C are sets, then $A-(B\cap C)=(A-B)\cap(A-C)$.

Disproof. This conjecture is false because of the following counterexample. Let $A=\{1,2,3\}$, $B=\{1,2\}$ and $C=\{2,3\}$. Notice that $A-(B\cap C)=\{1,3\}$ and $(A-B)\cap(A-C)=\emptyset$, so $A-(B\cap C)\neq(A-B)\cap(A-C)$. ■

(To see where this counterexample came from, draw Venn diagrams for $A-(B\cap C)$ and $(A-B)\cap(A-C)$. You will see that the diagrams are different. The numbers 1, 2 and 3 can then be inserted into the regions of the diagrams in such a way as to create the above counterexample.)

9.2 Disproving Existence Statements

We have seen that we can disprove a universally quantified statement or a conditional statement simply by finding a counterexample. Now let's turn to the problem of disproving an existence statement such as

$$\exists x \in S, P(x).$$

Proving this would involve simply finding an example of an x that makes $P(x)$ true. To *disprove* it, we have to prove its negation $\sim(\exists x \in S, P(x)) = \forall x \in S, \sim P(x)$. But this negation is universally quantified. Proving *it* involves showing that $\sim P(x)$ is true for *all* $x \in S$, and for this an example does not suffice. Instead we must use direct, contrapositive or contradiction proof to prove the conditional statement *"If $x \in S$, then $\sim P(x)$."* As an example, here is a conjecture to either prove or disprove.

Example 9.3 Either prove or disprove the following conjecture.

Conjecture: There is a real number x for which $x^4 < x < x^2$.

This may not seem like an unreasonable statement at first glance. After all, if the statement were asserting the existence of a real number for which $x^3 < x < x^2$, then it would be true: just take $x=-2$. But it asserts there is an x for which $x^4 < x < x^2$. When we apply some intelligent guessing to locate such an x we run into trouble. If $x=\frac{1}{2}$, then $x^4 < x$, but we don't have $x < x^2$; similarly if $x=2$, we have $x < x^2$ but not $x^4 < x$. Since finding an x with $x^4 < x < x^2$ seems problematic, we may begin to suspect that the given statement is false.

Let's see if we can disprove it. According to our strategy for disproof, to *disprove* it we must *prove* its negation. Symbolically, the statement is

$\exists x \in \mathbb{R}, x^4 < x < x^2$, so its negation is

$$\sim (\exists x \in \mathbb{R}, x^4 < x < x^2) \;=\; \forall x \in \mathbb{R}, \sim (x^4 < x < x^2).$$

Thus, in words the negation is:

For every real number x, it is not the case that $x^4 < x < x^2$.

This can be proved with contradiction, as follows. Suppose for the sake of contradiction that there **is** an x for which $x^4 < x < x^2$. Then x must be positive since it's greater than the non-negative number x^4. Dividing all parts of $x^4 < x < x^2$ by the positive number x produces $x^3 < 1 < x$. Now subtract 1 from all parts of $x^3 < 1 < x$ to obtain $x^3 - 1 < 0 < x - 1$ and reason as follows:

$$\begin{array}{rcl} x^3 - 1 & < 0 < & x - 1 \\ (x-1)(x^2 + x + 1) & < 0 < & (x-1) \\ x^2 + x + 1 & < 0 < & 1 \end{array}$$

(Division by $x - 1$ did not reverse the inequality $<$ because the second line above shows $0 < x - 1$, that is, $x - 1$ is positive.) Now we have $x^2 + x + 1 < 0$, which is a contradiction because x being positive forces $x^2 + x + 1 > 0$

We summarize our work as follows.

The statement *"There is a real number x for which $x^4 < x < x^2$"* is **false** because we have proved its negation *"For every real number x, it is not the case that $x^4 < x < x^2$."*

As you work the exercises, keep in mind that not every conjecture will be false. If one is true, then a disproof is impossible and you must produce a proof. Here is an example:

Example 9.4 Either prove or disprove the following conjecture.

Conjecture There exist three integers x, y, z, all greater than 1 and no two equal, for which $x^y = y^z$.

This conjecture is true. It is an existence statement, so to prove it we just need to give an example of three integers x, y, z, all greater than 1 and no two equal, so that $x^y = y^z$. A proof follows.

Proof. Note that if $x = 2$, $y = 16$ and $z = 4$, then $x^y = 2^{16} = (2^4)^4 = 16^4 = y^z$. ■

9.3 Disproof by Contradiction

Contradiction can be a very useful way to disprove a statement. To see how this works, suppose we wish to disprove a statement P. We know that to disprove P, we must *prove* $\sim P$. To prove $\sim P$ with contradiction, we assume $\sim\sim P$ is true and deduce a contradiction. But since $\sim\sim P = P$, this boils down to assuming P is true and deducing a contradiction. Here is an outline:

How to disprove P with contradiction:
Assume P is true, and deduce a contradiction.

To illustrate this, let's revisit Example 9.3 but do the disproof with contradiction. You will notice that the work duplicates much of what we did in Example 9.3, but is it much more streamlined because here we do not have to negate the conjecture.

Example 9.5 Disprove the following conjecture.

Conjecture: There is a real number x for which $x^4 < x < x^2$.

Disproof. Suppose for the sake of contradiction that this conjecture is true. Let x be a real number for which $x^4 < x < x^2$. Then x is positive, since it is greater than the non-negative number x^4. Dividing all parts of $x^4 < x < x^2$ by the positive number x produces $x^3 < 1 < x$. Now subtract 1 from all parts of $x^3 < 1 < x$ to obtain $x^3 - 1 < 0 < x - 1$ and reason as follows:

$$\begin{array}{rcl} x^3-1 & < 0 < & x-1 \\ (x-1)(x^2+x+1) & < 0 < & (x-1) \\ x^2+x+1 & < 0 < & 1 \end{array}$$

Now we have $x^2 + x + 1 < 0$, which is a contradiction because x is positive. Thus the conjecture must be false. ■

Exercises for Chapter 9

Each of the following statements is either true or false. If a statement is true, prove it. If a statement is false, disprove it. These exercises are cumulative, covering all topics addressed in Chapters 1–9.

1. If $x, y \in \mathbb{R}$, then $|x+y| = |x| + |y|$.
2. For every natural number n, the integer $2n^2 - 4n + 31$ is prime.

3. If $n \in \mathbb{Z}$ and $n^5 - n$ is even, then n is even.
4. For every natural number n, the integer $n^2 + 17n + 17$ is prime.
5. If A, B, C and D are sets, then $(A \times B) \cup (C \times D) = (A \cup C) \times (B \cup D)$.
6. If A, B, C and D are sets, then $(A \times B) \cap (C \times D) = (A \cap C) \times (B \cap D)$.
7. If A, B and C are sets, and $A \times C = B \times C$, then $A = B$.
8. If A, B and C are sets, then $A - (B \cup C) = (A - B) \cup (A - C)$.
9. If A and B are sets, then $\mathscr{P}(A) - \mathscr{P}(B) \subseteq \mathscr{P}(A - B)$.
10. If A and B are sets and $A \cap B = \emptyset$, then $\mathscr{P}(A) - \mathscr{P}(B) \subseteq \mathscr{P}(A - B)$.
11. If $a, b \in \mathbb{N}$, then $a + b < ab$.
12. If $a, b, c \in \mathbb{N}$ and ab, bc and ac all have the same parity, then a, b and c all have the same parity.
13. There exists a set X for which $\mathbb{R} \subseteq X$ and $\emptyset \in X$.
14. If A and B are sets, then $\mathscr{P}(A) \cap \mathscr{P}(B) = \mathscr{P}(A \cap B)$.
15. Every odd integer is the sum of three odd integers.
16. If A and B are finite sets, then $|A \cup B| = |A| + |B|$.
17. For all sets A and B, if $A - B = \emptyset$, then $B \neq \emptyset$.
18. If $a, b, c \in \mathbb{N}$, then at least one of $a - b$, $a + c$ and $b - c$ is even.
19. For every $r, s \in \mathbb{Q}$ with $r < s$, there is an irrational number u for which $r < u < s$.
20. There exist prime numbers p and q for which $p - q = 1000$.
21. There exist prime numbers p and q for which $p - q = 97$.
22. If p and q are prime numbers for which $p < q$, then $2p + q^2$ is odd.
23. If $x, y \in \mathbb{R}$ and $x^3 < y^3$, then $x < y$.
24. The inequality $2^x \geq x + 1$ is true for all positive real numbers x.
25. For all $a, b, c \in \mathbb{Z}$, if $a \mid bc$, then $a \mid b$ or $a \mid c$.
26. Suppose A, B and C are sets. If $A = B - C$, then $B = A \cup C$.
27. The equation $x^2 = 2^x$ has three real solutions.
28. Suppose $a, b \in \mathbb{Z}$. If $a \mid b$ and $b \mid a$, then $a = b$.
29. If $x, y \in \mathbb{R}$ and $|x + y| = |x - y|$, then $y = 0$.
30. There exist integers a and b for which $42a + 7b = 1$.
31. No number (other than 1) appears in Pascal's triangle more than four times.
32. If $n, k \in \mathbb{N}$ and $\binom{n}{k}$ is a prime number, then $k = 1$ or $k = n - 1$.
33. Suppose $f(x) = a_0 + a_1x + a_2x^2 + \cdots + a_nx^n$ is a polynomial of degree 1 or greater, and for which each coefficient a_i is in $\mathbb{N}$. Then there is an $n \in \mathbb{N}$ for which the integer $f(n)$ is not prime.
34. If $X \subseteq A \cup B$, then $X \subseteq A$ or $X \subseteq B$.

CHAPTER 10

Mathematical Induction

This chapter explains a powerful proof technique called **mathematical induction** (or just **induction** for short). To motivate the discussion, let's first examine the kinds of statements that induction is used to prove. Consider the following statement.

Conjecture. The sum of the first n odd natural numbers equals n^2.

The following table illustrates what this conjecture says. Each row is headed by a natural number n, followed by the sum of the first n odd natural numbers, followed by n^2.

n	sum of the first n odd natural numbers	n^2
1	$1 =$	1
2	$1+3 =$	4
3	$1+3+5 =$	9
4	$1+3+5+7 =$	16
5	$1+3+5+7+9 =$	25
$\vdots$	$\vdots$	$\vdots$
n	$1+3+5+7+9+11+\cdots+(2n-1) =$	n^2
$\vdots$	$\vdots$	$\vdots$

Note that in the first five lines of the table, the sum of the first n odd numbers really does add up to n^2. Notice also that these first five lines indicate that the nth odd natural number (the last number in each sum) is $2n-1$. (For instance, when $n = 2$, the second odd natural number is $2 \cdot 2 - 1 = 3$; when $n = 3$, the third odd natural number is $2 \cdot 3 - 1 = 5$, etc.)

The table raises a question. Does the sum $1+3+5+7+\cdots+(2n-1)$ really always equal n^2? In other words, is the conjecture true?

Let's rephrase this as follows. For each natural number n (i.e., for each line of the table), we have a statement S_n, as follows:

$$S_1 : 1 = 1^2$$
$$S_2 : 1 + 3 = 2^2$$
$$S_3 : 1 + 3 + 5 = 3^2$$
$$\vdots$$
$$S_n : 1 + 3 + 5 + 7 + \cdots + (2n - 1) = n^2$$
$$\vdots$$

Our question is: Are all of these statements true?

Mathematical induction is designed to answer just this kind of question. It is used when we have a set of statements $S_1, S_2, S_3, \ldots, S_n, \ldots$, and we need to prove that they are all true. The method is really quite simple. To visualize it, think of the statements as dominoes, lined up in a row. Imagine you can prove the first statement S_1, and symbolize this as domino S_1 being knocked down. Additionally, imagine that you can prove that any statement S_k being true (falling) forces the next statement S_{k+1} to be true (to fall). Then S_1 falls, and knocks down S_2. Next S_2 falls and knocks down S_3, then S_3 knocks down S_4, and so on. The inescapable conclusion is that all the statements are knocked down (proved true).

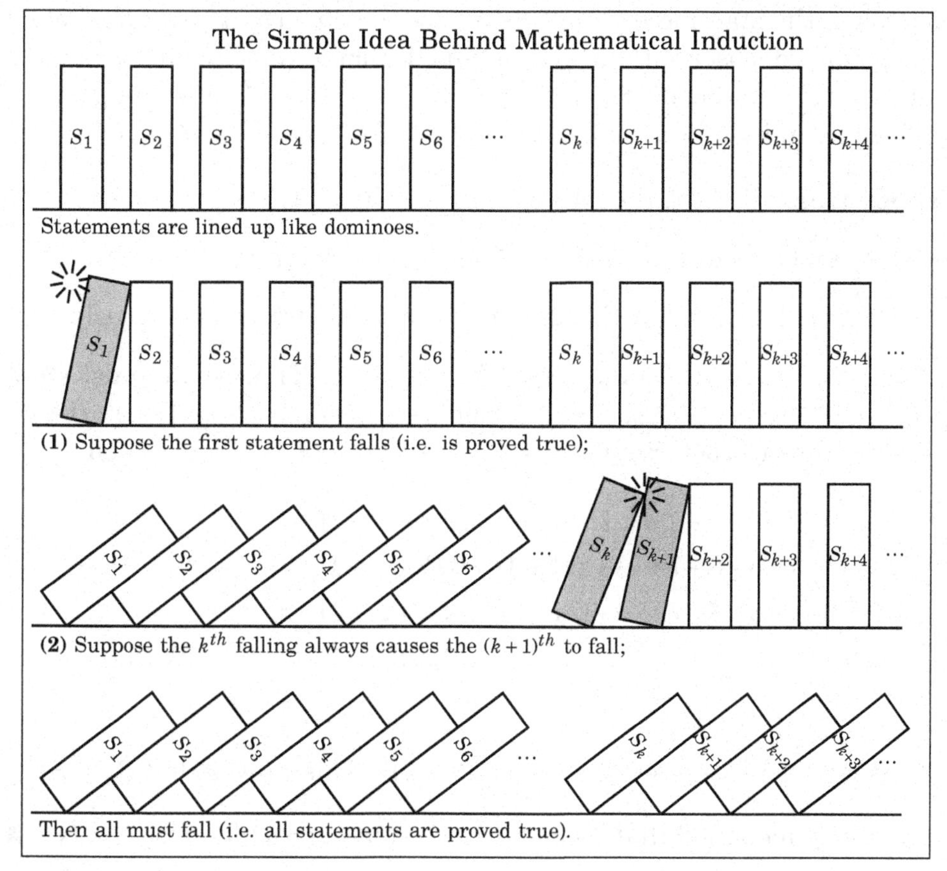

This picture gives our outline for *proof by mathematical induction.*

Outline for Proof by Induction

Proposition The statements $S_1, S_2, S_3, S_4, \ldots$ are all true.

Proof. (Induction)
(1) Prove that the first statement S_1 is true.
(2) Given any integer $k \geq 1$, prove that the statement $S_k \Rightarrow S_{k+1}$ is true.
It follows by mathematical induction that every S_n is true. ■

In this setup, the first step (1) is called the **basis step**. Because S_1 is usually a very simple statement, the basis step is often quite easy to do. The second step (2) is called the **inductive step**. In the inductive step direct proof is most often used to prove $S_k \Rightarrow S_{k+1}$, so this step is usually carried out by assuming S_k is true and showing this forces S_{k+1} to be true. The assumption that S_k is true is called the **inductive hypothesis**.

Now let's apply this technique to our original conjecture that the sum of the first n odd natural numbers equals n^2. Our goal is to show that for each $n \in \mathbb{N}$, the statement $S_n : 1+3+5+7+\cdots+(2n-1) = n^2$ is true. Before getting started, observe that S_k is obtained from S_n by plugging k in for n. Thus S_k is the statement $S_k : 1+3+5+7+\cdots+(2k-1) = k^2$. Also, we get S_{k+1} by plugging in $k+1$ for n, so that $S_{k+1} : 1+3+5+7+\cdots+(2(k+1)-1) = (k+1)^2$.

Proposition If $n \in \mathbb{N}$, then $1+3+5+7+\cdots+(2n-1) = n^2$.

Proof. We will prove this with mathematical induction.

(1) Observe that if $n = 1$, this statement is $1 = 1^2$, which is obviously true.

(2) We must now prove $S_k \Rightarrow S_{k+1}$ for any $k \geq 1$. That is, we must show that if $1+3+5+7+\cdots+(2k-1) = k^2$, then $1+3+5+7+\cdots+(2(k+1)-1) = (k+1)^2$. We use direct proof. Suppose $1+3+5+7+\cdots+(2k-1) = k^2$. Then

$$\begin{aligned}
1+3+5+7+\cdots\cdots\cdots\cdots+(2(k+1)-1) &= \\
1+3+5+7+\cdots+(2k-1)+(2(k+1)-1) &= \\
\big(1+3+5+7+\cdots+(2k-1)\big)+(2(k+1)-1) &= \\
k^2 \quad +(2(k+1)-1) &= k^2+2k+1 \\
&= (k+1)^2.
\end{aligned}$$

Thus $1+3+5+7+\cdots+(2(k+1)-1) = (k+1)^2$. This proves that $S_k \Rightarrow S_{k+1}$.

It follows by induction that $1+3+5+7+\cdots+(2n-1) = n^2$ for every $n \in \mathbb{N}$. ■

In induction proofs it is usually the case that the first statement S_1 is indexed by the natural number 1, but this need not always be so. Depending on the problem, the first statement could be S_0, or S_m for any other integer m. In the next example the statements are $S_0, S_1, S_2, S_3, \ldots$ The same outline is used, except that the basis step verifies S_0, not S_1.

Proposition If n is a non-negative integer, then $5 \mid (n^5 - n)$.

Proof. We will prove this with mathematical induction. Observe that the first non-negative integer is 0, so the basis step involves $n = 0$.

(1) If $n = 0$, this statement is $5 \mid (0^5 - 0)$ or $5 \mid 0$, which is obviously true.

(2) Let $k \geq 0$. We need to prove that if $5 \mid (k^5 - k)$, then $5 \mid ((k+1)^5 - (k+1))$. We use direct proof. Suppose $5 \mid (k^5 - k)$. Thus $k^5 - k = 5a$ for some $a \in \mathbb{Z}$. Observe that

$$\begin{aligned}
(k+1)^5 - (k+1) &= k^5 + 5k^4 + 10k^3 + 10k^2 + 5k + 1 - k - 1 \\
&= (k^5 - k) + 5k^4 + 10k^3 + 10k^2 + 5k \\
&= 5a + 5k^4 + 10k^3 + 10k^2 + 5k \\
&= 5(a + k^4 + 2k^3 + 2k^2 + k).
\end{aligned}$$

This shows $(k+1)^5 - (k+1)$ is an integer multiple of 5, so $5 \mid ((k+1)^5 - (k+1))$. We have now shown that $5 \mid (k^5 - k)$ implies $5 \mid ((k+1)^5 - (k+1))$.

It follows by induction that $5 \mid (n^5 - n)$ for all non-negative integers n. ■

As noted, induction is used to prove statements of the form $\forall n \in \mathbb{N}, S_n$. But notice the outline does *not* work for statements of form $\forall n \in \mathbb{Z}, S_n$ (where n is in $\mathbb{Z}$, not $\mathbb{N}$). The reason is that if you are trying to prove $\forall n \in \mathbb{Z}, S_n$ by induction, and you've shown S_1 is true and $S_k \Rightarrow S_{k+1}$, then it only follows from this that S_n is true for $n \geq 1$. You haven't proved that any of the statements $S_0, S_{-1}, S_{-2}, \ldots$ are true. If you ever want to prove $\forall n \in \mathbb{Z}, S_n$ by induction, you have to show that some S_a is true and $S_k \Rightarrow S_{k+1}$ **and** $S_k \Rightarrow S_{k-1}$.

Unfortunately, the term *mathematical induction* is sometimes confused with *inductive reasoning*, that is, the process of reaching the conclusion that something is likely to be true based on prior observations of similar circumstances. Please note that that mathematical induction, as introduced here, is a rigorous method that proves statements with absolute certainty.

To round out this section, we present four additional induction proofs.

Proposition If $n \in \mathbb{Z}$ and $n \geq 0$, then $\sum_{i=0}^{n} i \cdot i! = (n+1)! - 1$.

Proof. We will prove this with mathematical induction.

(1) If $n = 0$, this statement is

$$\sum_{i=0}^{0} i \cdot i! = (0+1)! - 1.$$

Since the left-hand side is $0 \cdot 0! = 0$, and the right-hand side is $1! - 1 = 0$, the equation $\sum_{i=0}^{0} i \cdot i! = (0+1)! - 1$ holds, as both sides are zero.

(2) Consider any integer $k \geq 0$. We must show that S_k implies S_{k+1}. That is, we must show that

$$\sum_{i=0}^{k} i \cdot i! = (k+1)! - 1$$

implies

$$\sum_{i=0}^{k+1} i \cdot i! = ((k+1)+1)! - 1.$$

We use direct proof. Suppose $\sum_{i=0}^{k} i \cdot i! = (k+1)! - 1$. Observe that

$$\begin{aligned}
\sum_{i=0}^{k+1} i \cdot i! &= \left(\sum_{i=0}^{k} i \cdot i!\right) + (k+1)(k+1)! \\
&= \big((k+1)! - 1\big) + (k+1)(k+1)! \\
&= (k+1)! + (k+1)(k+1)! - 1 \\
&= \big(1 + (k+1)\big)(k+1)! - 1 \\
&= (k+2)(k+1)! - 1 \\
&= (k+2)! - 1 \\
&= ((k+1)+1)! - 1.
\end{aligned}$$

Therefore $\sum_{i=0}^{k+1} i \cdot i! = ((k+1)+1)! - 1$.

It follows by induction that $\sum_{i=0}^{n} i \cdot i! = (n+1)! - 1$ for every integer $n \geq 0$. ■

The next example illustrates a trick that is occasionally useful. You know that you can add equal quantities to both sides of an equation without violating equality. But don't forget that you can add *unequal* quantities to both sides of an *inequality*, as long as the quantity added to the bigger side is bigger than the quantity added to the smaller side. For example, if $x \leq y$ and $a \leq b$, then $x+a \leq y+b$. Similarly, if $x \leq y$ and b is positive, then $x \leq y+b$. This oft-forgotten fact is used in the next proof.

Proposition For each $n \in \mathbb{N}$, it follows that $2^n \leq 2^{n+1}-2^{n-1}-1$.

Proof. We will prove this with mathematical induction.

(1) If $n = 1$, this statement is $2^1 \leq 2^{1+1}-2^{1-1}-1$, which simplifies to $2 \leq 4-1-1$, which is obviously true.

(2) Suppose $k \geq 1$. We need to show that $2^k \leq 2^{k+1}-2^{k-1}-1$ implies $2^{k+1} \leq 2^{(k+1)+1}-2^{(k+1)-1}-1$. We use direct proof. Suppose $2^k \leq 2^{k+1}-2^{k-1}-1$, and reason as follows:

$$\begin{array}{rcll}
2^k & \leq & 2^{k+1}-2^{k-1}-1 & \\
2(2^k) & \leq & 2(2^{k+1}-2^{k-1}-1) & \text{(multiply both sides by 2)} \\
2^{k+1} & \leq & 2^{k+2}-2^k-2 & \\
2^{k+1} & \leq & 2^{k+2}-2^k-2+1 & \text{(add 1 to the bigger side)} \\
2^{k+1} & \leq & 2^{k+2}-2^k-1 & \\
2^{k+1} & \leq & 2^{(k+1)+1}-2^{(k+1)-1}-1. &
\end{array}$$

It follows by induction that $2^n \leq 2^{n+1}-2^{n-1}-1$ for each $n \in \mathbb{N}$. ■

We next prove that if $n \in \mathbb{N}$, then the inequality $(1+x)^n \geq 1+nx$ holds for all $x \in \mathbb{R}$ with $x > -1$. Thus we will need to prove that the statement

$$S_n : (1+x)^n \geq 1+nx \text{ for every } x \in \mathbb{R} \text{ with } x > -1$$

is true for every natural number n. This is (only) slightly different from our other examples, which proved statements of the form $\forall n \in \mathbb{N}, P(n)$, where $P(n)$ is a statement about the number n. This time we are proving something of form

$$\forall n \in \mathbb{N}, P(n,x),$$

where the statement $P(n,x)$ involves not only n, but also a second variable x. (For the record, the inequality $(1+x)^n \geq 1+nx$ is known as *Bernoulli's inequality*.)

Proposition If $n \in \mathbb{N}$, then $(1+x)^n \geq 1+nx$ for all $x \in \mathbb{R}$ with $x > -1$.

Proof. We will prove this with mathematical induction.

(1) For the basis step, notice that when $n = 1$ the statement is $(1+x)^1 \geq 1+1\cdot x$, and this is true because both sides equal $1+x$.

(2) Assume that for some $k \geq 1$, the statement $(1+x)^k \geq 1+kx$ is true for all $x \in \mathbb{R}$ with $x > -1$. From this we need to prove $(1+x)^{k+1} \geq 1+(k+1)x$. Now, $1+x$ is positive because $x > -1$, so we can multiply both sides of $(1+x)^k \geq 1+kx$ by $(1+x)$ without changing the direction of the $\geq$.

$$\begin{aligned}(1+x)^k(1+x) &\geq (1+kx)(1+x)\\ (1+x)^{k+1} &\geq 1+x+kx+kx^2\\ (1+x)^{k+1} &\geq 1+(k+1)x+kx^2\end{aligned}$$

The above term kx^2 is positive, so removing it from the right-hand side will only make that side smaller. Thus we get $(1+x)^{k+1} \geq 1+(k+1)x$. ■

Next, an example where the basis step involves more than routine checking. (It will be used later, so it is numbered for reference.)

Proposition 10.1 Suppose $a_1, a_2, \ldots, a_n$ are n integers, where $n \geq 2$. If p is prime and $p \mid (a_1 \cdot a_2 \cdot a_3 \cdots a_n)$, then $p \mid a_i$ for at least one of the a_i.

Proof. The proof is induction on n.

(1) The basis step involves $n = 2$. Let p be prime and suppose $p \mid (a_1 a_2)$. We need to show that $p \mid a_1$ or $p \mid a_2$, or equivalently, if $p \nmid a_1$, then $p \mid a_2$. Thus suppose $p \nmid a_1$. Since p is prime, it follows that $\gcd(p, a_1) = 1$. By Proposition 7.1 (on page 126), there are integers k and ℓ for which $1 = pk + a_1\ell$. Multiplying this by a_2 gives

$$a_2 = pka_2 + a_1a_2\ell.$$

As we are assuming that p divides a_1a_2, it is clear that p divides the expression $pka_2 + a_1a_2\ell$ on the right; hence $p \mid a_2$. We've now proved that if $p \mid (a_1a_2)$, then $p \mid a_1$ or $p \mid a_2$. This completes the basis step.

(2) Suppose that $k \geq 2$, and $p \mid (a_1 \cdot a_2 \cdots a_k)$ implies then $p \mid a_i$ for some a_i. Now let $p \mid (a_1 \cdot a_2 \cdots a_k \cdot a_{k+1})$. Then $p \mid \big((a_1 \cdot a_2 \cdots a_k) \cdot a_{k+1}\big)$. By what we proved in the basis step, it follows that $p \mid (a_1 \cdot a_2 \cdots a_k)$ or $p \mid a_{k+1}$. This and the inductive hypothesis imply that p divides one of the a_i. ■

Please test your understanding now by working a few exercises.

10.1 Proof by Strong Induction

This section describes a useful variation on induction.

Occasionally it happens in induction proofs that it is difficult to show that S_k forces S_{k+1} to be true. Instead you may find that you need to use the fact that some "lower" statements S_m (with $m < k$) force S_{k+1} to be true. For these situations you can use a slight variant of induction called strong induction. Strong induction works just like regular induction, except that in Step (2) instead of assuming S_k is true and showing this forces S_{k+1} to be true, we assume that *all* the statements $S_1, S_2, \ldots, S_k$ are true and show this forces S_{k+1} to be true. The idea is that if it always happens that the first k dominoes falling makes the $(k+1)$th domino fall, then all the dominoes must fall. Here is the outline.

Outline for Proof by Strong Induction

Proposition The statements $S_1, S_2, S_3, S_4, \ldots$ are all true.

Proof. (Strong induction)
(1) Prove the first statement S_1. (Or the first several S_n.)
(2) Given any integer $k \geq 1$, prove $(S_1 \wedge S_2 \wedge S_3 \wedge \cdots \wedge S_k) \Rightarrow S_{k+1}$. ■

Strong induction can be useful in situations where assuming S_k is true does not neatly lend itself to forcing S_{k+1} to be true. You might be better served by showing some other statement (S_{k-1} or S_{k-2} for instance) forces S_k to be true. Strong induction says you are allowed to use any (or all) of the statements $S_1, S_2, \ldots, S_k$ to prove S_{k+1}.

As our first example of strong induction, we are going to prove that $12 \mid (n^4 - n^2)$ for any $n \in \mathbb{N}$. But first, let's look at how regular induction would be problematic. In regular induction we would start by showing $12 \mid (n^4 - n^2)$ is true for $n = 1$. This part is easy because it reduces to $12 \mid 0$, which is clearly true. Next we would assume that $12 \mid (k^4 - k^2)$ and try to show this implies $12 \mid ((k+1)^4 - (k+1)^2)$. Now, $12 \mid (k^4 - k^2)$ means $k^4 - k^2 = 12a$ for some $a \in \mathbb{Z}$. Next we use this to try to show $(k+1)^4 - (k+1)^2 = 12b$ for some integer b. Working out $(k+1)^4 - (k+1)^2$, we get

$$\begin{aligned}
(k+1)^4 - (k+1)^2 &= (k^4 + 4k^3 + 6k^2 + 4k + 1) - (k^2 + 2k + 1) \\
&= (k^4 - k^2) + 4k^3 + 6k^2 + 6k \\
&= 12a + 4k^3 + 6k^2 + 6k.
\end{aligned}$$

At this point we're stuck because we can't factor out a 12. Now let's see how strong induction can get us out of this bind.

Strong induction involves assuming each of statements $S_1, S_2, \ldots, S_k$ is true, and showing that this forces S_{k+1} to be true. In particular, if S_1 through S_k are true, then certainly S_{k-5} is true, provided that $1 \le k-5 < k$. The idea is then to show $S_{k-5} \Rightarrow S_{k+1}$ instead of $S_k \Rightarrow S_{k+1}$. For this to make sense, our basis step must involve checking that $S_1, S_2, S_3, S_4, S_5, S_6$ are all true. Once this is established, $S_{k-5} \Rightarrow S_{k+1}$ will imply that the other S_k are all true. For example, if $k = 6$, then $S_{k-5} \Rightarrow S_{k+1}$ is $S_1 \Rightarrow S_7$, so S_7 is true; for $k = 7$, then $S_{k-5} \Rightarrow S_{k+1}$ is $S_2 \Rightarrow S_8$, so S_8 is true, etc.

Proposition If $n \in \mathbb{N}$, then $12 \mid (n^4 - n^2)$.

Proof. We will prove this with strong induction.

(1) First note that the statement is true for the first six positive integers:

If $n = 1$, 12 divides $n^4 - n^2 = 1^4 - 1^2 = 0$.

If $n = 2$, 12 divides $n^4 - n^2 = 2^4 - 2^2 = 12$.

If $n = 3$, 12 divides $n^4 - n^2 = 3^4 - 3^2 = 72$.

If $n = 4$, 12 divides $n^4 - n^2 = 4^4 - 4^2 = 240$.

If $n = 5$, 12 divides $n^4 - n^2 = 5^4 - 5^2 = 600$.

If $n = 6$, 12 divides $n^4 - n^2 = 6^4 - 6^2 = 1260$.

(2) Let $k \ge 6$ and assume $12 \mid (m^4 - m^2)$ for $1 \le m \le k$. (That is, assume statements $S_1, S_2, \ldots, S_k$ are all true.) We must show $12 \mid \big((k+1)^4 - (k+1)^2\big)$. (That is, we must show that S_{k+1} is true.) Since S_{k-5} is true, we have $12 \mid ((k-5)^4 - (k-5)^2)$. For simplicity, let's set $\boxed{m = k-5,}$ so we know $12 \mid (m^4 - m^2)$, meaning $\boxed{m^4 - m^2 = 12a}$ for some integer a. Observe that:

$$\begin{aligned}
(k+1)^4 - (k+1)^2 &= (m+6)^4 - (m+6)^2 \\
&= m^4 + 24m^3 + 216m^2 + 864m + 1296 - (m^2 + 12m + 36) \\
&= (m^4 - m^2) + 24m^3 + 216m^2 + 852m + 1260 \\
&= 12a + 24m^3 + 216m^2 + 852m + 1260 \\
&= 12\big(a + 2m^3 + 18m^2 + 71m + 105\big).
\end{aligned}$$

As $(a + 2m^3 + 18m^2 + 71m + 105)$ is an integer, we get $12 \mid ((k+1)^4 - (k+1)^2)$.

This shows by strong induction that $12 \mid (n^4 - n^2)$ for every $n \in \mathbb{N}$. ■

Our next example involves mathematical objects called *graphs.* In mathematics, the word *graph* is used in two contexts. One context involves the graphs of equations and functions from algebra and calculus. In the other context, a **graph** is a configuration consisting of points (called **vertices**) and **edges** which are lines connecting the vertices. Following are some pictures of graphs. Let's agree that all of our graphs will be in "one piece," that is, you can travel from any vertex of a graph to any other vertex by traversing a route of edges from one vertex to the other.

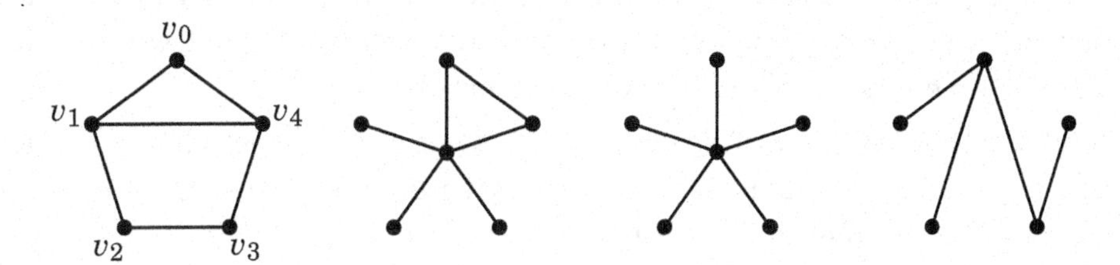

Figure 10.1. Examples of Graphs

A **cycle** in a graph is a sequence of distinct edges in the graph that form a route that ends where it began. For example, the graph on the far left of Figure 10.1 has a cycle that starts at vertex v_1, then goes to v_2, then to v_3, then v_4 and finally back to its starting point v_1. You can find cycles in both of the graphs on the left, but the two graphs on the right do not have cycles. There is a special name for a graph that has no cycles; it is called a **tree**. Thus the two graphs on the right of Figure 10.1 are trees, but the two graphs on the left are not trees.

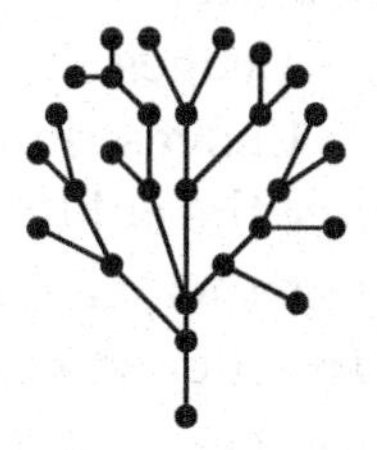

Figure 10.2. A tree

Note that the trees in Figure 10.1 both have one fewer edge than vertex. The tree on the far right has 5 vertices and 4 edges. The one next to it has 6 vertices and 5 edges. Draw any tree; you will find that if it has n vertices, then it has $n-1$ edges. We now prove that this is always true.

Proposition If a tree has n vertices, then it has $n-1$ edges.

Proof. Notice that this theorem asserts that for any $n \in \mathbb{N}$, the following statement is true: S_n : *A tree with n vertices has* $n-1$ *edges.* We use strong induction to prove this.

(1) Observe that if a tree has $n = 1$ vertex then it has no edges. Thus it has $n-1 = 0$ edges, so the theorem is true when $n = 1$.

(2) Now take an integer $k \geq 1$. We must show $(S_1 \wedge S_2 \wedge \cdots \wedge S_k) \Rightarrow S_{k+1}$. In words, we must show that if it is true that any tree with m vertices has $m-1$ edges, where $1 \leq m \leq k$, then any tree with $k+1$ vertices has $(k+1)-1 = k$ edges. We will use direct proof.

Suppose that for each integer m with $1 \leq m \leq k$, any tree with m vertices has $m-1$ edges. Now let T be a tree with $k+1$ vertices. Single out an edge of T and label it e, as illustrated below.

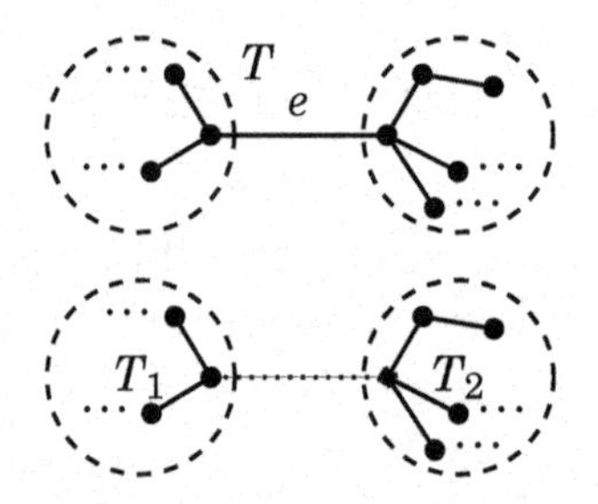

Now remove the edge e from T, but leave the two endpoints of e. This leaves two smaller trees that we call T_1 and T_2. Let's say T_1 has x vertices and T_2 has y vertices. As each of these two smaller trees has fewer than $k+1$ vertices, our inductive hypothesis guarantees that T_1 has $x-1$ edges, and T_2 has $y-1$ edges. Think about our original tree T. It has $x+y$ vertices. It has $x-1$ edges that belong to T_1 and $y-1$ edges that belong to T_2, *plus* it has the additional edge e that belongs to neither T_1 nor T_2. Thus, all together, the number of edges that T has is $(x-1)+(y-1)+1 = (x+y)-1$. In other words, T has one fewer edges than it has vertices. Thus it has $(k+1)-1 = k$ edges.

It follows by strong induction that a tree with n vertices has $n-1$ edges. ■

Notice that it was absolutely essential that we used strong induction in the above proof because the two trees T_1 and T_2 will not both have k vertices. At least one will have fewer than k vertices. Thus the statement S_k is not enough to imply S_{k+1}. We need to use the assumption that S_m will be true whenever $m \leq k$, and strong induction allows us to do this.

10.2 Proof by Smallest Counterexample

This section introduces yet another proof technique, called **proof by smallest counterexample**. It is a hybrid of induction and proof by contradiction. It has the nice feature that it leads you straight to a contradiction. It is therefore more "automatic" than the proof by contradiction that was introduced in Chapter 6. Here is the outline:

Outline for Proof by Smallest Counterexample

Proposition The statements $S_1, S_2, S_3, S_4, \ldots$ are all true.

Proof. (Smallest counterexample)
(1) Check that the first statement S_1 is true.
(2) For the sake of contradiction, suppose not every S_n is true.
(3) Let $k > 1$ be the smallest integer for which S_k is **false**.
(4) Then S_{k-1} is true and S_k is false. Use this to get a contradiction. ■

Notice that this setup leads you to a point where S_{k-1} is true and S_k is false. It is here, where true and false collide, that you will find a contradiction. Let's do an example.

Proposition If $n \in \mathbb{N}$, then $4 \mid (5^n - 1)$.

Proof. We use proof by smallest counterexample. (We will number the steps to match the outline, but that is not usually done in practice.)

(1) If $n = 1$, then the statement is $4 \mid (5^1 - 1)$, or $4 \mid 4$, which is true.

(2) For sake of contradiction, suppose it's not true that $4 \mid (5^n - 1)$ for all n.

(3) Let $k > 1$ be the smallest integer for which $4 \nmid (5^k - 1)$.

(4) Then $4 \mid (5^{k-1} - 1)$, so there is an integer a for which $5^{k-1} - 1 = 4a$. Then:

$$
\begin{aligned}
5^{k-1} - 1 &= 4a \\
5(5^{k-1} - 1) &= 5 \cdot 4a \\
5^k - 5 &= 20a \\
5^k - 1 &= 20a + 4 \\
5^k - 1 &= 4(5a + 1)
\end{aligned}
$$

This means $4 \mid (5^k - 1)$, a contradiction, because $4 \nmid (5^k - 1)$ in Step 3. Thus, we were wrong in Step 2 to assume that it is untrue that $4 \mid (5^n - 1)$ for every n. Therefore $4 \mid (5^n - 1)$ is true for every n. ■

We next prove the **fundamental theorem of arithmetic**, which says any integer greater than 1 has a unique prime factorization. For example, 12 factors into primes as $12 = 2\cdot 2\cdot 3$, and moreover *any* factorization of 12 into primes uses exactly the primes 2, 2 and 3. Our proof combines the techniques of induction, cases, minimum counterexample and the idea of uniqueness of existence outlined at the end of Section 7.3. We dignify this fundamental result with the label of "Theorem."

Theorem 10.1 (Fundamental Theorem of Arithmetic) Any integer $n > 1$ has a unique prime factorization. That is, if $n = p_1\cdot p_2\cdot p_3\cdots p_k$ and $n = a_1\cdot a_2\cdot a_3\cdots a_\ell$ are two prime factorizations of n, then $k = \ell$, and the primes p_i and a_i are the same, except that they may be in a different order.

Proof. Suppose $n > 1$. We first use strong induction to show that n has a prime factorization. For the basis step, if $n = 2$, it is prime, so it is already its own prime factorization. Let $n \geq 2$ and assume every integer between 2 and n (inclusive) has a prime factorization. Consider $n+1$. If it is prime, then it is its own prime factorization. If it is not prime, then it factors as $n+1 = ab$ with $a, b > 1$. Because a and b are both less than $n+1$ they have prime factorizations $a = p_1\cdot p_2\cdot p_3\cdots p_k$ and $b = p'_1\cdot p'_2\cdot p'_3\cdots p'_\ell$. Then

$$n+1 = ab = (p_1\cdot p_2\cdot p_3\cdots p_k)(p'_1\cdot p'_2\cdot p'_3\cdots p'_\ell)$$

is a prime factorization of $n+1$. This competes the proof by strong induction that every integer greater than 1 has a prime factorization.

Next we use proof by smallest counterexample to prove that the prime factorization of any $n \geq 2$ is unique. If $n = 2$, then n clearly has only one prime factorization, namely itself. Assume for the sake of contradiction that there is an $n > 2$ that has different prime factorizations $n = p_1\cdot p_2\cdot p_3\cdots p_k$ and $n = a_1\cdot a_2\cdot a_3\cdots a_\ell$. Assume n is the smallest number with this property. From $n = p_1\cdot p_2\cdot p_3\cdots p_k$, we see that $p_1 \mid n$, so $p_1 \mid (a_1\cdot a_2\cdot a_3\cdots a_\ell)$. By Proposition 10.1 (page 160), p_1 divides one of the primes a_i. As a_i is prime, we have $p_1 = a_i$. Dividing $n = p_1\cdot p_2\cdot p_3\cdots p_k = a_1\cdot a_2\cdot a_3\cdots a_\ell$ by $p_1 = a_i$ yields

$$p_2\cdot p_3\cdots p_k = a_1\cdot a_2\cdot a_3\cdots a_{i-1}\cdot a_{i+1}\cdots a_\ell.$$

These two factorizations are different, because the two prime factorizations of n were different. (Remember: the primes p_1 and a_i are equal, so the difference appears in the remaining factors, displayed above.) But also the above number $p_2\cdot p_3\cdots p_k$ is smaller than n, and this contradicts the fact that n was the smallest number with two different prime factorizations. ■

One word of warning about proof by smallest counterexample. In proofs in other textbooks or in mathematical papers, it often happens that the writer doesn't tell you up front that proof by smallest counterexample is being used. Instead, you will have to read through the proof to glean from context that this technique is being used. In fact, the same warning applies to *all* of our proof techniques. If you continue with mathematics, you will gradually gain through experience the ability to analyze a proof and understand exactly what approach is being used when it is not stated explicitly. Frustrations await you, but do not be discouraged by them. Frustration is a natural part of anything that's worth doing.

10.3 Fibonacci Numbers

Leonardo Pisano, now known as Fibonacci, was a mathematician born around 1175 in what is now Italy. His most significant work was a book *Liber Abaci*, which is recognized as a catalyst in medieval Europe's slow transition from Roman numbers to the Hindu-Arabic number system. But he is best known today for a number sequence that he described in his book and that bears his name. The **Fibonacci sequence** is

$$1,\ 1,\ 2,\ 3,\ 5,\ 8,\ 13,\ 21,\ 34,\ 55,\ 89,\ 144,\ 233,\ 377,\ldots$$

The numbers that appear in this sequence are called **Fibonacci numbers**. The first two numbers are 1 and 1, and thereafter any entry is the sum of the previous two entries. For example $3+5=8$, and $5+8=13$, etc. We denote the nth term of this sequence as F_n. Thus $F_1 = 1$, $F_2 = 1$, $F_3 = 2$, $F_4 = 3$, $F_7 = 13$ and so on. Notice that the Fibonacci Sequence is entirely determined by the rules $F_1 = 1$, $F_2 = 1$, and $F_n = F_{n-1} + F_{n-2}$.

We introduce Fibonacci's sequence here partly because it is something everyone should know about, but also because it is a great source of induction problems. This sequence, which appears with surprising frequency in nature, is filled with mysterious patterns and hidden structures. Some of these structures will be revealed to you in the examples and exercises.

We emphasize that the condition $F_n = F_{n-1}+F_{n-2}$ (or equivalently $F_{n+1} = F_n + F_{n-1}$) is the perfect setup for induction. It suggests that we can determine something about F_n by looking at earlier terms of the sequence. In using induction to prove something about the Fibonacci sequence, you should expect to use the equation $F_n = F_{n-1} + F_{n-2}$ somewhere.

For our first example we will prove that $F_{n+1}^2 - F_{n+1}F_n - F_n^2 = (-1)^n$ for any natural number n. For example, if $n = 5$ we have $F_6^2 - F_6F_5 - F_5^2 = 8^2 - 8\cdot 5 - 5^2 = 64 - 40 - 25 = -1 = (-1)^5$.

Proposition The Fibonacci sequence obeys $F_{n+1}^2 - F_{n+1}F_n - F_n^2 = (-1)^n$.

Proof. We will prove this with mathematical induction.

(1) If $n = 1$ we have $F_{n+1}^2 - F_{n+1}F_n - F_n^2 = F_2^2 - F_2F_1 - F_1^2 = 1^2 - 1 \cdot 1 - 1^2 = -1 = (-1)^1 = (-1)^n$, so indeed $F_{n+1}^2 - F_{n+1}F_n - F_n^2 = (-1)^n$ is true when $n = 1$.

(2) Take any integer $k \geq 1$. We must show that if $F_{k+1}^2 - F_{k+1}F_k - F_k^2 = (-1)^k$, then $F_{k+2}^2 - F_{k+2}F_{k+1} - F_{k+1}^2 = (-1)^{k+1}$. We use direct proof. Suppose $F_{k+1}^2 - F_{k+1}F_k - F_k^2 = (-1)^k$. Now we are going to carefully work out the expression $F_{k+2}^2 - F_{k+2}F_{k+1} - F_{k+1}^2$ and show that it really does equal $(-1)^{k+1}$. In so doing, we will use the fact that $F_{k+2} = F_{k+1} + F_k$.

$$\begin{aligned}
F_{k+2}^2 - F_{k+2}F_{k+1} - F_{k+1}^2 &= (F_{k+1} + F_k)^2 - (F_{k+1} + F_k)F_{k+1} - F_{k+1}^2 \\
&= F_{k+1}^2 + 2F_{k+1}F_k + F_k^2 - F_{k+1}^2 - F_kF_{k+1} - F_{k+1}^2 \\
&= -F_{k+1}^2 + F_{k+1}F_k + F_k^2 \\
&= -(F_{k+1}^2 - F_{k+1}F_k - F_k^2) \\
&= -(-1)^k \qquad \text{(inductive hypothesis)} \\
&= (-1)^1(-1)^k \\
&= (-1)^{k+1}
\end{aligned}$$

Therefore $F_{k+2}^2 - F_{k+2}F_{k+1} - F_{k+1}^2 = (-1)^{k+1}$.

It follows by induction that $F_{n+1}^2 - F_{n+1}F_n - F_n^2 = (-1)^n$ for every $n \in \mathbb{N}$. ■

Let's pause for a moment and think about what the result we just proved means. Dividing both sides of $F_{n+1}^2 - F_{n+1}F_n - F_n^2 = (-1)^n$ by F_n^2 gives

$$\left(\frac{F_{n+1}}{F_n}\right)^2 - \frac{F_{n+1}}{F_n} - 1 = \frac{(-1)^n}{F_n^2}.$$

For large values of n, the right-hand side is very close to zero, and the left-hand side is F_{n+1}/F_n plugged into the polynomial $x^2 - x - 1$. Thus, as n increases, the ratio F_{n+1}/F_n approaches a root of $x^2 - x - 1 = 0$. By the quadratic formula, the roots of $x^2 - x - 1$ are $\frac{1 \pm \sqrt{5}}{2}$. As $F_{n+1}/F_n > 1$, this ratio must be approaching the *positive* root $\frac{1+\sqrt{5}}{2}$. Therefore

$$\lim_{n \to \infty} \frac{F_{n+1}}{F_n} = \frac{1 + \sqrt{5}}{2}. \tag{10.1}$$

For a quick spot check, note that $F_{13}/F_{12} \approx 1.618025$, while $\frac{1+\sqrt{5}}{2} \approx 1.618033$. Even for the small value $n = 12$, the numbers match to four decimal places.

The number $\Phi = \frac{1+\sqrt{5}}{2}$ is sometimes called the **golden ratio**, and there has been much speculation about its occurrence in nature as well as in classical art and architecture. One theory holds that the Parthenon and the Great Pyramids of Egypt were designed in accordance with this number.

But we are here concerned with things that can be proved. We close by observing how the Fibonacci sequence in many ways resembles a geometric sequence. Recall that a **geometric sequence** with first term a and common ratio r has the form

$$a,\ ar,\ ar^2,\ ar^3,\ ar^4,\ ar^5,\ ar^6,\ ar^7,\ ar^8,\ldots$$

where any term is obtained by multiplying the previous term by r. In general its nth term is $G_n = ar^n$, and $G_{n+1}/G_n = r$. Equation (10.1) tells us that $F_{n+1}/F_n \approx \Phi$. Thus even though it is not a geometric sequence, the Fibonacci sequence tends to behave like a geometric sequence with common ratio Φ, and the further "out" you go, the higher the resemblance.

Exercises for Chapter 10

Prove the following statements with either induction, strong induction or proof by smallest counterexample.

1. For every integer $n \in \mathbb{N}$, it follows that $1+2+3+4+\cdots+n = \frac{n^2+n}{2}$.
2. For every integer $n \in \mathbb{N}$, it follows that $1^2+2^2+3^2+4^2+\cdots+n^2 = \frac{n(n+1)(2n+1)}{6}$.
3. For every integer $n \in \mathbb{N}$, it follows that $1^3+2^3+3^3+4^3+\cdots+n^3 = \frac{n^2(n+1)^2}{4}$.
4. If $n \in \mathbb{N}$, then $1\cdot2+2\cdot3+3\cdot4+4\cdot5+\cdots+n(n+1) = \frac{n(n+1)(n+2)}{3}$.
5. If $n \in \mathbb{N}$, then $2^1+2^2+2^3+\cdots+2^n = 2^{n+1}-2$.
6. For every natural number n, it follows that $\sum_{i=1}^{n}(8i-5) = 4n^2-n$.
7. If $n \in \mathbb{N}$, then $1\cdot3+2\cdot4+3\cdot5+4\cdot6+\cdots+n(n+2) = \frac{n(n+1)(2n+7)}{6}$.
8. If $n \in \mathbb{N}$, then $\frac{1}{2!}+\frac{2}{3!}+\frac{3}{4!}+\cdots+\frac{n}{(n+1)!} = 1-\frac{1}{(n+1)!}$
9. For any integer $n \geq 0$, it follows that $24 \mid (5^{2n}-1)$.
10. For any integer $n \geq 0$, it follows that $3 \mid (5^{2n}-1)$.
11. For any integer $n \geq 0$, it follows that $3 \mid (n^3+5n+6)$.
12. For any integer $n \geq 0$, it follows that $9 \mid (4^{3n}+8)$.

13. For any integer $n \geq 0$, it follows that $6 \mid (n^3 - n)$.

14. Suppose $a \in \mathbb{Z}$. Prove that $5 \mid 2^n a$ implies $5 \mid a$ for any $n \in \mathbb{N}$.

15. If $n \in \mathbb{N}$, then $\frac{1}{1\cdot 2} + \frac{1}{2\cdot 3} + \frac{1}{3\cdot 4} + \frac{1}{4\cdot 5} + \cdots + \frac{1}{n(n+1)} = 1 - \frac{1}{n+1}$.

16. For every natural number n, it follows that $2^n + 1 \leq 3^n$.

17. Suppose $A_1, A_2, \ldots A_n$ are sets in some universal set U, and $n \geq 2$. Prove that $\overline{A_1 \cap A_2 \cap \cdots \cap A_n} = \overline{A_1} \cup \overline{A_2} \cup \cdots \cup \overline{A_n}$.

18. Suppose $A_1, A_2, \ldots A_n$ are sets in some universal set U, and $n \geq 2$. Prove that $\overline{A_1 \cup A_2 \cup \cdots \cup A_n} = \overline{A_1} \cap \overline{A_2} \cap \cdots \cap \overline{A_n}$.

19. Prove that $\frac{1}{1} + \frac{1}{4} + \frac{1}{9} + \cdots + \frac{1}{n^2} \leq 2 - \frac{1}{n}$.

20. Prove that $(1+2+3+\cdots+n)^2 = 1^3 + 2^3 + 3^3 + \cdots + n^3$ for every $n \in \mathbb{N}$.

21. If $n \in \mathbb{N}$, then $\frac{1}{1} + \frac{1}{2} + \frac{1}{3} + \frac{1}{4} + \frac{1}{5} + \cdots + \frac{1}{2^n - 1} + \frac{1}{2^n} \geq 1 + \frac{n}{2}$.
(Note: This problem asserts that the sum of the first 2^n terms of the harmonic series is at least $1 + n/2$. It thus implies that the harmonic series diverges.)

22. If $n \in \mathbb{N}$, then $\left(1 - \frac{1}{2}\right)\left(1 - \frac{1}{4}\right)\left(1 - \frac{1}{8}\right)\left(1 - \frac{1}{16}\right) \cdots \left(1 - \frac{1}{2^n}\right) \geq \frac{1}{4} + \frac{1}{2^{n+1}}$.

23. Use mathematical induction to prove the binomial theorem (Theorem 3.1 on page 80). You may find that you need Equation (3.2) on page 78.

24. Prove that $\sum_{k=1}^{n} k\binom{n}{k} = n2^{n-1}$ for each natural number n.

25. Concerning the Fibonacci sequence, prove that $F_1 + F_2 + F_3 + F_4 + \ldots + F_n = F_{n+2} - 1$.

26. Concerning the Fibonacci sequence, prove that $\sum_{k=1}^{n} F_k^2 = F_n F_{n+1}$.

27. Concerning the Fibonacci sequence, prove that $F_1 + F_3 + F_5 + F_7 + \ldots + F_{2n-1} = F_{2n}$.

28. Concerning the Fibonacci sequence, prove that $F_2 + F_4 + F_6 + F_8 + \ldots + F_{2n} = F_{2n+1} - 1$.

29. In this problem $n \in \mathbb{N}$ and F_n is the nth Fibonacci number. Prove that

$$\binom{n}{0} + \binom{n-1}{1} + \binom{n-2}{2} + \binom{n-3}{3} + \cdots + \binom{0}{n} = F_{n+1}.$$

(For example, $\binom{6}{0} + \binom{5}{1} + \binom{4}{2} + \binom{3}{3} + \binom{2}{4} + \binom{1}{5} + \binom{0}{6} = 1+5+6+1+0+0+0 = 13 = F_{6+1}$.)

30. Here F_n is the nth Fibonacci number. Prove that

$$F_n = \frac{\left(\frac{1+\sqrt{5}}{2}\right)^n - \left(\frac{1-\sqrt{5}}{2}\right)^n}{\sqrt{5}}.$$

31. Prove that $\sum_{k=0}^{n} \binom{k}{r} = \binom{n+1}{r+1}$, where $1 \leq r \leq n$.

32. Prove that the number of n-digit binary numbers that have no consecutive 1's is the Fibonacci number F_{n+2}. For example, for $n = 2$ there are three such numbers (00, 01, and 10), and $3 = F_{2+2} = F_4$. Also, for $n = 3$ there are five such numbers (000, 001, 010, 100, 101), and $5 = F_{3+2} = F_5$.

33. Suppose n (infinitely long) straight lines lie on a plane in such a way that no two of the lines are parallel, and no three of the lines intersect at a single point. Show that this arrangement divides the plane into $\frac{n^2+n+2}{2}$ regions.

34. Prove that $3^1 + 3^2 + 3^3 + 3^4 + \cdots + 3^n = \dfrac{3^{n+1} - 3}{2}$ for every $n \in \mathbb{N}$.

35. Prove that if $n, k \in \mathbb{N}$, and n is even and k is odd, then $\binom{n}{k}$ is even.

36. Prove that if $n = 2^k - 1$ for some $k \in \mathbb{N}$, then every entry in the nth row of Pascal's triangle is odd.

The remaining odd-numbered exercises below are not solved in the back of the book.

37. Prove that if $m, n \in \mathbb{N}$, then $\sum_{k=0}^{n} k\binom{m+k}{m} = n\binom{m+n+1}{m+1} - \binom{m+n+1}{m+2}$.

38. Prove that if n is a positive integer, then $\binom{n}{0}^2 + \binom{n}{1}^2 + \binom{n}{2}^2 + \cdots + \binom{n}{n}^2 = \binom{2n}{n}$.

39. Prove that if n is a positive integer, then $\binom{n+0}{0} + \binom{n+1}{1} + \binom{n+2}{2} + \cdots + \binom{n+k}{k} = \binom{n+k+1}{k}$.

40. Prove that $\sum_{k=0}^{p} \binom{m}{k}\binom{n}{p-k} = \binom{m+n}{p}$ for positive integers m, n and p.

41. Prove that $\sum_{k=0}^{m} \binom{m}{k}\binom{n}{p+k} = \binom{m+n}{m+p}$ for positive integers m, n and p.

Part IV

Relations, Functions and Cardinality

CHAPTER 11

Relations

In mathematics there are endless ways that two entities can be related to each other. Consider the following mathematical statements.

$$\begin{array}{ccccccc} 5 < 10 & 5 \le 5 & 6 = \frac{30}{5} & 5 \mid 80 & 7 > 4 & x \neq y & 8 \nmid 3 \\ a \equiv b \pmod{n} & 6 \in \mathbb{Z} & X \subseteq Y & \pi \approx 3.14 & 0 \geq -1 & \sqrt{2} \notin \mathbb{Z} & \mathbb{Z} \not\subseteq \mathbb{N} \end{array}$$

In each case two entities appear on either side of a symbol, and we interpret the symbol as expressing some relationship between the two entities. Symbols such as $<, \le, =, \mid, \nmid, \geq, >$, $\in$ and $\subset$, etc., are called *relations* because they convey relationships among things.

Relations are significant. In fact, you would have to admit that there would be precious little left of mathematics if we took away all the relations. Therefore it is important to have a firm understanding of them, and this chapter is intended to develop that understanding.

Rather than focusing on each relation individually (an impossible task anyway since there are infinitely many different relations), we will develop a general theory that encompasses *all* relations. Understanding this general theory will give us the conceptual framework and language needed to understand and discuss any specific relation.

Before stating the theoretical definition of a relation, let's look at a motivational example. This example will lead naturally to our definition.

Consider the set $A = \{1,2,3,4,5\}$. (There's nothing special about this particular set; any set of numbers would do for this example.) Elements of A can be compared to each other by the symbol "$<$." For example, $1 < 4$, $2 < 3$, $2 < 4$, and so on. You have no trouble understanding this because the notion of numeric order is so ingrained. But imagine you had to explain it to an idiot savant, one with an obsession for detail but absolutely no understanding of the meaning of (or relationships between) integers. You might consider writing down for your student the following set:

$$R = \{(1,2),(1,3),(1,4),(1,5),(2,3),(2,4),(2,5),(3,4),(3,5),(4,5)\}.$$

The set R encodes the meaning of the $<$ relation for elements in A. An ordered pair (a,b) appears in the set if and only if $a<b$. If asked whether or not it is true that $3<4$, your student could look through R until he found the ordered pair $(3,4)$; then he would know $3<4$ is true. If asked about $5<2$, he would see that $(5,2)$ *does not* appear in R, so $5 \not< 2$. The set R, which is a subset of $A \times A$, completely describes the relation $<$ for A.

Though it may seem simple-minded at first, this is exactly the idea we will use for our main definition. This definition is general enough to describe not just the relation $<$ for the set $A = \{1,2,3,4,5\}$, but *any* relation for *any* set A.

Definition 11.1 A **relation** on a set A is a subset $R \subseteq A \times A$. We often abbreviate the statement $(x,y) \in R$ as xRy. The statement $(x,y) \notin R$ is abbreviated as $x \not R y$.

Notice that a relation is a set, so we can use what we know about sets to understand and explore relations. But before getting deeper into the theory of relations, let's look at some examples of Definition 11.1.

Example 11.1 Let $A = \{1,2,3,4\}$, and consider the following set:

$$R = \{(1,1),(2,1),(2,2),(3,3),(3,2),(3,1),(4,4),(4,3),(4,2),(4,1)\} \subseteq A \times A.$$

The set R is a relation on A, by Definition 11.1. Since $(1,1) \in R$, we have $1R1$. Similarly $2R1$ and $2R2$, and so on. However, notice that (for example) $(3,4) \notin R$, so $3 \not R 4$. Observe that R is the familiar relation $\geq$ for the set A.

Chapter 1 proclaimed that all of mathematics can be described with sets. Just look at how successful this program has been! The greater-than-or-equal-to relation is now a set R. (We might even express this in the rather cryptic form $\geq = R$.)

Example 11.2 Let $A = \{1,2,3,4\}$, and consider the following set:

$$S = \{(1,1),(1,3),(3,1),(3,3),(2,2),(2,4),(4,2),(4,4)\} \subseteq A \times A.$$

Here we have $1S1$, $1S3$, $4S2$, etc., but $3 \not S 4$ and $2 \not S 1$. What does S mean? Think of it as meaning "*has the same parity as*." Thus $1S1$ reads "1 *has the same parity as* 1," and $4S2$ reads "4 *has the same parity as* 2."

Example 11.3 Consider relations R and S of the previous two examples. Note that $R \cap S = \{(1,1),(2,2),(3,3),(3,1),(4,4),(4,2)\} \subseteq A \times A$ is a relation on A. The expression $x(R \cap S)y$ means "*$x \geq y$, and x,y have the same parity.*"

Example 11.4 Let $B = \{0,1,2,3,4,5\}$, and consider the following set:

$$U = \{(1,3),(3,3),(5,2),(2,5),(4,2)\} \subseteq B \times B.$$

Then U is a relation on B because $U \subseteq B \times B$. You may be hard-pressed to invent any "meaning" for this particular relation. A relation does not have to have any meaning. Any random subset of $B \times B$ is a relation on B, whether or not it describes anything familiar.

Some relations can be described with pictures. For example, we can depict the above relation U on B by drawing points labeled by elements of B. The statement $(x,y) \in U$ is then represented by an arrow pointing from x to y, a graphic symbol meaning "*x relates to y.*" Here's a picture of U:

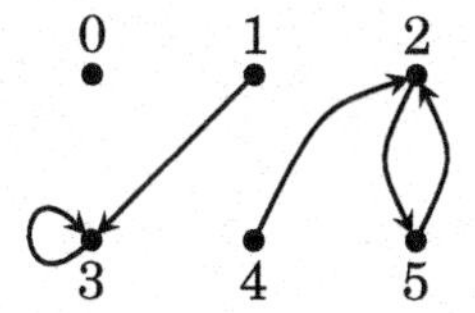

The next picture illustrates the relation R on the set $A = \{a,b,c,d\}$, where xRy means x comes before y in the alphabet. According to Definition 11.1, as a set this relation is $R = \{(a,b),(a,c),(a,d),(b,c),(b,d),(c,d)\}$. You may feel that the picture conveys the relation better than the set does. They are two different ways of expressing the same thing. In some instances pictures are more convenient than sets for discussing relations.

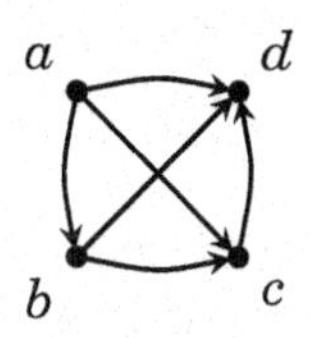

Although such diagrams can help us visualize relations, they do have their limitations. If A and R were infinite, then the diagram would be impossible to draw, but the set R might be easily expressed in set-builder notation. Here are some examples.

Example 11.5 Consider the set $R = \{(x,y) \in \mathbb{Z} \times \mathbb{Z} : x - y \in \mathbb{N}\} \subseteq \mathbb{Z} \times \mathbb{Z}$. This is the $>$ relation on the set $A = \mathbb{Z}$. It is infinite because there are infinitely many ways to have $x > y$ where x and y are integers.

Example 11.6 The set $R = \{(x,x) : x \in \mathbb{R}\} \subseteq \mathbb{R} \times \mathbb{R}$ is the relation $=$ on the set $\mathbb{R}$, because xRy means the same thing as $x = y$. Thus R is a set that expresses the notion of equality of real numbers.

Exercises for Section 11.0

1. Let $A = \{0,1,2,3,4,5\}$. Write out the relation R that expresses $>$ on A. Then illustrate it with a diagram.
2. Let $A = \{1,2,3,4,5,6\}$. Write out the relation R that expresses $\mid$ (divides) on A. Then illustrate it with a diagram.
3. Let $A = \{0,1,2,3,4,5\}$. Write out the relation R that expresses $\geq$ on A. Then illustrate it with a diagram.
4. Here is a diagram for a relation R on a set A. Write the sets A and R.

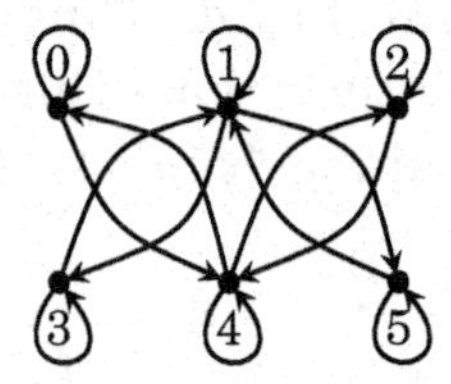

5. Here is a diagram for a relation R on a set A. Write the sets A and R.

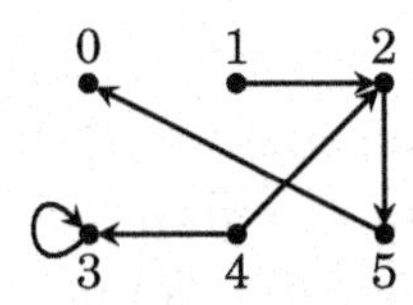

6. Congruence modulo 5 is a relation on the set $A = \mathbb{Z}$. In this relation xRy means $x \equiv y \pmod 5$. Write out the set R in set-builder notation.
7. Write the relation $<$ on the set $A = \mathbb{Z}$ as a subset R of $\mathbb{Z} \times \mathbb{Z}$. This is an infinite set, so you will have to use set-builder notation.
8. Let $A = \{1,2,3,4,5,6\}$. Observe that $\emptyset \subseteq A \times A$, so $R = \emptyset$ is a relation on A. Draw a diagram for this relation.
9. Let $A = \{1,2,3,4,5,6\}$. How many different relations are there on the set A?
10. Consider the subset $R = (\mathbb{R} \times \mathbb{R}) - \{(x,x) : x \in \mathbb{R}\} \subseteq \mathbb{R} \times \mathbb{R}$. What familiar relation on $\mathbb{R}$ is this? Explain.
11. Given a finite set A, how many different relations are there on A?

In the following exercises, subsets R of $\mathbb{R}^2 = \mathbb{R} \times \mathbb{R}$ or $\mathbb{Z}^2 = \mathbb{Z} \times \mathbb{Z}$ are indicated by gray shading. In each case, R is a familiar relation on $\mathbb{R}$ or $\mathbb{Z}$. State it.

12. 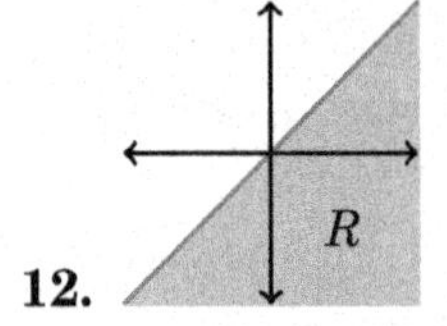

13. 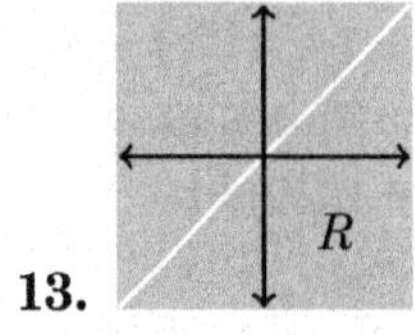

14. 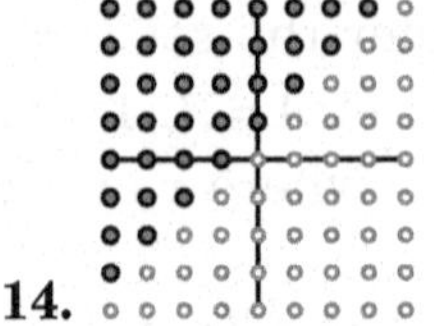

15. 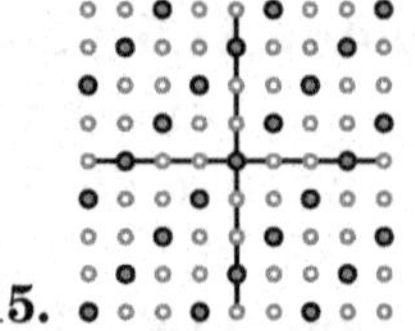

11.1 Properties of Relations

A relational expression xRy is a *statement* (or an *open sentence*); it is either true or false. For example, $5 < 10$ is true, and $10 < 5$ is false. (Thus an operation like + is not a relation, because, for instance, 5+10 has a numeric value, not a T/F value.) Since relational expressions have T/F values, we can combine them with logical operators; for example, $xRy \Rightarrow yRx$ is a statement or open sentence whose truth or falsity may depend on x and y.

With this in mind, note that some relations have properties that others don't have. For example, the relation $\leq$ on $\mathbb{Z}$ satisfies $x \leq x$ for every $x \in \mathbb{Z}$. But this is not so for $<$ because $x < x$ is never true. The next definition lays out three particularly significant properties that relations may have.

Definition 11.2 Suppose R is a relation on a set A.

1. Relation R is **reflexive** if xRx for every $x \in A$.
   That is, R is reflexive if $\forall x \in A, xRx$.
2. Relation R is **symmetric** if xRy implies yRx for all $x, y \in A$
   That is, R is symmetric if $\forall x, y \in A, xRy \Rightarrow yRx$.
3. Relation R is **transitive** if whenever xRy and yRz, then also xRz.
   That is, R is transitive if $\forall x, y, z \in A, \big((xRy) \wedge (yRz)\big) \Rightarrow xRz$.

To illustrate this, let's consider the set $A = \mathbb{Z}$. Examples of reflexive relations on $\mathbb{Z}$ include $\leq$, $=$, and $|$, because $x \leq x$, $x = x$ and $x|x$ are all true for any $x \in \mathbb{Z}$. On the other hand, $>$, $<$, $\neq$ and $\nmid$ are not reflexive for none of the statements $x < x$, $x > x$, $x \neq x$ and $x \nmid x$ is ever true.

The relation $\neq$ **is** symmetric, for if $x \neq y$, then surely $y \neq x$ also. Also, the relation $=$ is symmetric because $x = y$ always implies $y = x$.

The relation $\leq$ is **not** symmetric, as $x \leq y$ does not necessarily imply $y \leq x$. For instance $5 \leq 6$ is true, but $6 \leq 5$ is false. Notice $(x \leq y) \Rightarrow (y \leq x)$ *is* true for some x and y (for example, it is true when $x = 2$ and $y = 2$), but still $\leq$ is not symmetric because it is not the case that $(x \leq y) \Rightarrow (y \leq x)$ is true for *all* integers x and y.

The relation $\leq$ is transitive because whenever $x \leq y$ and $y \leq z$, it also is true that $x \leq z$. Likewise $<, \geq, >$ and $=$ are all transitive. Examine the following table and be sure you understand why it is labeled as it is.

Relations on $\mathbb{Z}$:	$<$	$\leq$	$=$	$\mid$	$\nmid$	$\neq$
Reflexive	no	yes	yes	yes	no	no
Symmetric	no	no	yes	no	no	yes
Transitive	yes	yes	yes	yes	no	no

Example 11.7 Here $A = \{b, c, d, e\}$, and R is the following relation on A: $R = \{(b,b),(b,c),(c,b),(c,c),(d,d),(b,d),(d,b),(c,d),(d,c)\}$.

This relation is **not** reflexive, for although bRb, cRc and dRd, it is **not** true that eRe. For a relation to be reflexive, xRx must be true for *all* $x \in A$.

The relation R **is** symmetric, because whenever we have xRy, it follows that yRx too. Observe that bRc and cRb; bRd and dRb; dRc and cRd. Take away the ordered pair (c,b) from R, and R is no longer symmetric.

The relation R is transitive, but it takes some work to check it. We must check that the statement $(xRy \wedge yRz) \Rightarrow xRz$ is true for all $x, y, z \in A$. For example, taking $x = b$, $y = c$ and $z = d$, we have $(bRc \wedge cRd) \Rightarrow bRd$, which is the true statement $(T \wedge T) \Rightarrow T$. Likewise, $(bRd \wedge dRc) \Rightarrow bRc$ is the true statement $(T \wedge T) \Rightarrow T$. Take note that if $x = b$, $y = e$ and $z = c$, then $(bRe \wedge eRc) \Rightarrow bRc$ becomes $(F \wedge F) \Rightarrow T$, which is *still* true. It's not much fun, but going through all the combinations, you can verify that $(xRy \wedge yRz) \Rightarrow xRz$ is true for all choices $x, y, z \in A$. (Try at least a few of them.)

The relation R from Example 11.7 has a meaning. You can think of xRy as meaning that x and y are both consonants. Thus bRc because b and c are both consonants; but $b \not R e$ because it's not true that b and e are both consonants. Once we look at it this way, it's immediately clear that R has to be transitive. If x and y are both consonants and y and z are both consonants, then surely x and z are both consonants. This illustrates a point that we will see again later in this section: Knowing the meaning of a relation can help us understand it and prove things about it.

Here is a picture of R. Notice that we can immediately spot several properties of R that may not have been so clear from its set description. For instance, we see that R is not reflexive because it lacks a loop at e, hence $e \not R e$.

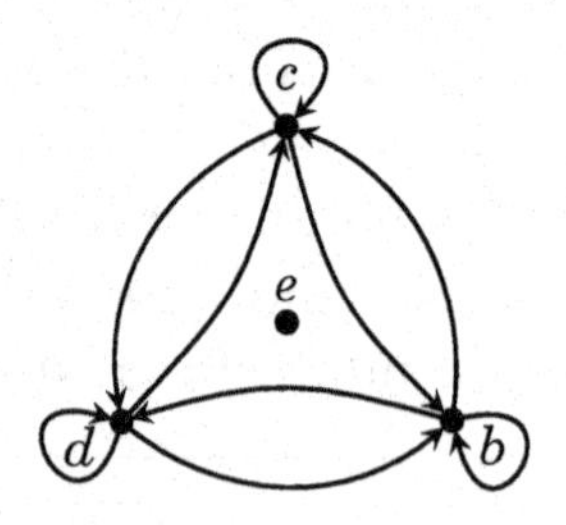

Figure 11.1. The relation R from Example 11.7

In what follows, we summarize how to spot the various properties of a relation from its diagram. Compare these with Figure 11.1.

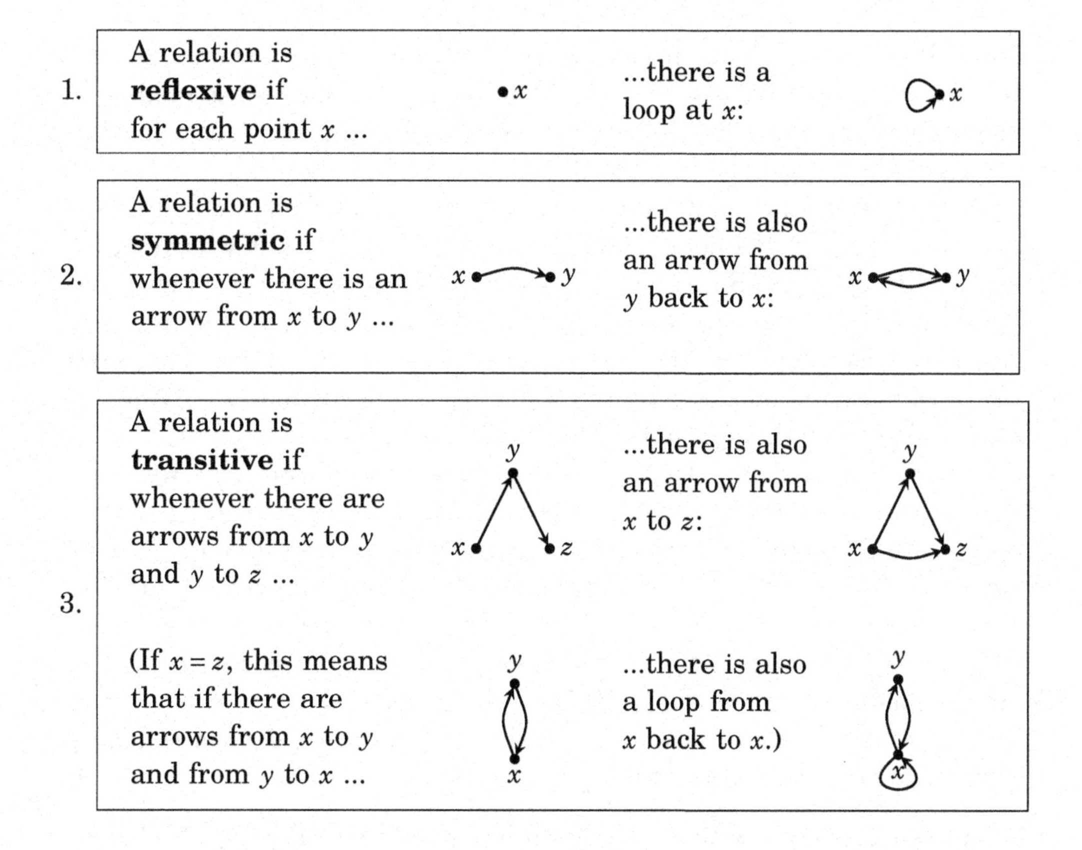

Consider the bottom diagram in Box 3, above. The transitive property demands $(xRy \wedge yRx) \Rightarrow xRx$. Thus, if xRy and yRx in a transitive relation, then also xRx, so there is a loop at x. In this case $(yRx \wedge xRy) \Rightarrow yRy$, so there will be a loop at y too.

Although these visual aids can be illuminating, their use is limited because many relations are too large and complex to be adequately described as diagrams. For example, it would be impossible to draw a diagram for the relation $\equiv \pmod{n}$, where $n \in \mathbb{N}$. Such a relation would best be explained in a more theoretical (and less visual) way.

We next prove that $\equiv \pmod{n}$ is reflexive, symmetric and transitive. Obviously we will not glean this from a drawing. Instead we will prove it from the properties of $\equiv \pmod{n}$ and Definition 11.2. Pay attention to this example. It illustrates how to **prove** things about relations.

Example 11.8 Prove the following proposition.

Proposition Let $n \in \mathbb{N}$. The relation $\equiv \pmod{n}$ on the set $\mathbb{Z}$ is reflexive, symmetric and transitive.

Proof. First we will show that $\equiv \pmod{n}$ is reflexive. Take any integer $x \in \mathbb{Z}$, and observe that $n \mid 0$, so $n \mid (x-x)$. By definition of congruence modulo n, we have $x \equiv x \pmod{n}$. This shows $x \equiv x \pmod{n}$ for every $x \in \mathbb{Z}$, so $\equiv \pmod{n}$ is reflexive.

Next, we will show that $\equiv \pmod{n}$ is symmetric. For this, we must show that for all $x, y \in \mathbb{Z}$, the condition $x \equiv y \pmod{n}$ implies that $y \equiv x \pmod{n}$. We use direct proof. Suppose $x \equiv y \pmod{n}$. Thus $n \mid (x-y)$ by definition of congruence modulo n. Then $x - y = na$ for some $a \in \mathbb{Z}$ by definition of divisibility. Multiplying both sides by -1 gives $y - x = n(-a)$. Therefore $n \mid (y-x)$, and this means $y \equiv x \pmod{n}$. We've shown that $x \equiv y \pmod{n}$ implies that $y \equiv x \pmod{n}$, and this means $\equiv \pmod{n}$ is symmetric.

Finally we will show that $\equiv \pmod{n}$ is transitive. For this we must show that if $x \equiv y \pmod{n}$ and $y \equiv z \pmod{n}$, then $x \equiv z \pmod{n}$. Again we use direct proof. Suppose $x \equiv y \pmod{n}$ and $y \equiv z \pmod{n}$. This means $n \mid (x-y)$ and $n \mid (y-z)$. Therefore there are integers a and b for which $x - y = na$ and $y - z = nb$. Adding these two equations, we obtain $x - z = na + nb$. Consequently, $x - z = n(a+b)$, so $n \mid (x-z)$, hence $x \equiv z \pmod{n}$. This completes the proof that $\equiv \pmod{n}$ is transitive.

The past three paragraphs have shown that $\equiv \pmod{n}$ is reflexive, symmetric and transitive, so the proof is complete. ■

As you continue with mathematics you will find that the reflexive, symmetric and transitive properties take on special significance in a variety of settings. In preparation for this, the next section explores further consequences of these properties. But first work some of the following exercises.

Exercises for Section 11.1

1. Consider the relation $R = \{(a,a),(b,b),(c,c),(d,d),(a,b),(b,a)\}$ on set $A = \{a,b,c,d\}$. Is R reflexive? Symmetric? Transitive? If a property does not hold, say why.
2. Consider the relation $R = \{(a,b),(a,c),(c,c),(b,b),(c,b),(b,c)\}$ on the set $A = \{a,b,c\}$. Is R reflexive? Symmetric? Transitive? If a property does not hold, say why.
3. Consider the relation $R = \{(a,b),(a,c),(c,b),(b,c)\}$ on the set $A = \{a,b,c\}$. Is R reflexive? Symmetric? Transitive? If a property does not hold, say why.

4. Let $A = \{a,b,c,d\}$. Suppose R is the relation

$$\begin{aligned} R &= \{(a,a),(b,b),(c,c),(d,d),(a,b),(b,a),(a,c),(c,a), \\ &\quad (a,d),(d,a),(b,c),(c,b),(b,d),(d,b),(c,d),(d,c)\}. \end{aligned}$$

   Is R reflexive? Symmetric? Transitive? If a property does not hold, say why.

5. Consider the relation $R = \{(0,0),(\sqrt{2},0),(0,\sqrt{2}),(\sqrt{2},\sqrt{2})\}$ on $\mathbb{R}$. Is R reflexive? Symmetric? Transitive? If a property does not hold, say why.
6. Consider the relation $R = \{(x,x) : x \in \mathbb{Z}\}$ on $\mathbb{Z}$. Is R reflexive? Symmetric? Transitive? If a property does not hold, say why. What familiar relation is this?
7. There are 16 possible different relations R on the set $A = \{a,b\}$. Describe all of them. (A picture for each one will suffice, but don't forget to label the nodes.) Which ones are reflexive? Symmetric? Transitive?
8. Define a relation on $\mathbb{Z}$ as xRy if $|x-y| < 1$. Is R reflexive? Symmetric? Transitive? If a property does not hold, say why. What familiar relation is this?
9. Define a relation on $\mathbb{Z}$ by declaring xRy if and only if x and y have the same parity. Is R reflexive? Symmetric? Transitive? If a property does not hold, say why. What familiar relation is this?
10. Suppose $A \neq \emptyset$. Since $\emptyset \subseteq A \times A$, the set $R = \emptyset$ is a relation on A. Is R reflexive? Symmetric? Transitive? If a property does not hold, say why.
11. Suppose $A = \{a,b,c,d\}$ and $R = \{(a,a),(b,b),(c,c),(d,d)\}$. Is R reflexive? Symmetric? Transitive? If a property does not hold, say why.
12. Prove that the relation $\mid$ (divides) on the set $\mathbb{Z}$ is reflexive and transitive. (Use Example 11.8 as a guide if you are unsure of how to proceed.)
13. Consider the relation $R = \{(x,y) \in \mathbb{R} \times \mathbb{R} : x - y \in \mathbb{Z}\}$ on $\mathbb{R}$. Prove that this relation is reflexive, symmetric and transitive.
14. Suppose R is a symmetric and transitive relation on a set A, and there is an element $a \in A$ for which aRx for every $x \in A$. Prove that R is reflexive.
15. Prove or disprove: If a relation is symmetric and transitive, then it is also reflexive.
16. Define a relation R on $\mathbb{Z}$ by declaring that xRy if and only if $x^2 \equiv y^2 \pmod{4}$. Prove that R is reflexive, symmetric and transitive.
17. Modifying the above Exercise 8 (above) slightly, define a relation $\sim$ on $\mathbb{Z}$ as $x \sim y$ if and only if $|x - y| \leq 1$. Say whether $\sim$ is reflexive. Is it symmetric? Transitive?
18. The table on page 179 shows that relations on $\mathbb{Z}$ may obey various combinations of the reflexive, symmetric and transitive properties. In all, there are $2^3 = 8$ possible combinations, and the table shows 5 of them. (There is some redundancy, as $\leq$ and $\mid$ have the same type.) Complete the table by finding examples of relations on $\mathbb{Z}$ for the three missing combinations.

11.2 Equivalence Relations

The relation = on the set $\mathbb{Z}$ (or on any set A) is reflexive, symmetric and transitive. There are many other relations that are also reflexive, symmetric and transitive. Relations that have all three of these properties occur very frequently in mathematics and often play quite significant roles. (For instance, this is certainly true of the relation =.) Such relations are given a special name. They are called *equivalence relations*.

Definition 11.3 A relation R on a set A is an **equivalence relation** if it is reflexive, symmetric and transitive.

As an example, Figure 11.2 shows four different equivalence relations R_1, R_2, R_3 and R_4 on the set $A = \{-1, 1, 2, 3, 4\}$. Each one has its own meaning, as labeled. For example, in the second row the relation R_2 literally means *"has the same parity as."* So $1R_2 3$ means "1 *has the same parity as* 3," etc.

Relation R	Diagram	Equivalence classes (see next page)
"is equal to" (=) $R_1 = \{(-1,-1),(1,1),(2,2),(3,3),(4,4)\}$		$\{-1\}$, $\{1\}$, $\{2\}$, $\{3\}$, $\{4\}$
"has same parity as" $R_2 = \{(-1,-1),(1,1),(2,2),(3,3),(4,4),$ $(-1,1),(1,-1),(-1,3),(3,-1),$ $(1,3),(3,1),(2,4),(4,2)\}$		$\{-1,1,3\}$, $\{2,4\}$
"has same sign as" $R_3 = \{(-1,-1),(1,1),(2,2),(3,3),(4,4),$ $(1,2),(2,1),(1,3),(3,1),(1,4),(4,1),(3,4),$ $(4,3),(2,3),(3,2),(2,4),(4,2),(1,3),(3,1)\}$		$\{-1\}$, $\{1,2,3,4\}$
"has same parity and sign as" $R_4 = \{(-1,-1),(1,1),(2,2),(3,3),(4,4),$ $(1,3),(3,1),(2,4),(4,2)\}$		$\{-1\}$, $\{1,3\}$, $\{2,4\}$

Figure 11.2. Examples of equivalence relations on the set $A = \{-1, 1, 2, 3, 4\}$

The above diagrams make it easy to check that each relation is reflexive, symmetric and transitive, i.e., that each is an equivalence relation. For example, R_1 is symmetric because $xR_1y \Rightarrow yR_1x$ is always true: When $x = y$ it becomes $T \Rightarrow T$ (true), and when $x \neq y$ it becomes $F \Rightarrow F$ (also true). In a similar fashion, R_1 is transitive because $(xR_1y \wedge yR_1z) \Rightarrow xR_1z$ is always true: It always works out to one of $T \Rightarrow T$, $F \Rightarrow T$ or $F \Rightarrow F$. (Check this.)

As you can see from the examples in Figure 11.2, equivalence relations on a set tend to express some measure of "sameness" among the elements of the set, whether it is true equality or something weaker (like having the same parity).

It's time to introduce an important definition. Whenever you have an equivalence relation R on a set A, it divides A into subsets called *equivalence classes*. Here is the definition:

Definition 11.4 Suppose R is an equivalence relation on a set A. Given any element $a \in A$, the **equivalence class containing** a is the subset $\{x \in A : xRa\}$ of A consisting of all the elements of A that relate to a. This set is denoted as $[a]$. Thus the equivalence class containing a is the set $[a] = \{x \in A : xRa\}$.

Example 11.9 Consider the relation R_1 in Figure 11.2. The equivalence class containing 2 is the set $[2] = \{x \in A : xR_12\}$. Because in this relation the only element that relates to 2 is 2 itself, we have $[2] = \{2\}$. Other equivalence classes for R_1 are $[-1] = \{-1\}$, $[1] = \{1\}$, $[3] = \{3\}$ and $[4] = \{4\}$. Thus this relation has five separate equivalence classes.

Example 11.10 Consider the relation R_2 in Figure 11.2. The equivalence class containing 2 is the set $[2] = \{x \in A : xR_22\}$. Because only 2 and 4 relate to 2, we have $[2] = \{2,4\}$. Observe that we also have $[4] = \{x \in A : xR_24\} = \{2,4\}$, so $[2] = [4]$. Another equivalence class for R_2 is $[1] = \{x \in A : xR_21\} = \{-1,1,3\}$. In addition, note that $[1] = [-1] = [3] = \{-1,1,3\}$. Thus this relation has just two equivalence classes, namely $\{2,4\}$ and $\{-1,1,3\}$.

Example 11.11 The relation R_4 in Figure 11.2 has three equivalence classes. They are $[-1] = \{-1\}$ and $[1] = [3] = \{1,3\}$ and $[2] = [4] = \{2,4\}$.

Don't be misled by Figure 11.2. It's important to realize that not every equivalence relation can be drawn as a diagram involving nodes and arrows. Even the simple relation $R = \{(x,x) : x \in \mathbb{R}\}$, which expresses equality in the set $\mathbb{R}$, is too big to be drawn. Its picture would involve a point for every real number and a loop at each point. Clearly that's too many points and loops to draw.

We close this section with several other examples of equivalence relations on infinite sets.

Example 11.12 Let P be the set of all polynomials with real coefficients. Define a relation R on P as follows. Given $f(x), g(x) \in P$, let $f(x) R\, g(x)$ mean that $f(x)$ and $g(x)$ have the same degree. Thus $(x^2+3x-4) R\, (3x^2-2)$ and $(x^3+3x^2-4) \not R\, (3x^2-2)$, for example. It takes just a quick mental check to see that R is an equivalence relation. (Do it.) It's easy to describe the equivalence classes of R. For example, $[3x^2+2]$ is the set of all polynomials that have the same degree as $3x^2+2$, that is, the set of all polynomials of degree 2. We can write this as $[3x^2+2] = \{ax^2+bx+c : a,b,c \in \mathbb{R}, a \neq 0\}$.

Example 11.8 proved that for a given $n \in \mathbb{N}$ the relation $\equiv \pmod{n}$ is reflexive, symmetric and transitive. Thus, in our new parlance, $\equiv \pmod{n}$ is an equivalence relation on $\mathbb{Z}$. Consider the case $n = 3$. Let's find the equivalence classes of the equivalence relation $\equiv \pmod{3}$. The equivalence class containing 0 seems like a reasonable place to start. Observe that

$$[0] = \{x \in \mathbb{Z} : x \equiv 0 \pmod{3}\} = \{x \in \mathbb{Z} : 3 \mid (x-0)\} = \{x \in \mathbb{Z} : 3 \mid x\} = \{\ldots,-3,0,3,6,9,\ldots\}.$$

Thus the class [0] consists of all the multiples of 3. (Or, said differently, [0] consists of all integers that have a remainder of 0 when divided by 3). Note that $[0] = [3] = [6] = [9]$, etc. The number 1 does not show up in the set [0] so let's next look at the equivalence class [1]:

$$[1] = \{x \in \mathbb{Z} : x \equiv 1 \pmod{3}\} = \{x \in \mathbb{Z} : 3 \mid (x-1)\} = \{\ldots,-5,-2,1,4,7,10,\ldots\}.$$

The equivalence class [1] consists of all integers that give a remainder of 1 when divided by 3. The number 2 is in neither of the sets [0] or [1], so we next look at the equivalence class [2]:

$$[2] = \{x \in \mathbb{Z} : x \equiv 2 \pmod{3}\} = \{x \in \mathbb{Z} : 3 \mid (x-2)\} = \{\ldots,-4,-1,2,5,8,11,\ldots\}.$$

The equivalence class [2] consists of all integers that give a remainder of 2 when divided by 3. Observe that any integer is in one of the sets [0], [1] or [2], so we have listed all of the equivalence classes. Thus $\equiv \pmod{3}$ has exactly three equivalence classes, as described above.

Similarly, you can show that the equivalence relation $\equiv \pmod{n}$ has n equivalence classes $[0],[1],[2],\ldots,[n-1]$.

Exercises for Section 11.2

1. Let $A = \{1,2,3,4,5,6\}$, and consider the following equivalence relation on A: $R = \{(1,1),(2,2),(3,3),(4,4),(5,5),(6,6),(2,3),(3,2),(4,5),(5,4),(4,6),(6,4),(5,6),(6,5)\}$ List the equivalence classes of R.
2. Let $A = \{a,b,c,d,e\}$. Suppose R is an equivalence relation on A. Suppose R has two equivalence classes. Also aRd, bRc and eRd. Write out R as a set.
3. Let $A = \{a,b,c,d,e\}$. Suppose R is an equivalence relation on A. Suppose R has three equivalence classes. Also aRd and bRc. Write out R as a set.
4. Let $A = \{a,b,c,d,e\}$. Suppose R is an equivalence relation on A. Suppose also that aRd and bRc, eRa and cRe. How many equivalence classes does R have?
5. There are two different equivalence relations on the set $A = \{a,b\}$. Describe them. Diagrams will suffice.
6. There are five different equivalence relations on the set $A = \{a,b,c\}$. Describe them all. Diagrams will suffice.
7. Define a relation R on $\mathbb{Z}$ as xRy if and only if $3x-5y$ is even. Prove R is an equivalence relation. Describe its equivalence classes.
8. Define a relation R on $\mathbb{Z}$ as xRy if and only if x^2+y^2 is even. Prove R is an equivalence relation. Describe its equivalence classes.
9. Define a relation R on $\mathbb{Z}$ as xRy if and only if $4 \mid (x+3y)$. Prove R is an equivalence relation. Describe its equivalence classes.
10. Suppose R and S are two equivalence relations on a set A. Prove that $R \cap S$ is also an equivalence relation. (For an example of this, look at Figure 11.2. Observe that for the equivalence relations R_2, R_3 and R_4, we have $R_2 \cap R_3 = R_4$.)
11. Prove or disprove: If R is an equivalence relation on an infinite set A, then R has infinitely many equivalence classes.
12. Prove or disprove: If R and S are two equivalence relations on a set A, then $R \cup S$ is also an equivalence relation on A.
13. Suppose R is an equivalence relation on a finite set A, and every equivalence class has the same cardinality m. Express $|R|$ in terms of m and $|A|$.
14. Suppose R is a reflexive and symmetric relation on a finite set A. Define a relation S on A by declaring xSy if and only if for some $n \in \mathbb{N}$ there are elements $x_1, x_2, \ldots, x_n \in A$ satisfying xRx_1, x_1Rx_2, x_2Rx_3, $x_3Rx_4, \ldots, x_{n-1}Rx_n$, and x_nRy. Show that S is an equivalence relation and $R \subseteq S$. Prove that S is the unique smallest equivalence relation on A containing R.
15. Suppose R is an equivalence relation on a set A, with four equivalence classes. How many different equivalence relations S on A are there for which $R \subseteq S$?

11.3 Equivalence Classes and Partitions

This section collects several properties of equivalence classes.

Our first result proves that $[a]=[b]$ if and only if aRb. This is useful because it assures us that whenever we are in a situation where $[a]=[b]$, we also have aRb, and vice versa. Being able to switch back and forth between these two pieces of information can be helpful in a variety of situations, and you may find yourself using this result a lot. Be sure to notice that the proof uses all three properties (reflexive, symmetric and transitive) of equivalence relations. Notice also that we have to use some Chapter 8 techniques in dealing with the sets $[a]$ and $[b]$.

Theorem 11.1 Suppose R is an equivalence relation on a set A. Suppose also that $a,b \in A$. Then $[a]=[b]$ if and only if aRb.

Proof. Suppose $[a]=[b]$. Note that aRa by the reflexive property of R, so $a \in \{x \in A : xRa\} = [a] = [b] = \{x \in A : xRb\}$. But a belonging to $\{x \in A : xRb\}$ means aRb. This completes the first part of the if-and-only-if proof.

Conversely, suppose aRb. We need to show $[a]=[b]$. We will do this by showing $[a] \subseteq [b]$ and $[b] \subseteq [a]$.

First we show $[a] \subseteq [b]$. Suppose $c \in [a]$. As $c \in [a] = \{x \in A : xRa\}$, we get cRa. Now we have cRa and aRb, so cRb because R is transitive. But cRb implies $c \in \{x \in A : xRb\} = [b]$. This demonstrates that $c \in [a]$ implies $c \in [b]$, so $[a] \subseteq [b]$.

Next we show $[b] \subseteq [a]$. Suppose $c \in [b]$. As $c \in [b] = \{x \in A : xRb\}$, we get cRb. Remember that we are assuming aRb, so bRa because R is symmetric. Now we have cRb and bRa, so cRa because R is transitive. But cRa implies $c \in \{x \in A : xRa\} = [a]$. This demonstrates that $c \in [b]$ implies $c \in [a]$; hence $[b] \subseteq [a]$.

The previous two paragraphs imply that $[a]=[b]$. ■

To illustrate Theorem 11.1, recall how we worked out the equivalence classes of $\equiv \pmod 3$ at the end of Section 11.2. We observed that

$$[-3] = [9] = \{\ldots, -3, 0, 3, 6, 9, \ldots\}.$$

Note that $[-3]=[9]$ and $-3 \equiv 9 \pmod 3$, just as Theorem 11.1 predicts. The theorem assures us that this will work for any equivalence relation. In the future you may find yourself using the result of Theorem 11.1 often. Over time it may become natural and familiar; you will use it automatically, without even thinking of it as a theorem.

Our next topic addresses the fact that an equivalence relation on a set A divides A into various equivalence classes. There is a special word for this kind of situation. We address it now, as you are likely to encounter it in subsequent mathematics classes.

Definition 11.5 A **partition** of a set A is a set of non-empty subsets of A, such that the union of all the subsets equals A, and the intersection of any two different subsets is $\emptyset$.

Example 11.13 Let $A = \{a,b,c,d\}$. One partition of A is $\{\{a,b\},\{c\},\{d\}\}$. This is a set of three subsets $\{a,b\}$, $\{c\}$ and $\{d\}$ of A. The union of the three subsets equals A; the intersection of any two subsets is $\emptyset$.

Other partitions of A are

$$\{\{a,b\},\{c,d\}\}, \qquad \{\{a,c\},\{b\},\{d\}\}, \qquad \{\{a\},\{b\},\{c\}\{d\}\}, \qquad \{\{a,b,c,d\}\},$$

to name a few. Intuitively, a partition is just a dividing up of A into pieces.

Example 11.14 Consider the equivalence relations in Figure 11.2. Each of these is a relation on the set $A = \{-1,1,2,3,4\}$. The equivalence classes of each relation are listed on the right side of the figure. Observe that, in each case, the set of equivalence classes forms a partition of A. For example, the relation R_1 yields the partition $\{\{-1\},\{1\},\{2\},\{3\},\{4\}\}$ of A. Likewise the equivalence classes of R_2 form the partition $\{\{-1,1,3\},\{2,4\}\}$.

Example 11.15 Recall that we worked out the equivalence classes of the equivalence relation $\equiv \pmod 3$ on the set $\mathbb{Z}$. These equivalence classes give the following partition of $\mathbb{Z}$:

$$\{\{\ldots,-3,0,3,6,9,\ldots\},\{\ldots,-2,1,4,7,10,\ldots\},\{\ldots,-1,2,5,8,11,\ldots\}\}.$$

We can write it more compactly as $\{[0],[1],[2]\}$.

Our examples and experience suggest that the equivalence classes of an equivalence relation on a set form a partition of that set. This is indeed the case, and we now prove it.

Theorem 11.2 Suppose R is an equivalence relation on a set A. Then the set $\{[a] : a \in A\}$ of equivalence classes of R forms a partition of A.

Proof. To show that $\{[a] : a \in A\}$ is a partition of A we need to show two things: We need to show that the union of all the sets $[a]$ equals A, and we need to show that if $[a] \neq [b]$, then $[a] \cap [b] = \emptyset$.

Notationally, the union of all the sets $[a]$ is $\bigcup_{a\in A}[a]$, so we need to prove $\bigcup_{a\in A}[a] = A$. Suppose $x \in \bigcup_{a\in A}[a]$. This means $x \in [a]$ for some $a \in A$. Since $[a] \subseteq A$, it then follows that $x \in A$. Thus $\bigcup_{a\in A}[a] \subseteq A$. On the other hand, suppose $x \in A$. As $x \in [x]$, we know $x \in [a]$ for some $a \in A$ (namely $a = x$). Therefore $x \in \bigcup_{a\in A}[a]$, and this shows $A \subseteq \bigcup_{a\in A}[a]$. Since $\bigcup_{a\in A}[a] \subseteq A$ and $A \subseteq \bigcup_{a\in A}[a]$, it follows that $\bigcup_{a\in A}[a] = A$.

Next we need to show that if $[a] \neq [b]$ then $[a]\cap[b] = \emptyset$. Let's use contrapositive proof. Suppose it's not the case that $[a]\cap[b] = \emptyset$, so there is some element c with $c \in [a]\cap[b]$. Thus $c \in [a]$ and $c \in [b]$. Now, $c \in [a]$ means cRa, and then aRc since R is symmetric. Also $c \in [b]$ means cRb. Now we have aRc and cRb, so aRb (because R is transitive). By Theorem 11.1, aRb implies $[a] = [b]$. Thus $[a] \neq [b]$ is not true.

We've now shown that the union of all the equivalence classes is A, and the intersection of two different equivalence classes is $\emptyset$. Therefore the set of equivalence classes is a partition of A. ■

Theorem 11.2 says the equivalence classes of any equivalence relation on a set A form a partition of A. Conversely, any partition of A describes an equivalence relation R where xRy if and only if x and y belong to the same set in the partition. (See Exercise 4 for this section, below.) Thus equivalence relations and partitions are really just two different ways of looking at the same thing. In your future mathematical studies, you may find yourself easily switching between these two points of view.

Exercises for Section 11.3

1. List all the partitions of the set $A = \{a,b\}$. Compare your answer to the answer to Exercise 5 of Section 11.2.
2. List all the partitions of the set $A = \{a,b,c\}$. Compare your answer to the answer to Exercise 6 of Section 11.2.
3. Describe the partition of $\mathbb{Z}$ resulting from the equivalence relation $\equiv \pmod 4$.
4. Suppose P is a partition of a set A. Define a relation R on A by declaring xRy if and only if $x,y \in X$ for some $X \in P$. Prove R is an equivalence relation on A. Then prove that P is the set of equivalence classes of R.
5. Consider the partition $P = \{\{\ldots,-4,-2,0,2,4,\ldots\},\{\ldots,-5,-3,-1,1,3,5,\ldots\}\}$ of $\mathbb{Z}$. Let R be the equivalence relation whose equivalence classes are the two elements of P. What familiar equivalence relation is R?

11.4 The Integers Modulo n

Example 11.8 proved that for a given $n \in \mathbb{N}$, the relation $\equiv \pmod{n}$ is reflexive, symmetric and transitive, so it is an equivalence relation. This is a particularly significant equivalence relation in mathematics, and in the present section we deduce some of its properties.

To make matters simpler, let's pick a concrete n, say $n = 5$. Let's begin by looking at the equivalence classes of the relation $\equiv \pmod{5}$. There are five equivalence classes, as follows:

$$\begin{array}{lllllll}
[0] & = & \{x \in \mathbb{Z} : x \equiv 0 \pmod{5})\} & = & \{x \in \mathbb{Z} : 5 \mid (x-0)\} & = & \{\ldots,-10,-5,0,5,10,15,\ldots\}, \\
[1] & = & \{x \in \mathbb{Z} : x \equiv 1 \pmod{5})\} & = & \{x \in \mathbb{Z} : 5 \mid (x-1)\} & = & \{\ldots,-9,-4,1,6,11,16,\ldots\}, \\
[2] & = & \{x \in \mathbb{Z} : x \equiv 2 \pmod{5})\} & = & \{x \in \mathbb{Z} : 5 \mid (x-2)\} & = & \{\ldots,-8,-3,2,7,12,17,\ldots\}, \\
[3] & = & \{x \in \mathbb{Z} : x \equiv 3 \pmod{5})\} & = & \{x \in \mathbb{Z} : 5 \mid (x-3)\} & = & \{\ldots,-7,-2,3,8,13,18,\ldots\}, \\
[4] & = & \{x \in \mathbb{Z} : x \equiv 4 \pmod{5})\} & = & \{x \in \mathbb{Z} : 5 \mid (x-4)\} & = & \{\ldots,-6,-1,4,9,14,19,\ldots\}.
\end{array}$$

Notice how these equivalence classes form a partition of the set $\mathbb{Z}$. We label the five equivalence classes as $[0],[1],[2],[3]$ and $[4]$, but you know of course that there are other ways to label them. For example, $[0] = [5] = [10] = [15]$, and so on; and $[1] = [6] = [-4]$, etc. Still, for this discussion we denote the five classes as $[0],[1],[2],[3]$ and $[4]$.

These five classes form a set, which we shall denote as $\mathbb{Z}_5$. Thus

$$\mathbb{Z}_5 = \{[0],[1],[2],[3],[4]\}$$

is a set of five sets. The interesting thing about $\mathbb{Z}_5$ is that even though its elements are sets (and not numbers), it is possible to add and multiply them. In fact, we can define the following rules that tell how elements of $\mathbb{Z}_5$ can be added and multiplied.

$$\begin{array}{rcl}
[a]+[b] & = & [a+b] \\
[a]\cdot[b] & = & [a\cdot b]
\end{array}$$

For example, $[2]+[1] = [2+1] = [3]$, and $[2]\cdot[2] = [2\cdot 2] = [4]$. We stress that in doing this we are adding and multiplying *sets* (more precisely equivalence classes), not numbers. We added (or multiplied) two elements of $\mathbb{Z}_5$ and obtained another element of $\mathbb{Z}_5$.

Here is a trickier example. Observe that $[2]+[3] = [5]$. This time we added elements $[2],[3] \in \mathbb{Z}_5$, and got the element $[5] \in \mathbb{Z}_5$. That was easy, except where is our answer $[5]$ in the set $\mathbb{Z}_5 = \{[0],[1],[2],[3],[4]\}$? Since $[5] = [0]$, it is more appropriate to write $[2]+[3] = [0]$.

In a similar vein, $[2]\cdot[3]=[6]$ would be written as $[2]\cdot[3]=[1]$ because $[6]=[1]$. Test your skill with this by verifying the following addition and multiplication tables for $\mathbb{Z}_5$.

+	[0]	[1]	[2]	[3]	[4]
[0]	[0]	[1]	[2]	[3]	[4]
[1]	[1]	[2]	[3]	[4]	[0]
[2]	[2]	[3]	[4]	[0]	[1]
[3]	[3]	[4]	[0]	[1]	[2]
[4]	[4]	[0]	[1]	[2]	[3]

·	[0]	[1]	[2]	[3]	[4]
[0]	[0]	[0]	[0]	[0]	[0]
[1]	[0]	[1]	[2]	[3]	[4]
[2]	[0]	[2]	[4]	[1]	[3]
[3]	[0]	[3]	[1]	[4]	[2]
[4]	[0]	[4]	[3]	[2]	[1]

We call the set $\mathbb{Z}_5=\{[0],[1],[2],[3],[4]\}$ the **integers modulo 5**. As our tables suggest, $\mathbb{Z}_5$ is more than just a set: It is a little number system with its own addition and multiplication. In this way it is like the familiar set $\mathbb{Z}$ which also comes equipped with an addition and a multiplication.

Of course, there is nothing special about the number 5. We can also define $\mathbb{Z}_n$ for any natural number n. Here is the definition:

Definition 11.6 Let $n\in\mathbb{N}$. The equivalence classes of the equivalence relation $\equiv \pmod{n}$ are $[0],[1],[2],\ldots,[n-1]$. The **integers modulo n** is the set $\mathbb{Z}_n=\{[0],[1],[2],\ldots,[n-1]\}$. Elements of $\mathbb{Z}_n$ can be added by the rule $[a]+[b]=[a+b]$ and multiplied by the rule $[a]\cdot[b]=[ab]$.

Given a natural number n, the set $\mathbb{Z}_n$ is a number system containing n elements. It has many of the algebraic properties that $\mathbb{Z}$, $\mathbb{R}$ and $\mathbb{Q}$ possess. For example, it is probably obvious to you already that elements of $\mathbb{Z}_n$ obey the commutative laws $[a]+[b]=[b]+[a]$ and $[a]\cdot[b]=[b]\cdot[a]$. You can also verify the distributive law $[a]\cdot([b]+[c])=[a]\cdot[b]+[a]\cdot[c]$, as follows:

$$\begin{aligned} [a]\cdot([b]+[c]) &= [a]\cdot[b+c] \\ &= [a(b+c)] \\ &= [ab+ac] \\ &= [ab]+[ac] \\ &= [a]\cdot[b]+[a]\cdot[c]. \end{aligned}$$

The integers modulo n are significant because they more closely fit certain applications than do other number systems such as $\mathbb{Z}$ or $\mathbb{R}$. If you go on to take a course in abstract algebra, then you will work extensively with $\mathbb{Z}_n$ as well as other, more exotic, number systems. (In such a course you will also use all of the proof techniques that we have discussed, as well as the ideas of equivalence relations.)

To close this section we take up an issue that may have bothered you earlier. It has to do with our definitions of addition $[a]+[b]=[a+b]$ and multiplication $[a]\cdot[b]=[ab]$. These definitions define addition and multiplication of equivalence classes in terms of representatives a and b in the equivalence classes. Since there are many different ways to choose such representatives, we may well wonder if addition and multiplication are consistently defined. For example, suppose two people, Alice and Bob, want to multiply the elements [2] and [3] in $\mathbb{Z}_5$. Alice does the calculation as $[2]\cdot[3]=[6]=[1]$, so her final answer is [1]. Bob does it differently. Since $[2]=[7]$ and $[3]=[8]$, he works out $[2]\cdot[3]$ as $[7]\cdot[8]=[56]$. Since $56\equiv 1$ (mod 5), Bob's answer is $[56]=[1]$, and that agrees with Alice's answer. Will their answers always agree or did they just get lucky (with the arithmetic)?

The fact is that no matter how they do the multiplication in $\mathbb{Z}_n$, their answers will agree. To see why, suppose Alice and Bob want to multiply the elements $[a],[b]\in\mathbb{Z}_n$, and suppose $[a]=[a']$ and $[b]=[b']$. Alice and Bob do the multiplication as follows:

$$\begin{array}{ll}\text{Alice:} & [a]\cdot[b]=[ab],\\ \text{Bob:} & [a']\cdot[b']=[a'b'].\end{array}$$

We need to show that their answers agree, that is, we need to show $[ab]=[a'b']$. Since $[a]=[a']$, we know by Theorem 11.1 that $a\equiv a' \pmod{n}$. Thus $n\mid(a-a')$, so $a-a'=nk$ for some integer k. Likewise, as $[b]=[b']$, we know $b\equiv b' \pmod{n}$, or $n\mid(b-b')$, so $b-b'=n\ell$ for some integer ℓ. Thus we get $a=a'+nk$ and $b=b'+n\ell$. Therefore:

$$\begin{array}{rcl} ab & = & (a'+nk)(b'+n\ell)\\ & = & a'b'+a'n\ell+nkb'+n^2k\ell,\\ \text{hence } ab-a'b' & = & n(a'\ell+kb'+nk\ell).\end{array}$$

This shows $n\mid(ab-a'b')$, so $ab\equiv a'b' \pmod{n}$, and from that we conclude $[ab]=[a'b']$. Consequently Alice and Bob really do get the same answer, so we can be assured that the definition of multiplication in $\mathbb{Z}_n$ is consistent.

Exercise 8 below asks you to show that addition in $\mathbb{Z}_n$ is similarly consistent.

Exercises for Section 11.4

1. Write the addition and multiplication tables for $\mathbb{Z}_2$.
2. Write the addition and multiplication tables for $\mathbb{Z}_3$.
3. Write the addition and multiplication tables for $\mathbb{Z}_4$.
4. Write the addition and multiplication tables for $\mathbb{Z}_6$.
5. Suppose $[a],[b] \in \mathbb{Z}_5$ and $[a]\cdot[b] = [0]$. Is it necessarily true that either $[a] = [0]$ or $[b] = [0]$?
6. Suppose $[a],[b] \in \mathbb{Z}_6$ and $[a]\cdot[b] = [0]$. Is it necessarily true that either $[a] = [0]$ or $[b] = [0]$?
7. Do the following calculations in $\mathbb{Z}_9$, in each case expressing your answer as $[a]$ with $0 \le a \le 8$.
   (a) $[8]+[8]$ **(b)** $[24]+[11]$ **(c)** $[21]\cdot[15]$ **(d)** $[8]\cdot[8]$
8. Suppose $[a],[b] \in \mathbb{Z}_n$, and $[a] = [a']$ and $[b] = [b']$. Alice adds $[a]$ and $[b]$ as $[a]+[b] = [a+b]$. Bob adds them as $[a']+[b'] = [a'+b']$. Show that their answers $[a+b]$ and $[a'+b']$ are the same.

11.5 Relations Between Sets

In the beginning of this chapter, we defined a relation on a set A to be a subset $R \subseteq A \times A$. This created a framework that could model any situation in which elements of A are compared to themselves. In this setting, the statement xRy has elements x and y from A on either side of the R because R compares elements from A. But there are other relational symbols that don't work this way. Consider $\in$. The statement $5 \in \mathbb{Z}$ expresses a relationship between 5 and $\mathbb{Z}$ (namely that the element 5 is in the set $\mathbb{Z}$) but 5 and $\mathbb{Z}$ are not in any way naturally regarded as both elements of some set A. To overcome this difficulty, we generalize the idea of a relation on A to a *relation from A to B*.

Definition 11.7 A **relation** from a set A to a set B is a subset $R \subseteq A \times B$. We often abbreviate the statement $(x,y) \in R$ as xRy. The statement $(x,y) \notin R$ is abbreviated as $x \not R\, y$.

Example 11.16 Suppose $A = \{1,2\}$ and $B = \mathscr{P}(A) = \{\emptyset,\{1\},\{2\},\{1,2\}\}$. Then $R = \{(1,\{1\}),(2,\{2\}),(1,\{1,2\}),(2,\{1,2\})\} \subseteq A \times B$ is a relation from A to B. Note that we have $1R\{1\}$, $2R\{2\}$, $1R\{1,2\}$ and $2R\{1,2\}$. The relation R is the familiar relation $\in$ for the set A, that is, xRX means exactly the same thing as $x \in X$.

Diagrams for relations from A to B differ from diagrams for relations on A. Since there are two sets A and B in a relation from A to B, we have to draw labeled nodes for each of the two sets. Then we draw arrows from x to y whenever xRy. The following figure illustrates this for Example 11.16.

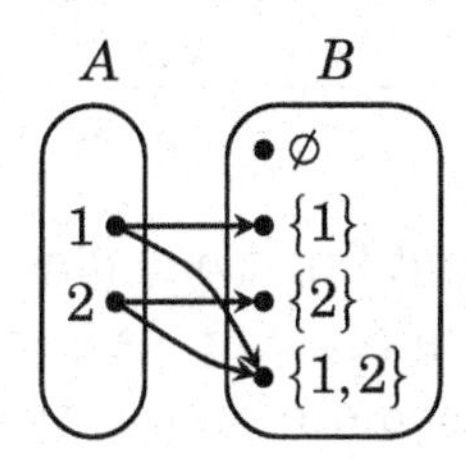

Figure 11.3. A relation from A to B

The ideas from this chapter show that any relation (whether it is a familiar one like $\geq$, $\leq$, $=$, $|$, $\in$ or $\subseteq$, or a more exotic one) is really just a set. Therefore the theory of relations is a part of the theory of sets. In the next chapter, we will see that this idea touches on another important mathematical construction, namely functions. We will define a function to be a special kind of relation from one set to another, and in this context we will see that any function is really just a set.

CHAPTER 12

Functions

You know from calculus that functions play a fundamental role in mathematics. You likely view a function as a kind of formula that describes a relationship between two (or more) quantities. You certainly understand and appreciate the fact that relationships between quantities are important in all scientific disciplines, so you do not need to be convinced that functions are important. Still, you may not be aware of the full significance of functions. Functions are more than merely descriptions of numeric relationships. In a more general sense, functions can compare and relate different kinds of mathematical structures. You will see this as your understanding of mathematics deepens. In preparation of this deepening, we will now explore a more general and versatile view of functions.

The concept of a relation between sets (Definition 11.7) plays a big role here, so you may want to quickly review it.

12.1 Functions

Let's start on familiar ground. Consider the function $f(x) = x^2$ from $\mathbb{R}$ to $\mathbb{R}$. Its graph is the set of points $R = \{(x, x^2) : x \in \mathbb{R}\} \subseteq \mathbb{R} \times \mathbb{R}$.

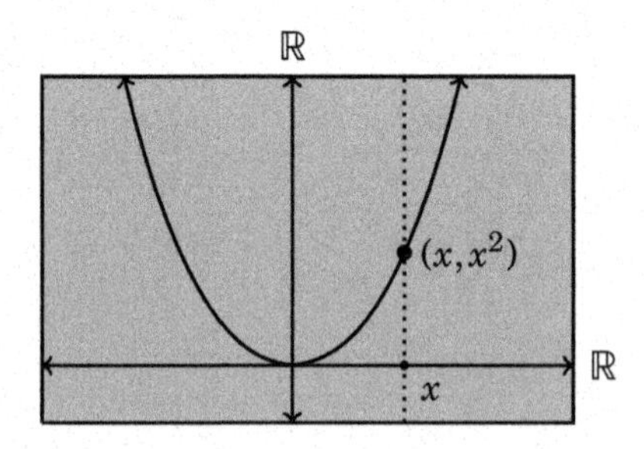

Figure 12.1. A familiar function

Having read Chapter 11, you may see f in a new light. Its graph $R \subseteq \mathbb{R} \times \mathbb{R}$ is a relation on the set $\mathbb{R}$. In fact, as we shall see, functions are just special kinds of relations. Before stating the exact definition, we look at another example. Consider the function $f(n) = |n| + 2$ that converts integers n into natural numbers $|n| + 2$. Its graph is $R = \{(n, |n| + 2) : n \in \mathbb{Z}\} \subseteq \mathbb{Z} \times \mathbb{N}$.

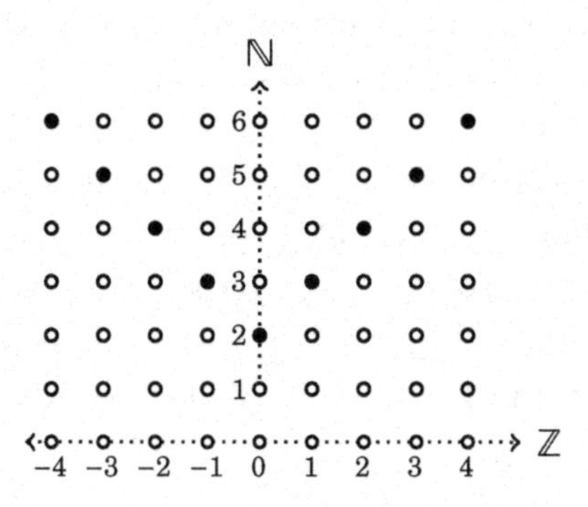

Figure 12.2. The function $f : \mathbb{Z} \to \mathbb{N}$, where $f(n) = |n| + 2$

Figure 12.2 shows the graph R as darkened dots in the grid of points $\mathbb{Z} \times \mathbb{N}$. Notice that in this example R is not a relation on a single set. The set of input values $\mathbb{Z}$ is different from the set $\mathbb{N}$ of output values, so the graph $R \subseteq \mathbb{Z} \times \mathbb{N}$ is *a relation from* $\mathbb{Z}$ *to* $\mathbb{N}$.

This example illustrates three things. First, a function can be viewed as sending elements from one set A to another set B. (In the case of f, $A = \mathbb{Z}$ and $B = \mathbb{N}$.) Second, such a function can be regarded as a relation from A to B. Third, for every input value n, there is *exactly one* output value $f(n)$. In your high school algebra course, this was expressed by the *vertical line test*: Any vertical line intersects a function's graph at most once. It means that for any input value x, the graph contains exactly one point of form $(x, f(x))$. Our main definition, given below, incorporates all of these ideas.

Definition 12.1 Suppose A and B are sets. A **function** f from A to B (denoted as $f : A \to B$) is a relation $f \subseteq A \times B$ from A to B, satisfying the property that for each $a \in A$ the relation f contains exactly one ordered pair of form (a, b). The statement $(a, b) \in f$ is abbreviated $f(a) = b$.

Example 12.1 Consider the function f graphed in Figure 12.2. According to Definition 12.1, we regard f as the set of points in its graph, that is, $f = \{(n, |n| + 2) : n \in \mathbb{Z}\} \subseteq \mathbb{Z} \times \mathbb{N}$. This is a relation from $\mathbb{Z}$ to $\mathbb{N}$, and indeed given any $a \in \mathbb{Z}$ the set f contains exactly one ordered pair $(a, |a| + 2)$ whose first coordinate is a. Since $(1, 3) \in f$, we write $f(1) = 3$; and since $(-3, 5) \in f$ we write $f(-3) = 5$, etc. In general, $(a, b) \in f$ means that f sends the input value a to the output value b, and we express this as $f(a)=b$. This function can be expressed by a formula: For each input value n, the output value is $|n|+2$, so we may write $f(n)=|n|+2$. All this agrees with the way we thought of functions in algebra and calculus; the only difference is that now we also think of a function as a relation.

Definition 12.2 For a function $f:A\to B$, the set A is called the **domain** of f. (Think of the domain as the set of possible "input values" for f.) The set B is called the **codomain** of f. The **range** of f is the set $\{f(a):a\in A\}=\{b:(a,b)\in f\}$. (Think of the range as the set of all possible "output values" for f. Think of the codomain as a sort of "target" for the outputs.)

Continuing Example 12.1, the domain of f is $\mathbb{Z}$ and its codomain is $\mathbb{N}$. Its range is $\{f(a):a\in\mathbb{Z}\}=\{|a|+2:a\in\mathbb{Z}\}=\{2,3,4,5,\ldots\}$. Notice that the range is a subset of the codomain, but it does not (in this case) equal the codomain.

In our examples so far, the domains and codomains are sets of numbers, but this needn't be the case in general, as the next example indicates.

Example 12.2 Let $A=\{p,q,r,s\}$ and $B=\{0,1,2\}$, and

$$f=\{(p,0),(q,1),(r,2),(s,2)\}\subseteq A\times B.$$

This is a function $f:A\to B$ because each element of A occurs exactly once as a first coordinate of an ordered pair in f. We have $f(p)=0$, $f(q)=1$, $f(r)=2$ and $f(s)=2$. The domain of f is $\{p,q,r,s\}$, and the codomain and range are both $\{0,1,2\}$.

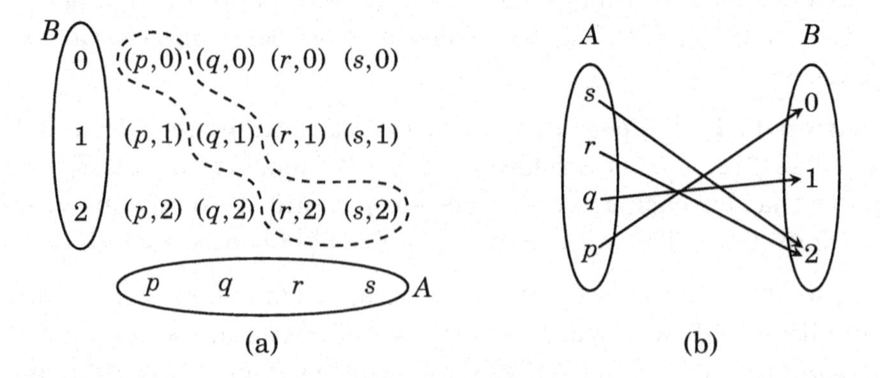

Figure 12.3. Two ways of drawing the function $f=\{(p,0),(q,1),(r,2),(s,2)\}$

If A and B are not both sets of numbers it can be difficult to draw a graph of $f : A \to B$ in the traditional sense. Figure 12.3(a) shows an attempt at a graph of f from Example 12.2. The sets A and B are aligned roughly as x- and y-axes, and the Cartesian product $A \times B$ is filled in accordingly. The subset $f \subseteq A \times B$ is indicated with dashed lines, and this can be regarded as a "graph" of f. A more natural visual description of f is shown in 12.3(b). The sets A and B are drawn side-by-side, and arrows point from a to b whenever $f(a) = b$.

In general, if $f : A \to B$ is the kind of function you may have encountered in algebra or calculus, then conventional graphing techniques offer the best visual description of it. On the other hand, if A and B are finite or if we are thinking of them as generic sets, then describing f with arrows is often a more appropriate way of visualizing it.

We emphasize that, according to Definition 12.1, a function is really just a special kind of set. Any function $f : A \to B$ is a subset of $A \times B$. By contrast, your calculus text probably defined a function as a certain kind of "rule." While that intuitive outlook is adequate for the first few semesters of calculus, it does not hold up well to the rigorous mathematical standards necessary for further progress. The problem is that words like "rule" are too vague. Defining a function as a set removes the ambiguity. It makes a function into a concrete mathematical object.

Still, in practice we tend to think of functions as rules. Given $f : \mathbb{Z} \to \mathbb{N}$ where $f(x) = |x| + 2$, we think of this as a rule that associates any number $n \in \mathbb{Z}$ to the number $|n| + 2$ in $\mathbb{N}$, rather than a set containing ordered pairs $(n, |n| + 2)$. It is only when we have to understand or interpret the theoretical nature of functions (as we do in this text) that Definition 12.1 comes to bear. The definition is a foundation that gives us license to think about functions in a more informal way.

The next example brings up a point about notation. Consider a function such as $f : \mathbb{Z}^2 \to \mathbb{Z}$, whose domain is a Cartesian product. This function takes as input an ordered pair $(m, n) \in \mathbb{Z}^2$ and sends it to a number $f((m, n)) \in \mathbb{Z}$. To simplify the notation, it is common to write $f(m, n)$ instead of $f((m, n))$, even though this is like writing fx instead of $f(x)$. We also remark that although we've been using the letters f, g and h to denote functions, any other reasonable symbol could be used. Greek letters such as φ and θ are common.

Example 12.3 Say a function $\varphi : \mathbb{Z}^2 \to \mathbb{Z}$ is defined as $\varphi(m, n) = 6m - 9n$. Note that as a set, this function is $\varphi = \{((m, n), 6m - 9n) : (m, n) \in \mathbb{Z}^2\} \subseteq \mathbb{Z}^2 \times \mathbb{Z}$. What is the range of φ?

To answer this, first observe that for any $(m,n) \in \mathbb{Z}^2$, the value $f(m,n) = 6m-9n = 3(2m-3n)$ is a multiple of 3. Thus every number in the range is a multiple of 3, so the range is a *subset* of the set of all multiples of 3. On the other hand if $b = 3k$ is a multiple of 3 we have $\varphi(-k,-k) = 6(-k)-9(-k) = 3k = b$, which means any multiple of 3 is in the range of φ. Therefore the range of φ is the set $\{3k : k \in \mathbb{Z}\}$ of all multiples of 3.

To conclude this section, let's use Definition 12.1 to help us understand what it means for two functions $f : A \to B$ and $g : C \to D$ to be equal. According to our definition, functions f and g are subsets $f \subseteq A \times B$ and $g \subseteq C \times D$. It makes sense to say that f and g are equal if $f = g$, that is, if they are equal as sets.

Thus the two functions $f = \{(1,a),(2,a),(3,b)\}$ and $g = \{(3,b),(2,a),(1,a)\}$ are equal because the sets f and g are equal. Notice that the domain of both functions is $A = \{1,2,3\}$, the set of first elements x in the ordered pairs $(x,y) \in f = g$. In general, equal functions must have equal domains.

Observe also that the equality $f = g$ means $f(x) = g(x)$ for every $x \in A$. We repackage these ideas in the following definition.

Definition 12.3 Two functions $f : A \to B$ and $g : A \to D$ are **equal** if $f(x) = g(x)$ for every $x \in A$.

Observe that f and g can have different codomains and still be equal. Consider the functions $f : \mathbb{Z} \to \mathbb{N}$ and $g : \mathbb{Z} \to \mathbb{Z}$ defined as $f(x) = |x|+2$ and $g(x) = |x|+2$. Even though their codomains are different, the functions are equal because $f(x) = g(x)$ for every x in the domain.

Exercises for Section 12.1

1. Suppose $A = \{0,1,2,3,4\}$, $B = \{2,3,4,5\}$ and $f = \{(0,3),(1,3),(2,4),(3,2),(4,2)\}$. State the domain and range of f. Find $f(2)$ and $f(1)$.
2. Suppose $A = \{a,b,c,d\}$, $B = \{2,3,4,5,6\}$ and $f = \{(a,2),(b,3),(c,4),(d,5)\}$. State the domain and range of f. Find $f(b)$ and $f(d)$.
3. There are four different functions $f : \{a,b\} \to \{0,1\}$. List them all. Diagrams will suffice.
4. There are eight different functions $f : \{a,b,c\} \to \{0,1\}$. List them all. Diagrams will suffice.
5. Give an example of a relation from $\{a,b,c,d\}$ to $\{d,e\}$ that is not a function.
6. Suppose $f : \mathbb{Z} \to \mathbb{Z}$ is defined as $f = \{(x,4x+5) : x \in \mathbb{Z}\}$. State the domain, codomain and range of f. Find $f(10)$.

7. Consider the set $f = \{(x,y) \in \mathbb{Z} \times \mathbb{Z} : 3x + y = 4\}$. Is this a function from $\mathbb{Z}$ to $\mathbb{Z}$? Explain.

8. Consider the set $f = \{(x,y) \in \mathbb{Z} \times \mathbb{Z} : x + 3y = 4\}$. Is this a function from $\mathbb{Z}$ to $\mathbb{Z}$? Explain.

9. Consider the set $f = \{(x^2, x) : x \in \mathbb{R}\}$. Is this a function from $\mathbb{R}$ to $\mathbb{R}$? Explain.

10. Consider the set $f = \{(x^3, x) : x \in \mathbb{R}\}$. Is this a function from $\mathbb{R}$ to $\mathbb{R}$? Explain.

11. Is the set $\theta = \{(X, |X|) : X \subseteq \mathbb{Z}_5\}$ a function? If so, what is its domain and range?

12. Is the set $\theta = \{((x,y),(3y, 2x, x+y)) : x, y \in \mathbb{R}\}$ a function? If so, what is its domain, codomain and range?

12.2 Injective and Surjective Functions

You may recall from algebra and calculus that a function may be *one-to-one* and *onto*, and these properties are related to whether or not the function is invertible. We now review these important ideas. In advanced mathematics, the word *injective* is often used instead of *one-to-one*, and *surjective* is used instead of *onto*. Here are the exact definitions:

Definition 12.4 A function $f : A \to B$ is:

1. **injective** (or one-to-one) if for every $x, y \in A$, $x \neq y$ implies $f(x) \neq f(y)$;
2. **surjective** (or onto) if for every $b \in B$ there is an $a \in A$ with $f(a) = b$;
3. **bijective** if f is both injective and surjective.

Below is a visual description of Definition 12.4. In essence, injective means that unequal elements in A always get sent to unequal elements in B. Surjective means that every element of B has an arrow pointing to it, that is, it equals $f(a)$ for some a in the domain of f.

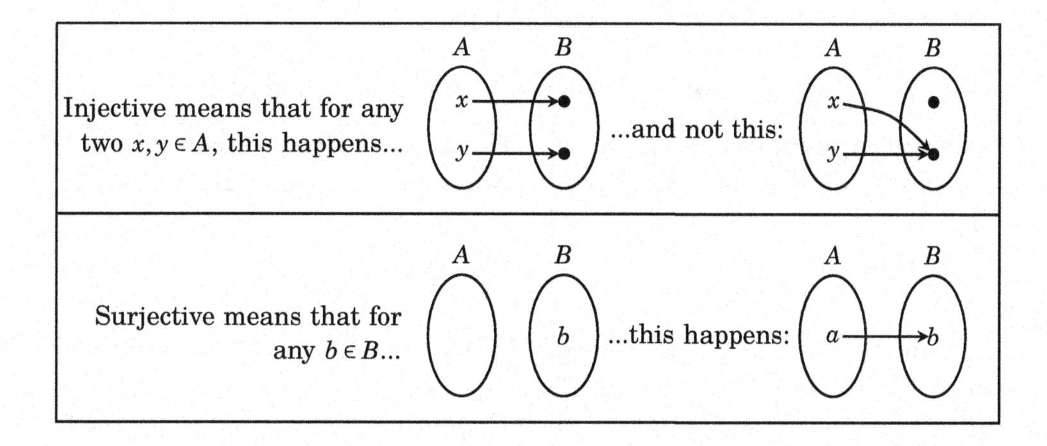

For more concrete examples, consider the following functions from $\mathbb{R}$ to $\mathbb{R}$. The function $f(x) = x^2$ is not injective because $-2 \neq 2$, but $f(-2) = f(2)$. Nor is it surjective, for if $b = -1$ (or if b is any negative number), then there is no $a \in \mathbb{R}$ with $f(a) = b$. On the other hand, $g(x) = x^3$ is both injective and surjective, so it is also bijective.

There are four possible injective/surjective combinations that a function may possess. This is illustrated in the following figure showing four functions from A to B. Functions in the first column are injective, those in the second column are not injective. Functions in the first row are surjective, those in the second row are not.

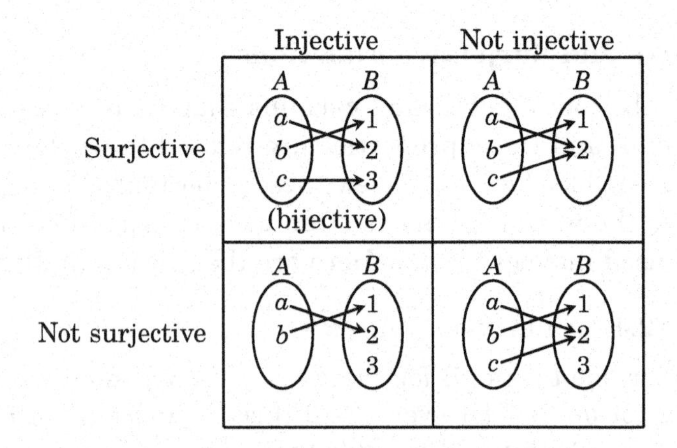

We note in passing that, according to the definitions, a function is surjective if and only if its codomain equals its range.

Often it is necessary to prove that a particular function $f : A \to B$ is injective. For this we must prove that for any two elements $x, y \in A$, the conditional statement $(x \neq y) \Rightarrow \big(f(x) \neq f(y)\big)$ is true. The two main approaches for this are summarized below.

How to show a function $f : A \to B$ is injective:

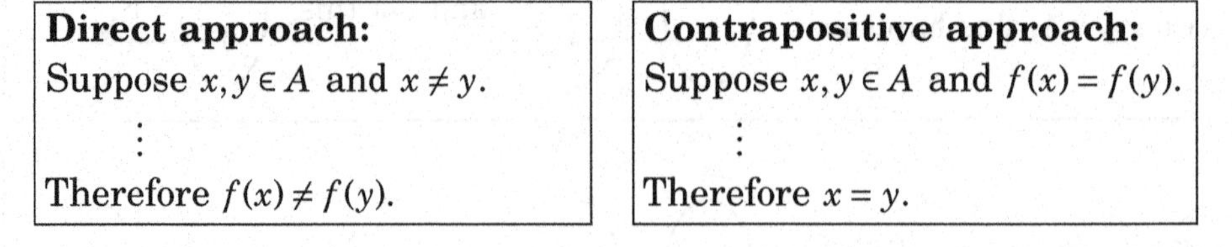

Direct approach:	**Contrapositive approach:**
Suppose $x, y \in A$ and $x \neq y$.	Suppose $x, y \in A$ and $f(x) = f(y)$.
$\vdots$	$\vdots$
Therefore $f(x) \neq f(y)$.	Therefore $x = y$.

Of these two approaches, the contrapositive is often the easiest to use, especially if f is defined by an algebraic formula. This is because the contrapositive approach starts with the *equation* $f(x) = f(y)$ and proceeds to the *equation* $x = y$. In algebra, as you know, it is usually easier to work with equations than inequalities.

To prove that a function is *not* injective, you must *disprove* the statement $(x \neq y) \Rightarrow \big(f(x) \neq f(y)\big)$. For this it suffices to find example of two elements $x, y \in A$ for which $x \neq y$ and $f(x) = f(y)$.

Next we examine how to prove that $f : A \to B$ is *surjective*. According to Definition 12.4, we must prove the statement $\forall b \in B, \exists a \in A, f(a) = b$. In words, we must show that for any $b \in B$, there is at least one $a \in A$ (which may depend on b) having the property that $f(a) = b$. Here is an outline.

How to show a function $f : A \to B$ is surjective:

Suppose $b \in B$. [Prove there exists $a \in A$ for which $f(a) = b$.]

In the second step, we have to prove the existence of an a for which $f(a) = b$. For this, just finding an example of such an a would suffice. (How to find such an example depends on how f is defined. If f is given as a formula, we may be able to find a by solving the equation $f(a) = b$ for a. Sometimes you can find a by just plain common sense.) To show f is *not* surjective, we must prove the negation of $\forall b \in B, \exists a \in A, f(a) = b$, that is, we must prove $\exists b \in B, \forall a \in A, f(a) \neq b$.

The following examples illustrate these ideas. (For the first example, note that the set $\mathbb{R} - \{0\}$ is $\mathbb{R}$ with the number 0 removed.)

Example 12.4 Show that the function $f : \mathbb{R} - \{0\} \to \mathbb{R}$ defined as $f(x) = \frac{1}{x} + 1$ is injective but not surjective.

We will use the contrapositive approach to show that f is injective. Suppose $x, y \in \mathbb{R} - \{0\}$ and $f(x) = f(y)$. This means $\frac{1}{x} + 1 = \frac{1}{y} + 1$. Subtracting 1 from both sides and inverting produces $x = y$. Therefore f is injective.

Function f is not surjective because there exists an element $b = 1 \in \mathbb{R}$ for which $f(x) = \frac{1}{x} + 1 \neq 1$ for every $x \in \mathbb{R} - \{0\}$.

Example 12.5 Show that the function $g : \mathbb{Z} \times \mathbb{Z} \to \mathbb{Z} \times \mathbb{Z}$ defined by the formula $g(m,n) = (m+n, m+2n)$, is both injective and surjective.

We will use the contrapositive approach to show that g is injective. Thus we need to show that $g(m,n) = g(k,\ell)$ implies $(m,n) = (k,\ell)$. Suppose $(m,n), (k,\ell) \in \mathbb{Z} \times \mathbb{Z}$ and $g(m,n) = g(k,\ell)$. Then $(m+n, m+2n) = (k+\ell, k+2\ell)$. It follows that $m+n = k+\ell$ and $m+2n = k+2\ell$. Subtracting the first equation from the second gives $n = \ell$. Next, subtract $n = \ell$ from $m+n = k+\ell$ to get $m = k$. Since $m = k$ and $n = \ell$, it follows that $(m,n) = (k,\ell)$. Therefore g is injective.

To see that g is surjective, consider an arbitrary element $(b,c) \in \mathbb{Z} \times \mathbb{Z}$. We need to show that there is some $(x,y) \in \mathbb{Z} \times \mathbb{Z}$ for which $g(x,y) = (b,c)$. To find (x,y), note that $g(x,y) = (b,c)$ means $(x+y, x+2y) = (b,c)$. This leads to the following system of equations:

$$\begin{array}{rcrcl} x & + & y & = & b \\ x & + & 2y & = & c. \end{array}$$

Solving gives $x = 2b - c$ and $y = c - b$. Then $(x,y) = (2b-c, c-b)$. We now have $g(2b-c, c-b) = (b,c)$, and it follows that g is surjective.

Example 12.6 Consider function $h : \mathbb{Z} \times \mathbb{Z} \to \mathbb{Q}$ defined as $h(m,n) = \dfrac{m}{|n|+1}$. Determine whether this is injective and whether it is surjective.

This function is *not* injective because of the unequal elements $(1,2)$ and $(1,-2)$ in $\mathbb{Z} \times \mathbb{Z}$ for which $h(1,2) = h(1,-2) = \frac{1}{3}$. However, h is surjective: Take any element $b \in \mathbb{Q}$. Then $b = \frac{c}{d}$ for some $c, d \in \mathbb{Z}$. Notice we may assume d is positive by making c negative, if necessary. Then $h(c, d-1) = \frac{c}{|d-1|+1} = \frac{c}{d} = b$.

Exercises for Section 12.2

1. Let $A = \{1,2,3,4\}$ and $B = \{a,b,c\}$. Give an example of a function $f : A \to B$ that is neither injective nor surjective.
2. Consider the logarithm function $\ln : (0,\infty) \to \mathbb{R}$. Decide whether this function is injective and whether it is surjective.
3. Consider the cosine function $\cos : \mathbb{R} \to \mathbb{R}$. Decide whether this function is injective and whether it is surjective. What if it had been defined as $\cos : \mathbb{R} \to [-1,1]$?
4. A function $f : \mathbb{Z} \to \mathbb{Z} \times \mathbb{Z}$ is defined as $f(n) = (2n, n+3)$. Verify whether this function is injective and whether it is surjective.
5. A function $f : \mathbb{Z} \to \mathbb{Z}$ is defined as $f(n) = 2n+1$. Verify whether this function is injective and whether it is surjective.
6. A function $f : \mathbb{Z} \times \mathbb{Z} \to \mathbb{Z}$ is defined as $f(m,n) = 3n - 4m$. Verify whether this function is injective and whether it is surjective.
7. A function $f : \mathbb{Z} \times \mathbb{Z} \to \mathbb{Z}$ is defined as $f(m,n) = 2n - 4m$. Verify whether this function is injective and whether it is surjective.
8. A function $f : \mathbb{Z} \times \mathbb{Z} \to \mathbb{Z} \times \mathbb{Z}$ is defined as $f(m,n) = (m+n, 2m+n)$. Verify whether this function is injective and whether it is surjective.
9. Prove that the function $f : \mathbb{R} - \{2\} \to \mathbb{R} - \{5\}$ defined by $f(x) = \dfrac{5x+1}{x-2}$ is bijective.
10. Prove the function $f : \mathbb{R} - \{1\} \to \mathbb{R} - \{1\}$ defined by $f(x) = \left(\dfrac{x+1}{x-1}\right)^3$ is bijective.
11. Consider the function $\theta : \{0,1\} \times \mathbb{N} \to \mathbb{Z}$ defined as $\theta(a,b) = (-1)^a b$. Is θ injective? Is it surjective? Bijective? Explain.

12. Consider the function $\theta : \{0,1\} \times \mathbb{N} \to \mathbb{Z}$ defined as $\theta(a,b) = a-2ab+b$. Is θ injective? Is it surjective? Bijective? Explain.

13. Consider the function $f : \mathbb{R}^2 \to \mathbb{R}^2$ defined by the formula $f(x,y) = (xy, x^3)$. Is f injective? Is it surjective? Bijective? Explain.

14. Consider the function $\theta : \mathscr{P}(\mathbb{Z}) \to \mathscr{P}(\mathbb{Z})$ defined as $\theta(X) = \overline{X}$. Is θ injective? Is it surjective? Bijective? Explain.

15. This question concerns functions $f : \{A,B,C,D,E,F,G\} \to \{1,2,3,4,5,6,7\}$. How many such functions are there? How many of these functions are injective? How many are surjective? How many are bijective?

16. This question concerns functions $f : \{A,B,C,D,E\} \to \{1,2,3,4,5,6,7\}$. How many such functions are there? How many of these functions are injective? How many are surjective? How many are bijective?

17. This question concerns functions $f : \{A,B,C,D,E,F,G\} \to \{1,2\}$. How many such functions are there? How many of these functions are injective? How many are surjective? How many are bijective?

18. Prove that the function $f : \mathbb{N} \to \mathbb{Z}$ defined as $f(n) = \frac{(-1)^n(2n-1)+1}{4}$ is bijective.

12.3 The Pigeonhole Principle

Here is a simple but useful idea. Imagine there is a set A of pigeons and a set B of pigeon-holes, and all the pigeons fly into the pigeon-holes. You can think of this as describing a function $f : A \to B$, where pigeon X flies into pigeon-hole $f(X)$. Figure 12.4 illustrates this.

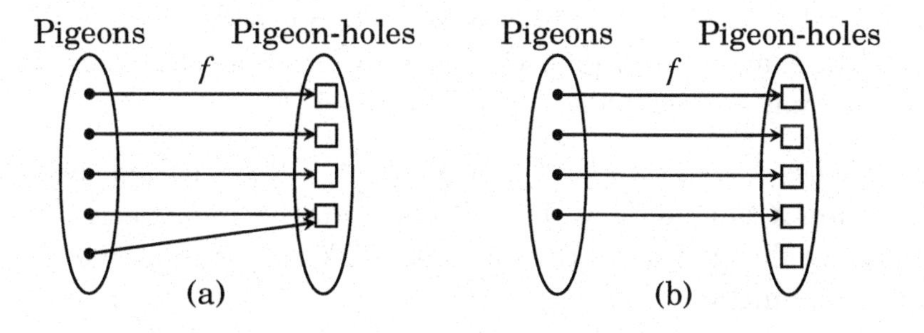

Figure 12.4. The pigeonhole principle

In Figure 12.4(a) there are more pigeons than pigeon-holes, and it is obvious that in such a case at least two pigeons have to fly into the same pigeon-hole, meaning that f is not injective. In Figure 12.4(b) there are fewer pigeons than pigeon-holes, so clearly at least one pigeon-hole remains empty, meaning that f is not surjective.

Although the underlying idea expressed by these figures has little to do with pigeons, it is nonetheless called the *pigeonhole principle:*

The Pigeonhole Principle
Suppose A and B are finite sets and $f : A \to B$ is any function. Then:

1. If $|A| > |B|$, then f is not injective.
2. If $|A| < |B|$, then f is not surjective.

Though the pigeonhole principle is obvious, it can be used to prove some things that are not so obvious.

Example 12.7 Prove the following proposition.

Proposition If A is any set of 10 integers between 1 and 100, then there exist two different subsets $X \subseteq A$ and $Y \subseteq A$ for which the sum of elements in X equals the sum of elements in Y.

To illustrate what this proposition is saying, consider the random set

$$A = \{5,7,12,11,17,50,51,80,90,100\}$$

of 10 integers between 1 and 100. Notice that A has subsets $X = \{5,80\}$ and $Y = \{7,11,17,50\}$ for which the sum of the elements in X equals the sum of those in Y. If we tried to "mess up" A by changing the 5 to a 6, we get

$$A = \{6,7,12,11,17,50,51,80,90,100\}$$

which has subsets $X = \{7,12,17,50\}$ and $Y = \{6,80\}$ both of whose elements add up to the same number (86). The proposition asserts that this is always possible, no matter what A is. Here is a proof:

Proof. Suppose $A \subseteq \{1,2,3,4,\ldots,99,100\}$ and $|A| = 10$, as stated. Notice that if $X \subseteq A$, then X has no more than 10 elements, each between 1 and 100, and therefore the sum of all the elements of X is less than $100 \cdot 10 = 1000$. Consider the function

$$f : \mathscr{P}(A) \to \{0,1,2,3,4,\ldots,1000\}$$

where $f(X)$ is the sum of the elements in X. (Examples: $f(\{3,7,50\}) = 60$; $f(\{1,70,80,95\}) = 246$.) As $|\mathscr{P}(A)| = 2^{10} = 1024 > 1001 = \big|\{0,1,2,3,\ldots,1000\}\big|$, it follows from the pigeonhole principle that f is not injective. Therefore there are two unequal sets $X, Y \in \mathscr{P}(A)$ for which $f(X) = f(Y)$. In other words, there are subsets $X \subseteq A$ and $Y \subseteq A$ for which the sum of elements in X equals the sum of elements in Y. ■

Example 12.8 Prove the following proposition.

Proposition There are at least two Texans with the same number of hairs on their heads.

Proof. We will use two facts. First, the population of Texas is more than twenty million. Second, it is a biological fact that every human head has fewer than one million hairs. Let A be the set of all Texans, and let $B = \{0,1,2,3,4,\ldots,1000000\}$. Let $f : A \to B$ be the function for which $f(x)$ equals the number of hairs on the head of x. Since $|A| > |B|$, the pigeonhole principle asserts that f is not injective. Thus there are two Texans x and y for whom $f(x) = f(y)$, meaning that they have the same number of hairs on their heads. ■

Proofs that use the pigeonhole principle tend to be inherently non-constructive, in the sense discussed in Section 7.4. For example, the above proof does not explicitly give us of two Texans with the same number of hairs on their heads; it only shows that two such people exist. If we were to make a constructive proof, we could find examples of two bald Texans. Then they have the same number of head hairs, namely zero.

Exercises for Section 12.3

1. Prove that if six numbers are chosen at random, then at least two of them will have the same remainder when divided by 5.
2. Prove that if a is a natural number, then there exist two unequal natural numbers k and ℓ for which $a^k - a^\ell$ is divisible by 10.
3. Prove that if six natural numbers are chosen at random, then the sum or difference of two of them is divisible by 9.
4. Consider a square whose side-length is one unit. Select any five points from inside this square. Prove that at least two of these points are within $\frac{\sqrt{2}}{2}$ units of each other.
5. Prove that any set of seven distinct natural numbers contains a pair of numbers whose sum or difference is divisible by 10.
6. Given a sphere S, a *great circle* of S is the intersection of S with a plane through its center. Every great circle divides S into two parts. A *hemisphere* is the union of the great circle and one of these two parts. Prove that if five points are placed arbitrarily on S, then there is a hemisphere that contains four of them.
7. Prove or disprove: Any subset $X \subseteq \{1,2,3,\ldots,2n\}$ with $|X| > n$ contains two (unequal) elements $a, b \in X$ for which $a \mid b$ or $b \mid a$.

12.4 Composition

You should be familiar with the notion of function composition from algebra and calculus. Still, it is worthwhile to revisit it now with our more sophisticated ideas about functions.

Definition 12.5 Suppose $f : A \to B$ and $g : B \to C$ are functions with the property that the codomain of f equals the domain of g. The **composition** of f with g is another function, denoted as $g \circ f$ and defined as follows: If $x \in A$, then $g \circ f(x) = g(f(x))$. Therefore $g \circ f$ sends elements of A to elements of C, so $g \circ f : A \to C$.

The following figure illustrates the definition. Here $f : A \to B$, $g : B \to C$, and $g \circ f : A \to C$. We have, for example, $g \circ f(0) = g(f(0)) = g(2) = 4$. Be very careful with the order of the symbols. Even though g comes first in the symbol $g \circ f$, we work out $g \circ f(x)$ as $g(f(x))$, with f acting on x first, followed by g acting on $f(x)$.

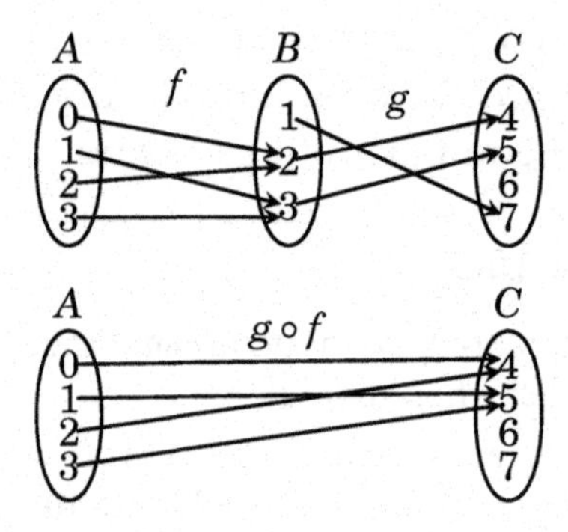

Figure 12.5. Composition of two functions

Notice that the composition $g \circ f$ also makes sense if the range of f is a *subset* of the domain of g. You should take note of this fact, but to keep matters simple we will continue to emphasize situations where the codomain of f equals the domain of g.

Example 12.9 Suppose $A = \{a, b, c\}$, $B = \{0, 1\}$, $C = \{1, 2, 3\}$. Let $f : A \to B$ be the function $f = \{(a, 0), (b, 1), (c, 0)\}$, and let $g : B \to C$ be the function $g = \{(0, 3), (1, 1)\}$. Then $g \circ f = \{(a, 3), (b, 1), (c, 3)\}$.

Example 12.10 Suppose $A = \{a, b, c\}$, $B = \{0, 1\}$, $C = \{1, 2, 3\}$. Let $f : A \to B$ be the function $f = \{(a, 0), (b, 1), (c, 0)\}$, and let $g : C \to B$ be the function $g = \{(1, 0), (2, 1), (3, 1)\}$. In this situation the composition $g \circ f$ is not defined because the codomain B of f is not the same set as the domain C of g.

Remember: In order for $g \circ f$ to make sense, the codomain of f must equal the domain of g. (Or at least be a subset of it.)

Example 12.11 Let $f : \mathbb{R} \to \mathbb{R}$ be defined as $f(x) = x^2 + x$, and $g : \mathbb{R} \to \mathbb{R}$ be defined as $g(x) = x + 1$. Then $g \circ f : \mathbb{R} \to \mathbb{R}$ is the function defined by the formula $g \circ f(x) = g(f(x)) = g(x^2 + x) = x^2 + x + 1$.

Since the domains and codomains of g and f are the same, we can in this case do a composition in the other order. Note that $f \circ g : \mathbb{R} \to \mathbb{R}$ is the function defined as $f \circ g(x) = f(g(x)) = f(x+1) = (x+1)^2 + (x+1) = x^2 + 3x + 2$.

This example illustrates that even when $g \circ f$ and $f \circ g$ are both defined, they are not necessarily equal. We can express this fact by saying *function composition is not commutative.*

We close this section by proving several facts about function composition that you are likely to encounter in your future study of mathematics. First, we note that, although it is not commutative, function composition *is* associative.

Theorem 12.1 Composition of functions is associative. That is if $f : A \to B$, $g : B \to C$ and $h : C \to D$, then $(h \circ g) \circ f = h \circ (g \circ f)$.

Proof. Suppose f, g, h are as stated. It follows from Definition 12.5 that both $(h \circ g) \circ f$ and $h \circ (g \circ f)$ are functions from A to D. To show that they are equal, we just need to show

$$\big((h \circ g) \circ f\big)(x) = \big(h \circ (g \circ f)\big)(x)$$

for every $x \in A$. Note that Definition 12.5 yields

$$\big((h \circ g) \circ f\big)(x) = (h \circ g)(f(x)) = h(g(f(x)).$$

Also

$$\big(h \circ (g \circ f)\big)(x) = h(g \circ f(x)) = h(g(f(x))).$$

Thus

$$\big((h \circ g) \circ f\big)(x) = \big(h \circ (g \circ f)\big)(x),$$

as both sides equal $h(g(f(x)))$. ■

Theorem 12.2 Suppose $f : A \to B$ and $g : B \to C$. If both f and g are injective, then $g \circ f$ is injective. If both f and g are surjective, then $g \circ f$ is surjective.

Proof. First suppose both f and g are injective. To see that $g \circ f$ is injective, we must show that $g \circ f(x) = g \circ f(y)$ implies $x = y$. Suppose $g \circ f(x) = g \circ f(y)$. This means $g(f(x)) = g(f(y))$. It follows that $f(x) = f(y)$. (For otherwise g wouldn't be injective.) But since $f(x) = f(y)$ and f is injective, it must be that $x = y$. Therefore $g \circ f$ is injective.

Next suppose both f and g are surjective. To see that $g \circ f$ is surjective, we must show that for any element $c \in C$, there is a corresponding element $a \in A$ for which $g \circ f(a) = c$. Thus consider an arbitrary $c \in C$. Because g is surjective, there is an element $b \in B$ for which $g(b) = c$. And because f is surjective, there is an element $a \in A$ for which $f(a) = b$. Therefore $g(f(a)) = g(b) = c$, which means $g \circ f(a) = c$. Thus $g \circ f$ is surjective. ■

Exercises for Section 12.4

1. Suppose $A = \{5,6,8\}$, $B = \{0,1\}$, $C = \{1,2,3\}$. Let $f : A \to B$ be the function $f = \{(5,1),(6,0),(8,1)\}$, and $g : B \to C$ be $g = \{(0,1),(1,1)\}$. Find $g \circ f$.
2. Suppose $A = \{1,2,3,4\}$, $B = \{0,1,2\}$, $C = \{1,2,3\}$. Let $f : A \to B$ be

$$f = \{(1,0),(2,1),(3,2),(4,0)\},$$

and $g : B \to C$ be $g = \{(0,1),(1,1),(2,3)\}$. Find $g \circ f$.
3. Suppose $A = \{1,2,3\}$. Let $f : A \to A$ be the function $f = \{(1,2),(2,2),(3,1)\}$, and let $g : A \to A$ be the function $g = \{(1,3),(2,1),(3,2)\}$. Find $g \circ f$ and $f \circ g$.
4. Suppose $A = \{a,b,c\}$. Let $f : A \to A$ be the function $f = \{(a,c),(b,c),(c,c)\}$, and let $g : A \to A$ be the function $g = \{(a,a),(b,b),(c,a)\}$. Find $g \circ f$ and $f \circ g$.
5. Consider the functions $f, g : \mathbb{R} \to \mathbb{R}$ defined as $f(x) = \sqrt[3]{x+1}$ and $g(x) = x^3$. Find the formulas for $g \circ f$ and $f \circ g$.
6. Consider the functions $f, g : \mathbb{R} \to \mathbb{R}$ defined as $f(x) = \frac{1}{x^2+1}$ and $g(x) = 3x + 2$. Find the formulas for $g \circ f$ and $f \circ g$.
7. Consider the functions $f, g : \mathbb{Z} \times \mathbb{Z} \to \mathbb{Z} \times \mathbb{Z}$ defined as $f(m,n) = (mn, m^2)$ and $g(m,n) = (m+1, m+n)$. Find the formulas for $g \circ f$ and $f \circ g$.
8. Consider the functions $f, g : \mathbb{Z} \times \mathbb{Z} \to \mathbb{Z} \times \mathbb{Z}$ defined as $f(m,n) = (3m - 4n, 2m + n)$ and $g(m,n) = (5m + n, m)$. Find the formulas for $g \circ f$ and $f \circ g$.
9. Consider the functions $f : \mathbb{Z} \times \mathbb{Z} \to \mathbb{Z}$ defined as $f(m,n) = m + n$ and $g : \mathbb{Z} \to \mathbb{Z} \times \mathbb{Z}$ defined as $g(m) = (m,m)$. Find the formulas for $g \circ f$ and $f \circ g$.
10. Consider the function $f : \mathbb{R}^2 \to \mathbb{R}^2$ defined by the formula $f(x,y) = (xy, x^3)$. Find a formula for $f \circ f$.

12.5 Inverse Functions

You may recall from calculus that if a function f is injective and surjective, then it has an inverse function f^{-1} that "undoes" the effect of f in the sense that $f^{-1}(f(x)) = x$ for every x in the domain. (For example, if $f(x) = x^3$, then $f^{-1}(x) = \sqrt[3]{x}$.) We now review these ideas. Our approach uses two ingredients, outlined in the following definitions.

Definition 12.6 Given a set A, the **identity function** on A is the function $i_A : A \to A$ defined as $i_A(x) = x$ for every $x \in A$.

Example: If $A = \{1,2,3\}$, then $i_A = \{(1,1),(2,2),(3,3)\}$. Also $i_{\mathbb{Z}} = \{(n,n) : n \in \mathbb{Z}\}$. The identity function on a set is the function that sends any element of the set to itself.

Notice that for any set A, the identity function i_A is bijective: It is injective because $i_A(x) = i_A(y)$ immediately reduces to $x = y$. It is surjective because if we take any element b in the codomain A, then b is also in the domain A, and $i_A(b) = b$.

Definition 12.7 Given a relation R from A to B, the **inverse relation of** R is the relation from B to A defined as $R^{-1} = \{(y,x) : (x,y) \in R\}$. In other words, the inverse of R is the relation R^{-1} obtained by interchanging the elements in every ordered pair in R.

For example, let $A = \{a,b,c\}$ and $B = \{1,2,3\}$, and suppose f is the relation $f = \{(a,2),(b,3),(c,1)\}$ from A to B. Then $f^{-1} = \{(2,a),(3,b),(1,c)\}$ and this is a relation from B to A. Notice that f is actually a function from A to B, and f^{-1} is a function from B to A. These two relations are drawn below. Notice the drawing for relation f^{-1} is just the drawing for f with arrows reversed.

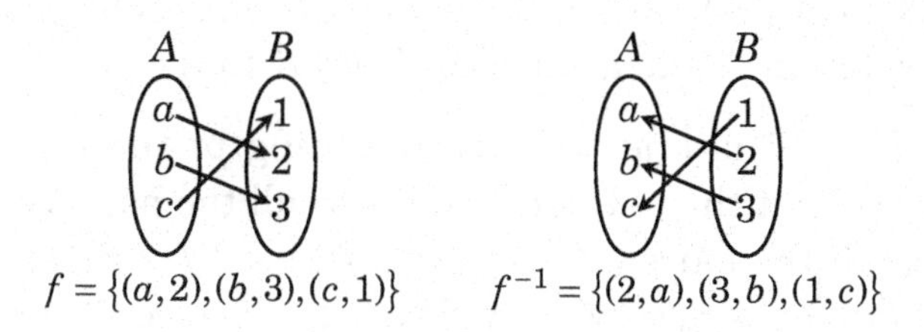

For another example, let A and B be the same sets as above, but consider the relation $g = \{(a,2),(b,3),(c,3)\}$ from A to B. Then $g^{-1} = \{(2,a),(3,b),(3,c)\}$ is a relation from B to A. These two relations are sketched below.

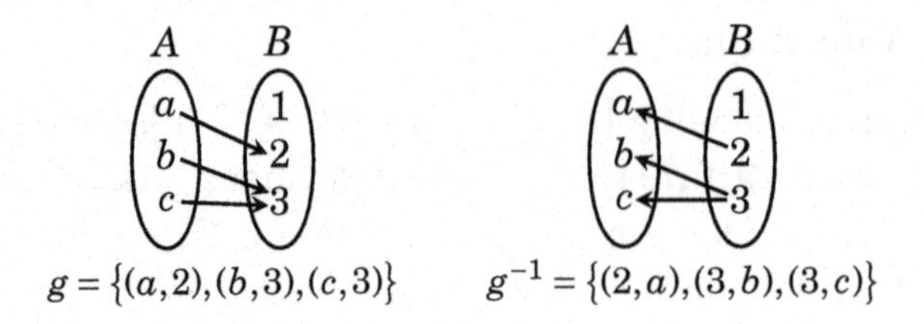

$g = \{(a,2),(b,3),(c,3)\}$ $\qquad$ $g^{-1} = \{(2,a),(3,b),(3,c)\}$

This time, even though the relation g is a function, its inverse g^{-1} is not a function because the element 3 occurs twice as a first coordinate of an ordered pair in g^{-1}.

In the above examples, relations f and g are both functions, and f^{-1} is a function and g^{-1} is not. This raises a question: What properties does f have and g lack that makes f^{-1} a function and g^{-1} not a function? The answer is not hard to see. Function g is not injective because $g(b) = g(c) = 3$, and thus $(b,3)$ and $(c,3)$ are both in g. This causes a problem with g^{-1} because it means $(3,b)$ and $(3,c)$ are both in g^{-1}, so g^{-1} can't be a function. Thus, in order for g^{-1} to be a function, it would be necessary that g be injective.

But that is not enough. Function g also fails to be surjective because no element of A is sent to the element $1 \in B$. This means g^{-1} contains no ordered pair whose first coordinate is 1, so it can't be a function from B to A. If g^{-1} were to be a function it would be necessary that g be surjective.

The previous two paragraphs suggest that if g is a function, then it must be bijective in order for its inverse relation g^{-1} to be a function. Indeed, this is easy to verify. Conversely, if a function is bijective, then its inverse relation is easily seen to be a function. We summarize this in the following theorem.

Theorem 12.3 Let $f : A \to B$ be a function. Then f is bijective if and only if the inverse relation f^{-1} is a function from B to A.

Suppose $f : A \to B$ is bijective, so according to the theorem f^{-1} is a function. Observe that the relation f contains all the pairs $(x, f(x))$ for $x \in A$, so f^{-1} contains all the pairs $(f(x), x)$. But $(f(x), x) \in f^{-1}$ means $f^{-1}(f(x)) = x$. Therefore $f^{-1} \circ f(x) = x$ for every $x \in A$. From this we get $f^{-1} \circ f = i_A$. Similar reasoning produces $f \circ f^{-1} = i_B$. This leads to the following definitions.

Definition 12.8 If $f : A \to B$ is bijective then its **inverse** is the function $f^{-1} : B \to A$. Functions f and f^{-1} obey the equations $f^{-1} \circ f = i_A$ and $f \circ f^{-1} = i_B$.

You probably recall from algebra and calculus at least one technique for computing the inverse of a bijective function f: to find f^{-1}, start with the equation $y = f(x)$. Then interchange variables to get $x = f(y)$. Solving this equation for y (if possible) produces $y = f^{-1}(x)$. The next two examples illustrate this.

Example 12.12 The function $f : \mathbb{R} \to \mathbb{R}$ defined as $f(x) = x^3 + 1$ is bijective. Find its inverse.

We begin by writing $y = x^3 + 1$. Now interchange variables to obtain $x = y^3 + 1$. Solving for y produces $y = \sqrt[3]{x-1}$. Thus

$$f^{-1}(x) = \sqrt[3]{x-1}.$$

(You can check your answer by computing

$$f^{-1}(f(x)) = \sqrt[3]{f(x)-1} = \sqrt[3]{x^3+1-1} = x.$$

Therefore $f^{-1}(f(x)) = x$. Any answer other than x indicates a mistake.)

We close with one final example. Example 12.5 showed that the function $g : \mathbb{Z} \times \mathbb{Z} \to \mathbb{Z} \times \mathbb{Z}$ defined by the formula $g(m,n) = (m+n, m+2n)$ is bijective. Let's find its inverse. The approach outlined above should work, but we need to be careful to keep track of coordinates in $\mathbb{Z} \times \mathbb{Z}$. We begin by writing $(x,y) = g(m,n)$, then interchanging the variables (x,y) and (m,n) to get $(m,n) = g(x,y)$. This gives

$$(m,n) = (x+y, x+2y),$$

from which we get the following system of equations:

$$\begin{array}{rcrcl} x & + & y & = & m \\ x & + & 2y & = & n. \end{array}$$

Solving this system using techniques from algebra with which you are familiar, we get

$$\begin{array}{rcrcl} x & & & = & 2m-n \\ & & y & = & n-m. \end{array}$$

Then $(x,y) = (2m-n, n-m)$, so $\boxed{g^{-1}(m,n) = (2m-n, n-m).}$

We can check our work by confirming that $g^{-1}(g(m,n)) = (m,n)$. Doing the math,

$$\begin{aligned} g^{-1}(g(m,n)) &= g^{-1}(m+n, m+2n) \\ &= \big(2(m+n)-(m+2n), (m+2n)-(m+n)\big) \\ &= (m,n). \end{aligned}$$

Exercises for Section 12.5

1. Check that the function $f : \mathbb{Z} \to \mathbb{Z}$ defined by $f(n) = 6 - n$ is bijective. Then compute f^{-1}.
2. In Exercise 9 of Section 12.2 you proved that $f : \mathbb{R} - \{2\} \to \mathbb{R} - \{5\}$ defined by $f(x) = \frac{5x+1}{x-2}$ is bijective. Now find its inverse.
3. Let $B = \{2^n : n \in \mathbb{Z}\} = \{\ldots, \frac{1}{4}, \frac{1}{2}, 1, 2, 4, 8, \ldots\}$. Show that the function $f : \mathbb{Z} \to B$ defined as $f(n) = 2^n$ is bijective. Then find f^{-1}.
4. The function $f : \mathbb{R} \to (0,\infty)$ defined as $f(x) = e^{x^3+1}$ is bijective. Find its inverse.
5. The function $f : \mathbb{R} \to \mathbb{R}$ defined as $f(x) = \pi x - e$ is bijective. Find its inverse.
6. The function $f : \mathbb{Z} \times \mathbb{Z} \to \mathbb{Z} \times \mathbb{Z}$ defined by the formula $f(m,n) = (5m+4n, 4m+3n)$ is bijective. Find its inverse.
7. Show that the function $f : \mathbb{R}^2 \to \mathbb{R}^2$ defined by the formula $f(x,y) = ((x^2+1)y, x^3)$ is bijective. Then find its inverse.
8. Is the function $\theta : \mathscr{P}(\mathbb{Z}) \to \mathscr{P}(\mathbb{Z})$ defined as $\theta(X) = \overline{X}$ bijective? If so, what is its inverse?
9. Consider the function $f : \mathbb{R} \times \mathbb{N} \to \mathbb{N} \times \mathbb{R}$ defined as $f(x,y) = (y, 3xy)$. Check that this is bijective; find its inverse.
10. Consider $f : \mathbb{N} \to \mathbb{Z}$ defined as $f(n) = \frac{(-1)^n(2n-1)+1}{4}$. This function is bijective by Exercise 18 in Section 12.2. Find its inverse.

12.6 Image and Preimage

It is time to take up a matter of notation that you will encounter in future mathematics classes. Suppose we have a function $f : A \to B$. If $X \subseteq A$, the expression $f(X)$ has a special meaning. It stands for the set $\{f(x) : x \in X\}$. Similarly, if $Y \subseteq B$ then $f^{-1}(Y)$ has a meaning *even if f is not invertible.* The expression $f^{-1}(Y)$ stands for the set $\{x \in A : f(x) \in Y\}$. Here are the precise definitions.

Definition 12.9 Suppose $f: A \to B$ is a function.

1. If $X \subseteq A$, the **image** of X is the set $f(X) = \{f(x) : x \in X\} \subseteq B$.
2. If $Y \subseteq B$, the **preimage** of Y is the set $f^{-1}(Y) = \{x \in A : f(x) \in Y\} \subseteq A$.

In words, the image $f(X)$ of X is the set of all things in B that f sends elements of X to. (Roughly speaking, you might think of $f(X)$ as a kind of distorted "copy" or "image" of X in B.) The preimage $f^{-1}(Y)$ of Y is the set of all things in A that f sends into Y.

Maybe you have already encountered these ideas in linear algebra, in a setting involving a linear transformation $T: V \to W$ between two vector spaces. If $X \subseteq V$ is a subspace of V, then its image $T(X)$ is a subspace of W. If $Y \subseteq W$ is a subspace of W, then its preimage $T^{-1}(Y)$ is a subspace of V. (If this does not sound familiar, then ignore it.)

Example 12.13 Let $f: \{s,t,u,v,w,x,y,z\} \to \{0,1,2,3,4,5,6,7,8,9\}$, where

$$f = \{(s,4),(t,8),(u,8),(v,1),(w,2),(x,4),(y,6),(z,4)\}.$$

Notice that f is neither injective nor surjective, so it certainly is not invertible. Be sure you understand the following statements.

1. $f(\{s,t,u,z\}) = \{8,4\}$
2. $f(\{s,x,z\}) = \{4\}$
3. $f(\{s,v,w,y\}) = \{1,2,4,6\}$
4. $f^{-1}(\{4\}) = \{s,x,z\}$
5. $f^{-1}(\{4,9\}) = \{s,x,z\}$
6. $f^{-1}(\{9\}) = \emptyset$
7. $f^{-1}(\{1,4,8\}) = \{s,t,u,v,x,z\}$

It is important to realize that the X and Y in Definition 12.9 are subsets (not elements!) of A and B. Note that in the above example we had $f^{-1}(\{4\}) = \{s,x,z\}$, while $f^{-1}(4)$ has absolutely no meaning because the inverse function f^{-1} does not exist. Likewise, there is a subtle difference between $f(\{s\}) = \{4\}$ and $f(s) = 4$. Be careful.

Example 12.14 Consider the function $f: \mathbb{R} \to \mathbb{R}$ defined as $f(x) = x^2$. Note that $f(\{0,1,2\}) = \{0,1,4\}$ and $f^{-1}(\{0,1,4\}) = \{-2,-1,0,1,2\}$. This shows $f^{-1}(f(X)) \neq X$ in general.

Using the same f, now check your understanding of the following statements involving images and preimages of intervals: $f([-2,3]) = [0,9]$, and $f^{-1}([0,9]) = [-3,3]$. Also $f(\mathbb{R}) = [0,\infty)$ and $f^{-1}([-2,-1]) = \emptyset$.

If you continue with mathematics you are likely to encounter the following results. For now, you are asked to prove them in the exercises.

Theorem 12.4 Suppose $f : A \to B$ is a function. Let $W, X \subseteq A$, and $Y, Z \subseteq B$. Then:

1. $f(W \cap X) \subseteq f(W) \cap f(X)$
2. $f(W \cup X) = f(W) \cup f(X)$
3. $f^{-1}(Y \cap Z) = f^{-1}(Y) \cap f^{-1}(Z)$
4. $f^{-1}(Y \cup Z) = f^{-1}(Y) \cup f^{-1}(Z)$
5. $X \subseteq f^{-1}(f(X))$
6. $f(f^{-1}(Y)) \subseteq Y$.

Exercises for Section 12.6

1. Consider the function $f : \mathbb{R} \to \mathbb{R}$ defined as $f(x) = x^2 + 3$. Find $f([-3,5])$ and $f^{-1}([12,19])$.

2. Consider the function $f : \{1,2,3,4,5,6,7\} \to \{0,1,2,3,4,5,6,7,8,9\}$ given as

$$f = \{(1,3),(2,8),(3,3),(4,1),(5,2),(6,4),(7,6)\}.$$

Find: $f(\{1,2,3\})$, $f(\{4,5,6,7\})$, $f(\emptyset)$, $f^{-1}(\{0,5,9\})$ and $f^{-1}(\{0,3,5,9\})$.

3. This problem concerns functions $f : \{1,2,3,4,5,6,7\} \to \{0,1,2,3,4\}$. How many such functions have the property that $|f^{-1}(\{3\})| = 3$?

4. This problem concerns functions $f : \{1,2,3,4,5,6,7,8\} \to \{0,1,2,3,4,5,6\}$. How many such functions have the property that $|f^{-1}(\{2\})| = 4$?

5. Consider a function $f : A \to B$ and a subset $X \subseteq A$. We observed in Section 12.6 that $f^{-1}(f(X)) \neq X$ in general. However $X \subseteq f^{-1}(f(X))$ is always true. Prove this.

6. Given a function $f : A \to B$ and a subset $Y \subseteq B$, is $f(f^{-1}(Y)) = Y$ always true? Prove or give a counterexample.

7. Given a function $f : A \to B$ and subsets $W, X \subseteq A$, prove $f(W \cap X) \subseteq f(W) \cap f(X)$.

8. Given a function $f : A \to B$ and subsets $W, X \subseteq A$, then $f(W \cap X) = f(W) \cap f(X)$ is *false* in general. Produce a counterexample.

9. Given a function $f : A \to B$ and subsets $W, X \subseteq A$, prove $f(W \cup X) = f(W) \cup f(X)$.

10. Given $f : A \to B$ and subsets $Y, Z \subseteq B$, prove $f^{-1}(Y \cap Z) = f^{-1}(Y) \cap f^{-1}(Z)$.

11. Given $f : A \to B$ and subsets $Y, Z \subseteq B$, prove $f^{-1}(Y \cup Z) = f^{-1}(Y) \cup f^{-1}(Z)$.

12. Consider $f : A \to B$. Prove that f is injective if and only if $X = f^{-1}(f(X))$ for all $X \subseteq A$. Prove that f is surjective if and only if $f(f^{-1}(Y)) = Y$ for all $Y \subseteq B$.

13. Let $f : A \to B$ be a function, and $X \subseteq A$. Prove or disprove: $f\big(f^{-1}(f(X))\big) = f(X)$.

14. Let$f : A \to B$ be a function, and $Y \subseteq B$. Prove or disprove: $f^{-1}\big(f(f^{-1}(Y))\big) = f^{-1}(Y)$.

CHAPTER 13

Cardinality of Sets

This chapter is all about cardinality of sets. At first this looks like a very simple concept. To find the cardinality of a set, just count its elements. If $A = \{a,b,c,d\}$, then $|A| = 4$; if $B = \{n \in \mathbb{Z} : -5 \le n \le 5\}$, then $|B| = 11$. In this case $|A| < |B|$. What could be simpler than that?

Actually, the idea of cardinality becomes quite subtle when the sets are infinite. The main point of this chapter is to explain how there are numerous different kinds of infinity, and some infinities are bigger than others. Two sets A and B can both have infinite cardinality, yet $|A| < |B|$.

13.1 Sets with Equal Cardinalities

We begin with a discussion of what it means for two sets to have the same cardinality. Up until this point we've said $|A| = |B|$ if A and B have the same number of elements: Count the elements of A, then count the elements of B. If you get the same number, then $|A| = |B|$.

Although this is a fine strategy if the sets are finite (and not too big!), it doesn't apply to infinite sets because we'd never be done counting their elements. We need a new approach that applies to both finite and infinite sets. Here it is:

Definition 13.1 Two sets A and B have the **same cardinality**, written $|A| = |B|$, if there exists a bijective function $f : A \to B$. If no such bijective function exists, then the sets have **unequal cardinalities**, that is, $|A| \neq |B|$.

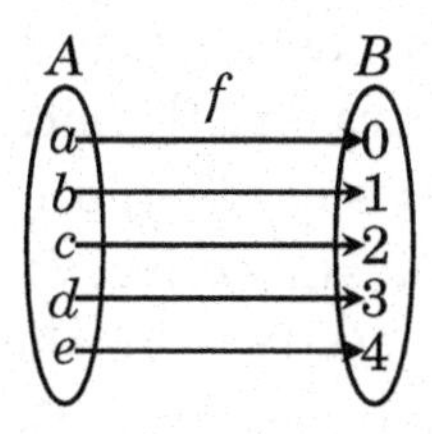

The above picture illustrates our definition. There is a bijective function $f : A \to B$, so $|A| = |B|$. The function f matches up A with B. Think of f as describing how to overlay A onto B so that they fit together perfectly.

On the other hand, if A and B are as indicated in either of the following figures, then there can be no bijection $f : A \to B$. (The best we can do is a function that is either injective or surjective, but not both). Therefore the definition says $|A| \neq |B|$ in these cases.

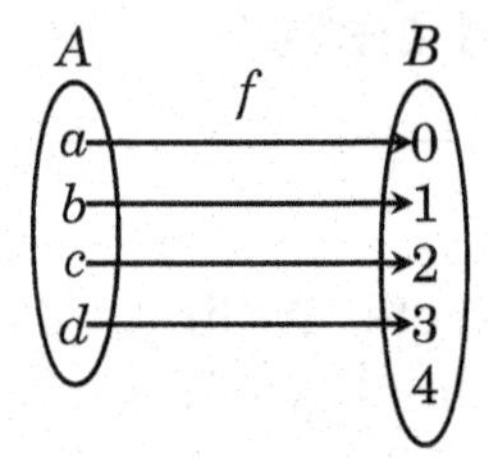

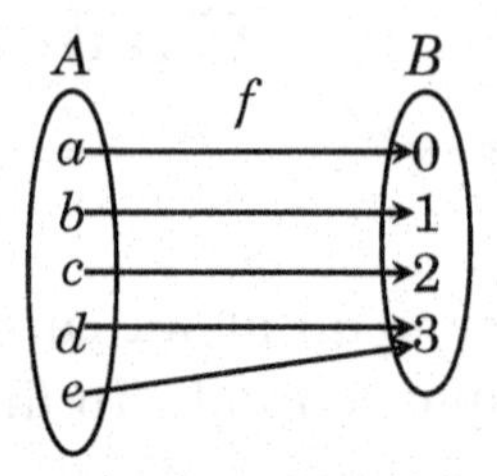

Example 13.1 The sets $A = \{n \in \mathbb{Z} : 0 \leq n \leq 5\}$ and $B = \{n \in \mathbb{Z} : -5 \leq n \leq 0\}$ have the same cardinality because there is a bijective function $f : A \to B$ given by the rule $f(n) = -n$.

Several comments are in order. First, if $|A| = |B|$, there can be *lots* of bijective functions from A to B. We only need to find one of them in order to conclude $|A| = |B|$. Second, as bijective functions play such a big role here, we use the word **bijection** to mean *bijective function*. Thus the function $f(n) = -n$ from Example 13.1 is a bijection. Also, an injective function is called an **injection** and a surjective function is called a **surjection**.

We emphasize and reiterate that Definition 13.1 applies to finite as well as infinite sets. If A and B are infinite, then $|A| = |B|$ provided there exists a bijection $f : A \to B$. If no such bijection exists, then $|A| \neq |B|$.

Example 13.2 This example shows that $|\mathbb{N}| = |\mathbb{Z}|$. To see why this is true, notice that the following table describes a bijection $f : \mathbb{N} \to \mathbb{Z}$.

n	1	2	3	4	5	6	7	8	9	10	11	12	13	14	15	...
$f(n)$	0	1	−1	2	−2	3	−3	4	−4	5	−5	6	−6	7	−7	...

Notice that f is described in such a way that it is both injective and surjective. Every integer appears exactly once on the infinitely long second row. Thus, according to the table, given any $b \in \mathbb{Z}$ there is some natural number n with $f(n) = b$, so f is surjective. It is injective because the way the table is constructed forces $f(m) \neq f(n)$ whenever $m \neq n$. Because of this bijection $f : \mathbb{N} \to \mathbb{Z}$, we must conclude from Definition 13.1 that $|\mathbb{N}| = |\mathbb{Z}|$.

Example 13.2 may seem slightly unsettling. On one hand it makes sense that $|\mathbb{N}| = |\mathbb{Z}|$ because $\mathbb{N}$ and $\mathbb{Z}$ are both infinite, so their cardinalities are both "infinity." On the other hand, $\mathbb{Z}$ may seem twice as large as

$\mathbb{N}$ because $\mathbb{Z}$ has all the negative integers as well as the positive ones. Definition 13.1 settles the issue. Because the bijection $f:\mathbb{N}\to\mathbb{Z}$ matches up $\mathbb{N}$ with $\mathbb{Z}$, it follows that $|\mathbb{N}|=|\mathbb{Z}|$. We summarize this with a theorem.

Theorem 13.1 There exists a bijection $f:\mathbb{N}\to\mathbb{Z}$. Therefore $|\mathbb{N}|=|\mathbb{Z}|$.

The fact that $\mathbb{N}$ and $\mathbb{Z}$ have the same cardinality might prompt us compare the cardinalities of other infinite sets. How, for example, do $\mathbb{N}$ and $\mathbb{R}$ compare? Let's turn our attention to this.

In fact, $|\mathbb{N}|\neq|\mathbb{R}|$. This was first recognized by Georg Cantor (1845–1918), who devised an ingenious argument to show that there are no surjective functions $f:\mathbb{N}\to\mathbb{R}$. (This in turn implies that there can be no bijections $f:\mathbb{N}\to\mathbb{R}$, so $|\mathbb{N}|\neq|\mathbb{R}|$ by Definition 13.1.)

We now describe Cantor's argument for why there are no surjections $f:\mathbb{N}\to\mathbb{R}$. We will reason informally, rather than writing out an exact proof. Take any arbitrary function $f:\mathbb{N}\to\mathbb{R}$. Here's why f can't be surjective:

Imagine making a table for f, where values of n in $\mathbb{N}$ are in the left-hand column and the corresponding values $f(n)$ are on the right. The first few entries might look something as follows. In this table, the real numbers $f(n)$ are written with all their decimal places trailing off to the right. Thus, even though $f(1)$ happens to be the real number 0.4, we write it as 0.40000000…., etc.

n	$f(n)$
1	0 . 4 0 0 0 0 0 0 0 0 0 0 0 0 0...
2	8 . 5 0 0 6 0 7 0 8 6 6 6 9 0 0...
3	7 . 5 0 5 0 0 9 4 0 0 4 4 1 0 1...
4	5 . 5 0 7 0 4 0 0 8 0 4 8 0 5 0...
5	6 . 9 0 0 2 6 0 0 0 0 0 0 5 0 6...
6	6 . 8 2 8 0 9 5 8 2 0 5 0 0 2 0...
7	6 . 5 0 5 0 5 5 5 0 6 5 5 8 0 8...
8	8 . 7 2 0 8 0 6 4 0 0 0 0 4 4 8...
9	0 . 5 5 0 0 0 0 8 8 8 8 0 0 7 7...
10	0 . 5 0 0 2 0 7 2 2 0 7 8 0 5 1...
11	2 . 9 0 0 0 0 8 8 0 0 0 0 9 0 0...
12	6 . 5 0 2 8 0 0 0 8 0 0 9 6 7 1...
13	8 . 8 9 0 0 8 0 2 4 0 0 8 0 5 0...
14	8 . 5 0 0 0 8 7 4 2 0 8 0 2 2 6...
⋮	⋮

There is a diagonal shaded band in the table. For each $n \in \mathbb{N}$, this band covers the nth decimal place of $f(n)$:

The 1st decimal place of $f(1)$ is the 1st entry on the diagonal.
The 2nd decimal place of $f(2)$ is the 2nd entry on the diagonal.
The 3rd decimal place of $f(3)$ is the 3rd entry on the diagonal.
The 4th decimal place of $f(4)$ is the 4th entry on the diagonal, etc.

The diagonal helps us construct a number $b \in \mathbb{R}$ that is unequal to any $f(n)$. Just let the nth decimal place of b differ from the nth entry of the diagonal. Then the nth decimal place of b differs from the nth decimal place of $f(n)$. In order to be definite, define b to be the positive number less than 1 whose nth decimal place is 0 if the nth decimal place of $f(n)$ is not 0, and whose nth decimal place is 1 if the nth decimal place of $f(n)$ *equals* 0. Thus, for the function f illustrated in the above table, we have

$$b = 0.01010001001000\ldots$$

and b has been defined so that, for any $n \in \mathbb{N}$, its nth decimal place is unequal to the nth decimal place of $f(n)$. Therefore $f(n) \neq b$ for every natural number n, meaning f is not surjective.

Since this argument applies to *any* function $f : \mathbb{N} \to \mathbb{R}$ (not just the one in the above example) we conclude that there exist no bijections $f : \mathbb{N} \to \mathbb{R}$, so $|\mathbb{N}| \neq |\mathbb{R}|$ by Definition 13.1. We summarize this as a theorem.

Theorem 13.2 There exists no bijection $f : \mathbb{N} \to \mathbb{R}$. Therefore $|\mathbb{N}| \neq |\mathbb{R}|$.

This is our first indication of how there are different kinds of infinities. Both $\mathbb{N}$ and $\mathbb{R}$ are infinite sets, yet $|\mathbb{N}| \neq |\mathbb{R}|$. We will continue to develop this theme throughout this chapter. The next example shows that the intervals $(0,\infty)$ and $(0,1)$ on $\mathbb{R}$ have the same cardinality.

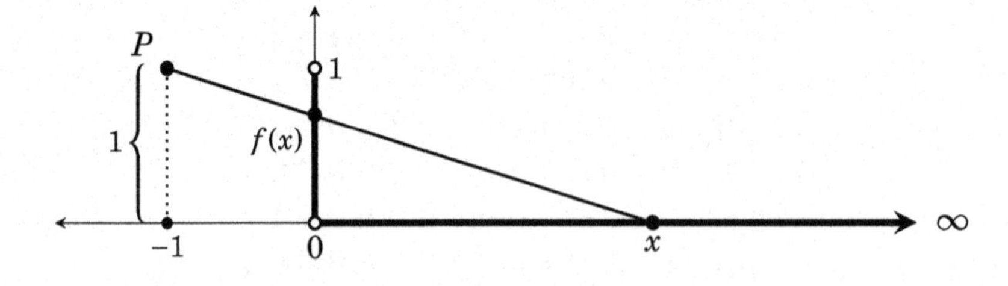

Figure 13.1. A bijection $f : (0,\infty) \to (0,1)$

Example 13.3 Show that $|(0,\infty)| = |(0,1)|$.

To accomplish this, we need to show that there is a bijection $f:(0,\infty)\to(0,1)$. We describe this function geometrically. Consider the interval $(0,\infty)$ as the positive x-axis of $\mathbb{R}^2$. Let the interval $(0,1)$ be on the y-axis as illustrated in Figure 13.1, so that $(0,\infty)$ and $(0,1)$ are perpendicular to each other.

The figure also shows a point $P=(-1,1)$. Define $f(x)$ to be the point on $(0,1)$ where the line from P to $x\in(0,\infty)$ intersects the y-axis. By similar triangles, we have

$$\frac{1}{x+1}=\frac{f(x)}{x},$$

and therefore

$$f(x)=\frac{x}{x+1}.$$

If it is not clear from the figure that $f:(0,\infty)\to(0,1)$ is bijective, then you can verify it using the techniques from Section 12.2. (Exercise 16, below.)

It is important to note that equality of cardinalities is an equivalence relation on sets: it is reflexive, symmetric and transitive. Let us confirm this. Given a set A, the identity function $A\to A$ is a bijection, so $|A|=|A|$. (This is the reflexive property.) For the symmetric property, if $|A|=|B|$, then there is a bijection $f:A\to B$, and its inverse is a bijection $f^{-1}:B\to A$, so $|B|=|A|$. For transitivity, suppose $|A|=|B|$ and $|B|=|C|$. Then there are bijections $f:A\to B$ and $g:B\to C$. The composition $g\circ f:A\to C$ is a bijection (Theorem 12.2), so $|A|=|C|$.

The transitive property can be useful. If, in trying to show two sets A and C have the same cardinality, we can produce a third set B for which $|A|=|B|$ and $|B|=|C|$, then transitivity assures us that indeed $|A|=|C|$. The next example uses this idea.

Example 13.4 Show that $|\mathbb{R}|=|(0,1)|$.

Because of the bijection $g:\mathbb{R}\to(0,\infty)$ where $g(x)=2^x$, we have $|\mathbb{R}|=|(0,\infty)|$. Also, Example 13.3 shows that $|(0,\infty)|=|(0,1)|$. Therefore $|\mathbb{R}|=|(0,1)|$.

So far in this chapter we have declared that two sets have "the same cardinality" if there is a bijection between them. They have "different cardinalities" if there exists no bijection between them. Using this idea, we showed that $|\mathbb{Z}|=|\mathbb{N}|\neq|\mathbb{R}|=|(0,\infty)|=|(0,1)|$. So, we have a means of determining when two sets have the same or different cardinalities. But we have neatly avoided saying exactly what cardinality *is*. For example, we can say that $|\mathbb{Z}|=|\mathbb{N}|$, but what exactly *is* $|\mathbb{Z}|$, or $|\mathbb{N}|$? What exactly *are* these things that are equal? Certainly not numbers, for they are too big.

And saying they are "infinity" is not accurate, because we now know that there are different types of infinity. So just what kind of mathematical entity is $|\mathbb{Z}|$? In general, given a set X, exactly what *is* its cardinality $|X|$?

This is a lot like asking what a number is. A number, say 5, is an abstraction, not a physical thing. Early in life we instinctively grouped together certain sets of things (five apples, five oranges, etc.) and conceived of 5 as the thing common to all such sets. In a very real sense, the number 5 is an abstraction of the fact that any two of these sets can be matched up via a bijection. That is, it can be identified with a certain equivalence class of sets under the "*has the same cardinality as*" relation. (Recall that this is an equivalence relation.) This is easy to grasp because our sense of numeric quantity is so innate. But in exactly the same way we can say that the cardinality of a set X is what is common to all sets that can be matched to X via a bijection. This may be harder to grasp, but it is really no different from the idea of the magnitude of a (finite) number.

In fact, we could be concrete and define $|X|$ to be the equivalence class of all sets whose cardinality is the same as that of X. This has the advantage of giving an explicit meaning to $|X|$. But there is no harm in taking the intuitive approach and just interpreting the cardinality $|X|$ of a set X to be a measure the "size" of X. The point of this section is that we have a means of deciding whether two sets have the same size or different sizes.

Exercises for Section 13.1

A. Show that the two given sets have equal cardinality by describing a bijection from one to the other. Describe your bijection with a formula (not as a table).

1. $\mathbb{R}$ and $(0,\infty)$

2. $\mathbb{R}$ and $(\sqrt{2},\infty)$

3. $\mathbb{R}$ and $(0,1)$

4. The set of even integers and the set of odd integers

5. $A = \{3k : k \in \mathbb{Z}\}$ and $B = \{7k : k \in \mathbb{Z}\}$

6. $\mathbb{N}$ and $S = \{\frac{\sqrt{2}}{n} : n \in \mathbb{N}\}$

7. $\mathbb{Z}$ and $S = \{\ldots, \frac{1}{8}, \frac{1}{4}, \frac{1}{2}, 1, 2, 4, 8, 16, \ldots\}$

8. $\mathbb{Z}$ and $S = \{x \in \mathbb{R} : \sin x = 1\}$

9. $\{0,1\} \times \mathbb{N}$ and $\mathbb{N}$

10. $\{0,1\} \times \mathbb{N}$ and $\mathbb{Z}$

11. $[0,1]$ and $(0,1)$

12. $\mathbb{N}$ and $\mathbb{Z}$ (Suggestion: use Exercise 18 of Section 12.2.)

13. $\mathscr{P}(\mathbb{N})$ and $\mathscr{P}(\mathbb{Z})$ (Suggestion: use Exercise 12, above.)

14. $\mathbb{N} \times \mathbb{N}$ and $\{(n,m) \in \mathbb{N} \times \mathbb{N} : n \le m\}$

B. Answer the following questions concerning bijections from this section.

15. Find a formula for the bijection f in Example 13.2 (page 218).

16. Verify that the function f in Example 13.3 is a bijection.

13.2 Countable and Uncountable Sets

Let's summarize the main points from the previous section.

1. $|A| = |B|$ if and only if there exists a bijection $A \to B$.
2. $|\mathbb{N}| = |\mathbb{Z}|$ because there exists a bijection $\mathbb{N} \to \mathbb{Z}$.
3. $|\mathbb{N}| \neq |\mathbb{R}|$ because there exists *no* bijection $\mathbb{N} \to \mathbb{R}$.

Thus, even though $\mathbb{N}$, $\mathbb{Z}$ and $\mathbb{R}$ are all infinite sets, their cardinalities are not all the same. The sets $\mathbb{N}$ and $\mathbb{Z}$ have the same cardinality, but $\mathbb{R}$'s cardinality is different from that of both the other sets. This means infinite sets can have different sizes. We now make some definitions to put words and symbols to this phenomenon.

In a certain sense you can count the elements of $\mathbb{N}$; you can count its elements off as $1,2,3,4,\ldots$, but you'd have to continue this process forever to count the whole set. Thus we will call $\mathbb{N}$ a *countably infinite set*, and the same term is used for any set whose cardinality equals that of $\mathbb{N}$.

Definition 13.2 Suppose A is a set. Then A is **countably infinite** if $|\mathbb{N}| = |A|$, that is, if there exists a bijection $\mathbb{N} \to A$. The set A is **uncountable** if A is infinite and $|\mathbb{N}| \neq |A|$, that is, if A is infinite and there exists *no* bijection $\mathbb{N} \to A$.

Thus $\mathbb{Z}$ is countably infinite but $\mathbb{R}$ is uncountable. This section deals mainly with countably infinite sets. Uncountable sets are treated later.

If A is countably infinite, then $|\mathbb{N}| = |A|$, so there is a bijection $f : \mathbb{N} \to A$. You can think of f as "counting" the elements of A. The first element of A is $f(1)$, followed by $f(2)$, then $f(3)$ and so on. It makes sense to think of a countably infinite set as the smallest type of infinite set, because if the counting process stopped, the set would be finite, not infinite; a countably infinite set has the fewest elements that a set can have and still be infinite. It is common to reserve the special symbol $\aleph_0$ to stand for the cardinality of countably infinite sets.

Definition 13.3 The cardinality of the natural numbers is denoted as $\aleph_0$. That is, $|\mathbb{N}| = \aleph_0$. Thus any countably infinite set has cardinality $\aleph_0$.

(The symbol $\aleph$ is the first letter in the Hebrew alphabet, and is pronounced "aleph." The symbol $\aleph_0$ is pronounced "aleph naught.") The summary of facts at the beginning of this section shows $|\mathbb{Z}| = \aleph_0$ and $|\mathbb{R}| \neq \aleph_0$.

Example 13.5 Let $E = \{2k : k \in \mathbb{Z}\}$ be the set of even integers. The function $f : \mathbb{Z} \to E$ defined as $f(n) = 2n$ is easily seen to be a bijection, so we have $|\mathbb{Z}| = |E|$. Thus, as $|\mathbb{N}| = |\mathbb{Z}| = |E|$, the set E is countably infinite and $|E| = \aleph_0$.

Here is a significant fact: The elements of any countably infinite set A can be written in an infinitely long list $a_1, a_2, a_3, a_4, \ldots$ that begins with some element $a_1 \in A$ and includes every element of A. For example, the set E in the above example can be written in list form as $0, 2, -2, 4, -4, 6, -6, 8, -8, \ldots$ The reason that this can be done is as follows. Since A is countably infinite, Definition 13.2 says there is a bijection $f : \mathbb{N} \to A$. This allows us to list out the set A as an infinite list $f(1), f(2), f(3), f(4), \ldots$ Conversely, if the elements of A can be written in list form as $a_1, a_2, a_3, \ldots$, then the function $f : \mathbb{N} \to A$ defined as $f(n) = a_n$ is a bijection, so A is countably infinite. We summarize this as follows.

Theorem 13.3 A set A is countably infinite if and only if its elements can be arranged in an infinite list $a_1, a_2, a_3, a_4, \ldots$

As an example of how this theorem might be used, let P denote the set of all prime numbers. Since we can list its elements as $2, 3, 5, 7, 11, 13, \ldots$, it follows that the set P is countably infinite.

As another consequence of Theorem 13.3, note that we can interpret the fact that the set $\mathbb{R}$ is not countably infinite as meaning that it is impossible to write out all the elements of $\mathbb{R}$ in an infinite list. (After all, we tried to do that in the table on page 219, and failed!)

This raises a question. Is it also impossible to write out all the elements of $\mathbb{Q}$ in an infinite list? In other words, is the set $\mathbb{Q}$ of rational numbers countably infinite or uncountable? If you start plotting the rational numbers on the number line, they seem to mostly fill up $\mathbb{R}$. Sure, some numbers such as $\sqrt{2}$, π and e will not be plotted, but the dots representing rational numbers seem to predominate. We might thus expect $\mathbb{Q}$ to be uncountable. However, it is a surprising fact that $\mathbb{Q}$ is countable. The proof presented below arranges all the rational numbers in an infinitely long list.

Theorem 13.4 The set $\mathbb{Q}$ of rational numbers is countably infinite.

Proof. To prove this, we just need to show how to write the set $\mathbb{Q}$ in list form. Begin by arranging all rational numbers in an infinite array. This is done by making the following chart. The top row has a list of all integers, beginning with 0, then alternating signs as they increase. Each column headed by an integer k contains all the fractions (in reduced form) with numerator k. For example, the column headed by 2 contains the fractions $\frac{2}{1}, \frac{2}{3}, \frac{2}{5}, \frac{2}{7}, \ldots$, and so on. It does not contain $\frac{2}{2}$, $\frac{2}{4}$, $\frac{2}{6}$, etc., because those are not reduced, and in fact their reduced forms appear in the column headed by 1. You should examine this table and convince yourself that it contains all rational numbers in $\mathbb{Q}$.

0	1	−1	2	−2	3	−3	4	−4	5	−5	⋯
$\frac{0}{1}$	$\frac{1}{1}$	$\frac{-1}{1}$	$\frac{2}{1}$	$\frac{-2}{1}$	$\frac{3}{1}$	$\frac{-3}{1}$	$\frac{4}{1}$	$\frac{-4}{1}$	$\frac{5}{1}$	$\frac{-5}{1}$	⋯
	$\frac{1}{2}$	$\frac{-1}{2}$	$\frac{2}{3}$	$\frac{-2}{3}$	$\frac{3}{2}$	$\frac{-3}{2}$	$\frac{4}{3}$	$\frac{-4}{3}$	$\frac{5}{2}$	$\frac{-5}{2}$	⋯
	$\frac{1}{3}$	$\frac{-1}{3}$	$\frac{2}{5}$	$\frac{-2}{5}$	$\frac{3}{4}$	$\frac{-3}{4}$	$\frac{4}{5}$	$\frac{-4}{5}$	$\frac{5}{3}$	$\frac{-5}{3}$	⋯
	$\frac{1}{4}$	$\frac{-1}{4}$	$\frac{2}{7}$	$\frac{-2}{7}$	$\frac{3}{5}$	$\frac{-3}{5}$	$\frac{4}{7}$	$\frac{-4}{7}$	$\frac{5}{4}$	$\frac{-5}{4}$	⋯
	$\frac{1}{5}$	$\frac{-1}{5}$	$\frac{2}{9}$	$\frac{-2}{9}$	$\frac{3}{7}$	$\frac{-3}{7}$	$\frac{4}{9}$	$\frac{-4}{9}$	$\frac{5}{6}$	$\frac{-5}{6}$	⋯
	$\frac{1}{6}$	$\frac{-1}{6}$	$\frac{2}{11}$	$\frac{-2}{11}$	$\frac{3}{8}$	$\frac{-3}{8}$	$\frac{4}{11}$	$\frac{-4}{11}$	$\frac{5}{7}$	$\frac{-5}{7}$	⋯
	$\frac{1}{7}$	$\frac{-1}{7}$	$\frac{2}{13}$	$\frac{-2}{13}$	$\frac{3}{10}$	$\frac{-3}{10}$	$\frac{4}{13}$	$\frac{-4}{13}$	$\frac{5}{8}$	$\frac{-5}{8}$	⋯
	⋮	⋮	⋮	⋮	⋮	⋮	⋮	⋮	⋮	⋮	⋱

Next, draw an infinite path in this array, beginning at $\frac{0}{1}$ and snaking back and forth as indicated below. Every rational number is on this path.

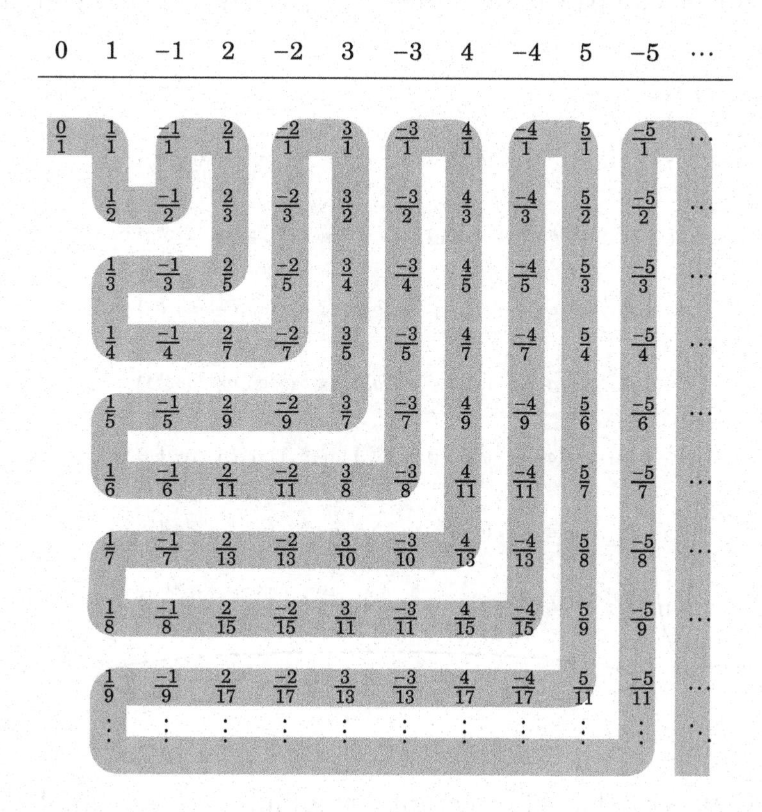

Beginning at $\frac{0}{1}$ and following the path, we get an infinite list of all rational numbers:

$$0,\ 1,\ \frac{1}{2},\ -\frac{1}{2},\ -1,\ 2,\ \frac{2}{3},\ \frac{2}{5},\ -\frac{1}{3},\ \frac{1}{3},\ \frac{1}{4},\ -\frac{1}{4},\ \frac{2}{7},\ -\frac{2}{7},\ -\frac{2}{5},\ -\frac{2}{3},\ -\frac{2}{3},\ -2,\ 3,\ \frac{3}{2},\ \ldots$$

By Theorem 13.3, it follows that $\mathbb{Q}$ is countably infinite, that is, $|\mathbb{Q}| = |\mathbb{N}|$. ∎

It is also true that the Cartesian product of two countably infinite sets is itself countably infinite, as our next theorem states.

Theorem 13.5 If A and B are both countably infinite, then so is $A \times B$.

Proof. Suppose A and B are both countably infinite. By Theorem 13.3, we know we can write A and B in list form as

$$\begin{aligned} A &= \{a_1, a_2, a_3, a_4, \ldots\}, \\ B &= \{b_1, b_2, b_3, b_4, \ldots\}. \end{aligned}$$

Figure 13.2 shows how to form an infinite path winding through all of $A \times B$. Therefore $A \times B$ can be written in list form, so it is countably infinite. ∎

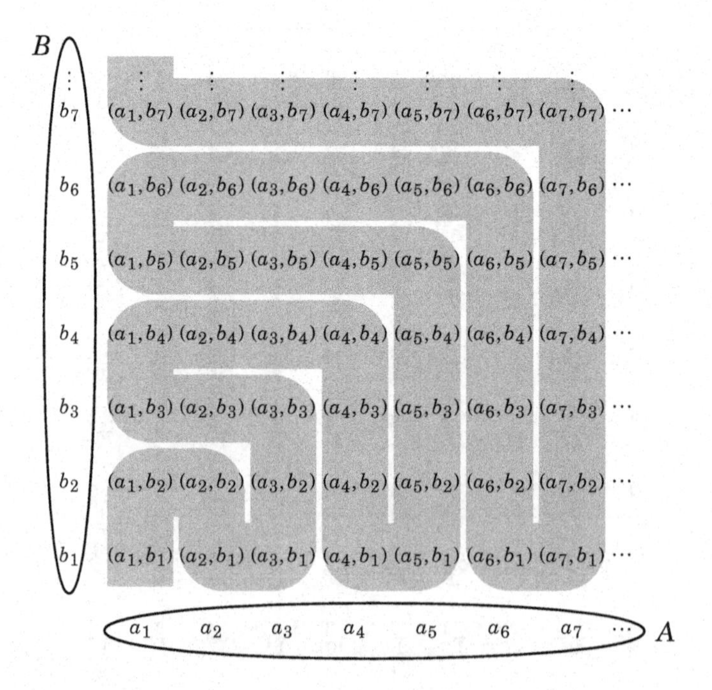

Figure 13.2. A product of two countably infinite sets is countably infinite

As an example of a consequence of this theorem, notice that since $\mathbb{Q}$ is countably infinite, the set $\mathbb{Q}\times\mathbb{Q}$ is also countably infinite.

Recall that the word "corollary" means a result that follows easily from some other result. We have the following corollary of Theorem 13.5.

Corollary 13.1 Given n countably infinite sets $A_1, A_2, A_3, \ldots, A_n$, with $n \geq 2$, the Cartesian product $A_1 \times A_2 \times A_3 \times \cdots \times A_n$ is also countably infinite.

Proof. The proof is by induction on n. For the basis step, notice that when $n = 2$ the statement asserts that for countably infinite sets A_1 and A_2, the product $A_1 \times A_2$ is countably infinite, and this is true by Theorem 13.5.

Assume that for $k \geq 2$, any product $A_1 \times A_2 \times A_3 \times \cdots \times A_k$ of countably infinite sets is countably infinite. Consider a product $A_1 \times A_2 \times A_3 \times \cdots \times A_{k+1}$ of $k+1$ countably infinite sets. It is easily confirmed that the function

$$\begin{aligned} f : A_1 \times A_2 \times A_3 \times \cdots \times A_k \times A_{k+1} &\longrightarrow (A_1 \times A_2 \times A_3 \times \cdots \times A_k) \times A_{k+1} \\ f(x_1, x_2, \ldots, x_k, x_{k+1}) &= \big((x_1, x_2, \ldots, x_k),\, x_{k+1}\big) \end{aligned}$$

is bijective, so $|A_1 \times A_2 \times A_3 \times \cdots \times A_k \times A_{k+1}| = |(A_1 \times A_2 \times A_3 \times \cdots \times A_k) \times A_{k+1}|$. By the induction hypothesis, $(A_1 \times A_2 \times A_3 \times \cdots \times A_k) \times A_{k+1}$ is a product of two countably infinite sets, so it is countably infinite by Theorem 13.5. As noted above, $A_1 \times A_2 \times A_3 \times \cdots \times A_k \times A_{k+1}$ has the same cardinality, so it too is countably infinite. ■

Theorem 13.6 If A and B are both countably infinite, then $A \cup B$ is countably infinite.

Proof. Suppose A and B are both countably infinite. By Theorem 13.3, we know we can write A and B in list form as

$$\begin{aligned} A &= \{a_1, a_2, a_3, a_4, \ldots\}, \\ B &= \{b_1, b_2, b_3, b_4, \ldots\}. \end{aligned}$$

We can "shuffle" A and B into one infinite list for $A \cup B$ as follows.

$$A \cup B = \{a_1, b_1, a_2, b_2, a_3, b_3, a_4, b_4, \ldots\}.$$

(We agree not to list an element twice if it belongs to both A and B.) Therefore, by Theorem 13.3, it follows that $A \cup B$ is countably infinite. ■

Exercises for Section 13.2

1. Prove that the set $A = \{\ln(n) : n \in \mathbb{N}\} \subseteq \mathbb{R}$ is countably infinite.
2. Prove that the set $A = \{(m,n) \in \mathbb{N} \times \mathbb{N} : m \le n\}$ is countably infinite.
3. Prove that the set $A = \{(5n, -3n) : n \in \mathbb{Z}\}$ is countably infinite.
4. Prove that the set of all irrational numbers is uncountable. (Suggestion: Consider proof by contradiction using Theorems 13.4 and 13.6.)
5. Prove or disprove: There exists a countably infinite subset of the set of irrational numbers.
6. Prove or disprove: There exists a bijective function $f : \mathbb{Q} \to \mathbb{R}$.
7. Prove or disprove: The set $\mathbb{Q}^{100}$ is countably infinite.
8. Prove or disprove: The set $\mathbb{Z} \times \mathbb{Q}$ is countably infinite.
9. Prove or disprove: The set $\{0,1\} \times \mathbb{N}$ is countably infinite.
10. Prove or disprove: The set $A = \{\frac{\sqrt{2}}{n} : n \in \mathbb{N}\}$ countably infinite.
11. Describe a partition of $\mathbb{N}$ that divides $\mathbb{N}$ into eight countably infinite subsets.
12. Describe a partition of $\mathbb{N}$ that divides $\mathbb{N}$ into $\aleph_0$ countably infinite subsets.
13. Prove or disprove: If $A = \{X \subseteq \mathbb{N} : X \text{ is finite}\}$, then $|A| = \aleph_0$.
14. Suppose $A = \{(m,n) \in \mathbb{N} \times \mathbb{R} : n = \pi m\}$. Is it true that $|\mathbb{N}| = |A|$?
15. Theorem 13.5 implies that $\mathbb{N} \times \mathbb{N}$ is countably infinite. Construct an alternate proof of this fact by showing that the function $\varphi : \mathbb{N} \times \mathbb{N} \to \mathbb{N}$ defined as $\varphi(m,n) = 2^{n-1}(2m-1)$ is bijective.

13.3 Comparing Cardinalities

At this point we know that there are at least two different kinds of infinity. On one hand, there are countably infinite sets such as $\mathbb{N}$, of cardinality $\aleph_0$. Then there is the uncountable set $\mathbb{R}$. Are there other kinds of infinity beyond these two kinds? The answer is "yes," but to see why we first need to introduce some new definitions and theorems.

Our first task will be to formulate a definition for what we mean by $|A| < |B|$. Of course if A and B are finite we know exactly what this means: $|A| < |B|$ means that when the elements of A and B are counted, A is found to have fewer elements than B. But this process breaks down if A or B is infinite, for then the elements can't be counted.

The language of functions helps us overcome this difficulty. Notice that for finite sets A and B it is intuitively clear that $|A| < |B|$ if and only if there exists an injective function $f : A \to B$ but there are no surjective functions $f : A \to B$. The following diagram illustrates this:

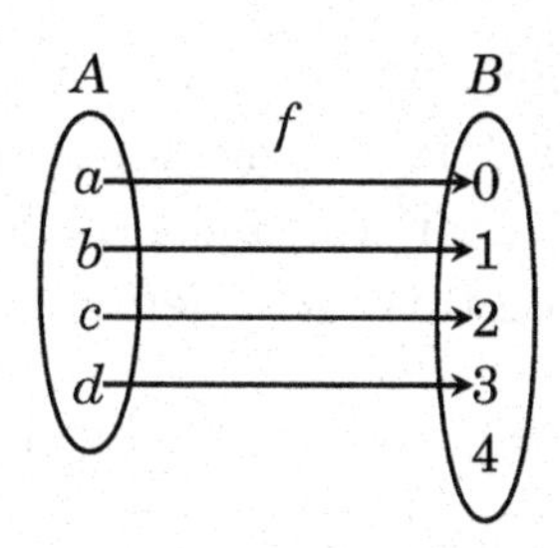

We will use this idea to define what is meant by $|A| < |B|$ and $|A| \leq |B|$. For emphasis, the following definition also restates what is meant by $|A| = |B|$.

Definition 13.4 Suppose A and B are sets.
(1) $|A| = |B|$ means there is a bijection $A \to B$.
(2) $|A| < |B|$ means there is an injection $A \to B$, but no surjection $A \to B$.
(3) $|A| \leq |B|$ means $|A| < |B|$ or $|A| = |B|$.

For example, consider $\mathbb{N}$ and $\mathbb{R}$. The function $f : \mathbb{N} \to \mathbb{R}$ defined as $f(n) = n$ is clearly injective, but it is not surjective because given the element $\frac{1}{2} \in \mathbb{R}$, we have $f(n) \neq \frac{1}{2}$ for every $n \in \mathbb{N}$. In fact, Theorem 13.2 of Section 13.1 asserts that there is no surjection $\mathbb{N} \to \mathbb{R}$. Definition 13.4 yields

$$|\mathbb{N}| < |\mathbb{R}|. \tag{13.1}$$

Said differently, $\aleph_0 < |\mathbb{R}|$.

Is there a set X for which $|\mathbb{R}| < |X|$? The answer is "yes," and the next theorem explains why. It implies $|\mathbb{R}| < |\mathscr{P}(\mathbb{R})|$. (Recall that $\mathscr{P}(A)$ denotes the power set of A.)

Theorem 13.7 If A is any set, then $|A| < |\mathscr{P}(A)|$.

Proof. Before beginning the proof, we remark that this statement is obvious if A is finite, for then $|A| < 2^{|A|} = |\mathscr{P}(A)|$. But our proof must apply to *all* sets A, both finite and infinite, so it must use Definition 13.4.

We prove the theorem with direct proof. Let A be an arbitrary set. According to Definition 13.4, to prove $|A| < |\mathscr{P}(A)|$ we must show that there is an injection $f : A \to \mathscr{P}(A)$, but no surjection $f : A \to \mathscr{P}(A)$.

To see that there is an injection $f : A \to \mathscr{P}(A)$, define f by the rule $f(x) = \{x\}$. In words, f sends any element x of A to the one-element set $\{x\} \in \mathscr{P}(A)$. Then $f : A \to \mathscr{P}(A)$ is injective, as follows. Suppose $f(x) = f(y)$. Then $\{x\} = \{y\}$. Now, the only way that $\{x\}$ and $\{y\}$ can be equal is if $x = y$, so it follows that $x = y$. Thus f is injective.

Next we need to show that there exists no surjection $f : A \to \mathscr{P}(A)$. Suppose for the sake of contradiction that there does exist a surjection

$f : A \to \mathscr{P}(A)$. Notice that for any element $x \in A$, we have $f(x) \in \mathscr{P}(A)$, so $f(x)$ is a subset of A. Thus f is a function that sends elements of A to subsets of A. It follows that for any $x \in A$, either x is an element of the subset $f(x)$ or it is not. Using this idea, define the following subset B of A:

$$B = \{x \in A : x \notin f(x)\} \subseteq A.$$

Now since $B \subseteq A$ we have $B \in \mathscr{P}(A)$, and since f is surjective there is an $a \in A$ for which $f(a) = B$. Now, either $a \in B$ or $a \notin B$. We will consider these two cases separately, and show that each leads to a contradiction.

Case 1. If $a \in B$, then the definition of B implies $a \notin f(a)$, and since $f(a) = B$ we have $a \notin B$, which is a contradiction.

Case 2. If $a \notin B$, then the definition of B implies $a \in f(a)$, and since $f(a) = B$ we have $a \in B$, again a contradiction.

Since the assumption that there is a surjection $f : A \to \mathscr{P}(A)$ leads to a contradiction, we conclude that there are no such surjective functions.

In conclusion, we have seen that there exists an injection $A \to \mathscr{P}(A)$ but no surjection $A \to \mathscr{P}(A)$, so Definition 13.4 implies that $|A| < |\mathscr{P}(A)|$. ■

Beginning with the set $A = \mathbb{N}$ and applying Theorem 13.7 over and over again, we get the following chain of infinite cardinalities.

$$\aleph_0 = |\mathbb{N}| < |\mathscr{P}(\mathbb{N})| < |\mathscr{P}(\mathscr{P}(\mathbb{N}))| < |\mathscr{P}(\mathscr{P}(\mathscr{P}(\mathbb{N})))| < \cdots \qquad (13.2)$$

Thus there is an infinite sequence of different types of infinity, starting with $\aleph_0$ and becoming ever larger. The set $\mathbb{N}$ is countable, and all the sets $\mathscr{P}(\mathbb{N})$, $\mathscr{P}(\mathscr{P}(\mathbb{N}))$, etc., are uncountable.

In the next section we will prove that $|\mathscr{P}(\mathbb{N})| = |\mathbb{R}|$. Thus $|\mathbb{N}|$ and $|\mathbb{R}|$ are the first two entries in the chain (13.2) above. They are are just two relatively tame infinities in a long list of other wild and exotic infinities.

Unless you plan on studying advanced set theory or the foundations of mathematics, you are unlikely to ever encounter any types of infinity beyond $\aleph_0$ and $|\mathbb{R}|$. Still you will in future mathematics courses need to distinguish between countably infinite and uncountable sets, so we close with two final theorems that can help you do this.

Theorem 13.8 An infinite subset of a countably infinite set is countably infinite.

Proof. Suppose A is an infinite subset of the countably infinite set B. Because B is countably infinite, its elements can be written in a list

$b_1, b_2, b_3, b_4, \ldots$ Then we can also write A's elements in list form by proceeding through the elements of B, in order, and selecting those that belong to A. Thus A can be written in list form, and since A is infinite, its list will be infinite. Consequently A is countably infinite. ■

Theorem 13.9 If $U \subseteq A$, and U is uncountable, then A is uncountable.

Proof. Suppose for the sake of contradiction that $U \subseteq A$, and U is uncountable but A is not uncountable. Then since $U \subseteq A$ and U is infinite, then A must be infinite too. Since A is infinite, and not uncountable, it must be countably infinite. Then U is an infinite subset of a countably infinite set A, so U is countably infinite by Theorem 13.8. Thus U is both uncountable and countably infinite, a contradiction. ■

Theorems 13.8 and 13.9 can be useful when we need to decide whether a set is countably infinite or uncountable. They sometimes allow us to decide its cardinality by comparing it to a set whose cardinality is known.

For example, suppose we want to decide whether or not the set $A = \mathbb{R}^2$ is uncountable. Since the x-axis $U = \{(x,0) : x \in \mathbb{R}\} \subseteq \mathbb{R}^2$ has the same cardinality as $\mathbb{R}$, it is uncountable. Theorem 13.9 implies that $\mathbb{R}^2$ is uncountable. Other examples can be found in the exercises.

Exercises for Section 13.3

1. Suppose B is an uncountable set and A is a set. Given that there is a surjective function $f : A \to B$, what can be said about the cardinality of A?
2. Prove that the set $\mathbb{C}$ of complex numbers is uncountable.
3. Prove or disprove: If A is uncountable, then $|A| = |\mathbb{R}|$.
4. Prove or disprove: If $A \subseteq B \subseteq C$ and A and C are countably infinite, then B is countably infinite.
5. Prove or disprove: The set $\{0,1\} \times \mathbb{R}$ is uncountable.
6. Prove or disprove: Every infinite set is a subset of a countably infinite set.
7. Prove or disprove: If $A \subseteq B$ and A is countably infinite and B is uncountable, then $B - A$ is uncountable.
8. Prove or disprove: The set $\{(a_1, a_2, a_3, \ldots) : a_i \in \mathbb{Z}\}$ of infinite sequences of integers is countably infinite.
9. Prove that if A and B are finite sets with $|A| = |B|$, then any injection $f : A \to B$ is also a surjection. Show this is not necessarily true if A and B are not finite.
10. Prove that if A and B are finite sets with $|A| = |B|$, then any surjection $f : A \to B$ is also an injection. Show this is not necessarily true if A and B are not finite.

13.4 The Cantor-Bernstein-Schröeder Theorem

An often used property of numbers is that if $a \leq b$ and $b \leq a$, then $a = b$. It is reasonable to ask if the same property applies to cardinality. If $|A| \leq |B|$ and $|B| \leq |A|$, is it true that $|A| = |B|$? This is in fact true, and this section's goal is to prove it. This will yield an alternate (and highly effective) method of proving that two sets have the same cardianlity.

Recall (Definition 13.4) that $|A| \leq |B|$ means that $|A| < |B|$ or $|A| = |B|$. If $|A| < |B|$ then (by Definition 13.4) there is an injection $A \to B$. On the other hand, if $|A| = |B|$, then there is a bijection (hence also an injection) $A \to B$. Thus $|A| \leq |B|$ implies that there is an injection $f : A \to B$.

Likewise, $|B| \leq |A|$ implies that there is an injection $g : B \to A$.

Our aim is to show that if $|A| \leq |B|$ and $|B| \leq |A|$, then $|A| = |B|$. In other words, we aim to show that if there are injections $f : A \to B$ and $g : B \to A$, then there is a bijection $h : A \to B$. The proof of this fact, though not particularly difficult, is not entirely trivial, either. The fact that f and g guarantee that such an h exists is called the **the Cantor-Bernstein-Schröeder theorem**. This theorem is very useful for proving two sets A and B have the same cardinality: it says that instead of finding a bijection $A \to B$, it suffices to find injections $A \to B$ and $B \to A$. This is useful because injections are often easier to find than bijections.

We will prove the Cantor-Bernstein-Schröeder theorem, but before doing so let's work through an informal visual argument that will guide us through (and illustrate) the proof.

Suppose there are injections $f : A \to B$ and $g : B \to A$. We want to use them to produce a bijection $h : A \to B$. Sets A and B are sketched below. For clarity, each has the shape of the letter that denotes it, and to help distinguish them the set A is shaded.

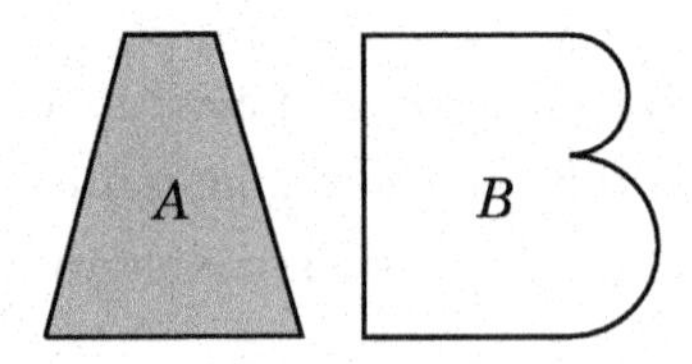

Figure 13.3. The sets A and B

The injections $f : A \to B$ and $g : B \to A$ are illustrated in Figure 13.4. Think of f as putting a "copy" $f(A) = \{f(x) : x \in A\}$ of A into B, as illustrated. This copy, the range of f, does not fill up all of B (unless f happens to be surjective). Likewise, g puts a "copy" $g(B)$ of B into A. Because they are not necessarily bijective, neither f nor g is guaranteed to have an inverse. But the map $g : B \to g(B)$ from B to $g(B) = \{g(x) : x \in B\}$ *is* bijective, so there is an inverse $g^{-1} : g(B) \to B$. (We will need this inverse soon.)

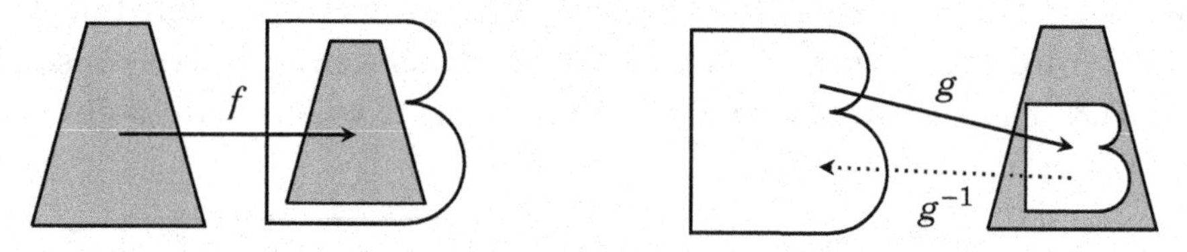

Figure 13.4. The injections $f : A \to B$ and $g : B \to A$

Consider the chain of injections illustrated in Figure 13.5. On the left, g puts a copy of B into A. Then f puts a copy of A (containing the copy of B) into B. Next, g puts a copy of this B-containing-A-containing-B into A, and so on, always alternating g and f.

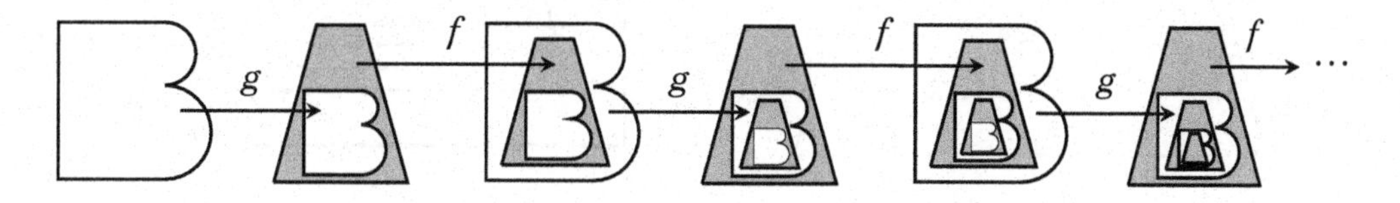

Figure 13.5. An infinite chain of injections

The first time A occurs in this sequence, it has a shaded region $A - g(B)$. In the second occurrence of A, the shaded region is $(A-g(B)) \cup (g \circ f)(A-g(B))$. In the third occurrence of A, the shaded region is

$$(A - g(B)) \cup (g \circ f)(A - g(B)) \cup (g \circ f \circ g \circ f)(A - g(B)).$$

To tame the notation, let's say $(g \circ f)^2 = (g \circ f) \circ (g \circ f)$, and $(g \circ f)^3 = (g \circ f) \circ (g \circ f) \circ (g \circ f)$, and so on. Let's also agree that $(g \circ f)^0 = \iota_A$, that is, it is the identity function on A. Then the shaded region of the nth occurrence of A in the sequence is

$$\bigcup_{k=0}^{n-1} (g \circ f)^k (A - g(B)).$$

This process divides A into gray and white regions: the gray region is

$$G = \bigcup_{k=0}^{\infty} (g \circ f)^k (A - g(B)),$$

and the white region is $A - G$. (See Figure 13.6.)

Figure 13.6 suggests our desired bijection $h : A \to B$. The injection f sends the gray areas on the left bijectively to the gray areas on the right. The injection $g^{-1} : g(B) \to B$ sends the white areas on the left bijectively to the white areas on the right. We can thus define $h : A \to B$ so that $h(x) = f(x)$ if x is a gray point, and $h(x) = g^{-1}(x)$ if x is a white point.

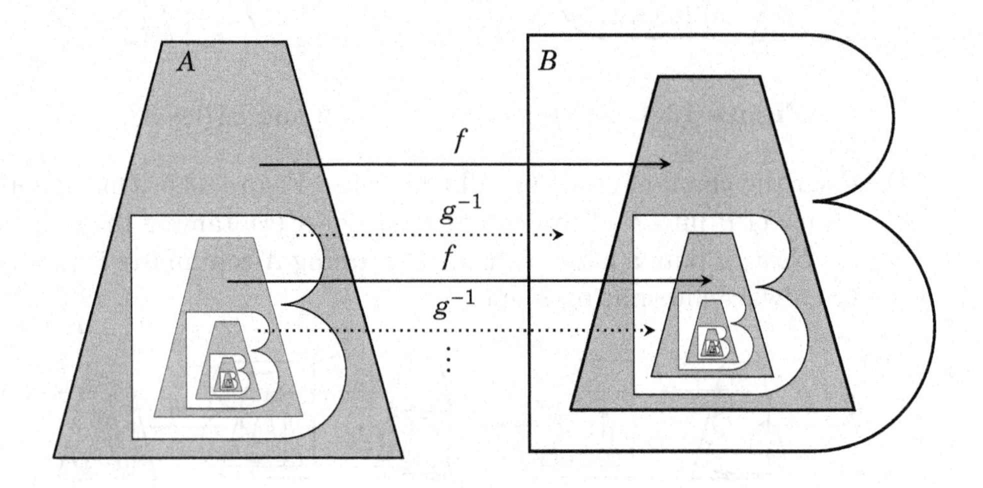

Figure 13.6. The bijection $h : A \to B$

This informal argument suggests that given injections $f : A \to B$ and $g : B \to A$, there is a bijection $h : A \to B$. But it is not a proof. We now present this as a theorem and tighten up our reasoning in a careful proof, with the above diagrams and ideas as a guide.

Theorem 13.10 (The Cantor-Bernstein-Schröeder Theorem)
If $|A| \le |B|$ and $|B| \le |A|$, then $|A| = |B|$. In other words, if there are injections $f : A \to B$ and $g : B \to A$, then there is a bijection $h : A \to B$.

Proof. (Direct) Suppose there are injections $f : A \to B$ and $g : B \to A$. Then, in particular, $g : B \to g(B)$ is a bijection from B onto the range of g, so it has an inverse $g^{-1} : g(B) \to B$. (Note that $g : B \to A$ itself has no inverse $g^{-1} : A \to B$ unless g is surjective.) Consider the subset

$$G = \bigcup_{k=0}^{\infty} (g \circ f)^k (A - g(B)) \subseteq A.$$

Let $W = A - G$, so $A = G \cup W$ is partitioned into two sets G (think gray) and W (think white). Define a function $h : A \to B$ as

$$h(x) = \begin{cases} f(x) & \text{if } x \in G \\ g^{-1}(x) & \text{if } x \in W. \end{cases}$$

Notice that this makes sense: if $x \in W$, then $x \notin G$, so $x \notin A - g(B) \subseteq G$, hence $x \in g(B)$, so $g^{-1}(x)$ is defined.

To finish the proof, we must show that h is both injective and surjective.

For injective, we assume $h(x) = h(y)$, and deduce $x = y$. There are three cases to consider. First, if x and y are both in G, then $h(x) = h(y)$ means $f(x) = f(y)$, so $x = y$ because f is injective. Second, if x and y are both in W, then $h(x) = h(y)$ means $g^{-1}(x) = g^{-1}(y)$, and applying g to both sides gives $x = y$. In the third case, one of x and y is in G and the other is in W. Say $x \in G$ and $y \in W$. The definition of G gives $x = (g \circ f)^k(z)$ for some $k \geq 0$ and $z \in A - g(B)$. Note $h(x) = h(y)$ now implies $f(x) = g^{-1}(y)$, that is, $f((g \circ f)^k(z)) = g^{-1}(y)$. Applying g to both sides gives $(g \circ f)^{k+1}(z) = y$, which means $y \in G$. But this is impossible, as $y \in W$. Thus this third case cannot happen. But in the first two cases $h(x) = h(y)$ implies $x = y$, so h is injective.

To see that h is surjective, take any $b \in B$. We will find an $x \in A$ with $h(x) = b$. Note that $g(b) \in A$, so either $g(b) \in W$ or $g(b) \in G$. In the first case, $h(g(b)) = g^{-1}(g(b)) = b$, so we have an $x = g(b) \in A$ for which $h(x) = b$. In the second case, $g(b) \in G$. The definition of G shows

$$g(b) = (g \circ f)^k(z)$$

for some $k > 0$, and $z \in A - g(B)$. Thus

$$g(b) = (g \circ f) \circ (g \circ f)^{k-1}(z).$$

Rewriting this,

$$g(b) = g\Big(f\big((g \circ f)^{k-1}(z)\big)\Big).$$

Because g is injective, this implies

$$b = f\big((g \circ f)^{k-1}(z)\big).$$

Let $x = (g \circ f)^{k-1}(z)$, so $x \in G$ by definition of G. Observe that $h(x) = f(x) = f\big((g \circ f)^{k-1}(z)\big) = b$. We have now seen that for any $b \in B$, there is an $x \in A$ for which $h(x) = b$. Thus h is surjective.

Since $h : A \to B$ is both injective and surjective, it is also bijective. ■

Here are some examples illustrating how the Cantor-Bernstein-Schröeder theorem can be used. This includes a proof that $|\mathbb{R}| = |\mathscr{P}(\mathbb{N})|$.

Example 13.6 The intervals $[0,1)$ and $(0,1)$ in $\mathbb{R}$ have equal cardinalities.

Surely this fact is plausible, for the two intervals are identical except for the endpoint 0. Yet concocting a bijection $[0,1) \to (0,1)$ is tricky. (Though not particularly difficult: see the solution of Exercise 11 of Section 13.1.)

For a simpler approach, note that $f(x) = \frac{1}{4} + \frac{1}{2}x$ is an injection $[0,1) \to (0,1)$. Also, $g(x) = x$ is an injection $(0,1) \to [0,1)$. The Cantor-Bernstein-Schröeder theorem guarantees a bijection $h : [0,1) \to (0,1)$, so $|[0,1)| = |(0,1)|$.

Theorem 13.11 The sets $\mathbb{R}$ and $\mathscr{P}(\mathbb{N})$ have the same cardinality.

Proof. Example 13.4 shows that $|\mathbb{R}| = |(0,1)|$, and Example 13.6 shows $|(0,1)| = |[0,1)|$. Thus $|\mathbb{R}| = |[0,1)|$, so to prove the theorem we just need to show that $|[0,1)| = |\mathscr{P}(\mathbb{N})|$. By the Cantor-Bernstein-Schröeder theorem, it suffices to find injections $f : [0,1) \to \mathscr{P}(\mathbb{N})$ and $g : \mathscr{P}(\mathbb{N}) \to [0,1)$.

To define $f : [0,1) \to \mathscr{P}(\mathbb{N})$, we use the fact that any number in $[0,1)$ has a unique decimal representation $0.b_1b_2b_3b_4\ldots$, where each b_i one of the digits $0,1,2,\ldots,9$, and there is not a repeating sequence of 9's at the end. (Recall that, e.g., $0.35999\overline{9} = 0.36\overline{0}$, etc.) Define $f : [0,1) \to \mathscr{P}(\mathbb{N})$ as

$$f\big(0.b_1b_2b_3b_4\ldots\big) = \big\{10b_1,\ 10^2b_2,\ 10^3b_3,\ \ldots\big\}.$$

For example, $f(0.1212\overline{12}) = \{10,200,1000,20000,100000,\ldots\}$, and $f(0.05) = \{0,500\}$. Also $f(0.5) = f(0.5\overline{0}) = \{0,50\}$. To see that f is injective, take two unequal numbers $0.b_1b_2b_3b_4\ldots$ and $0.d_1d_2d_3d_4\ldots$ in $[0,1)$. Then $b_i \neq d_i$ for some index i. Hence $b_i10^i \in f(0.b_1b_2b_3b_4\ldots)$ but $b_i10^i \notin f(0.d_1d_2d_3d_4\ldots)$, so $f(0.b_1b_2b_3b_4\ldots) \neq f(0.d_1d_2d_3d_4\ldots)$. Consequently f is injective.

Next, define $g : \mathscr{P}(\mathbb{N}) \to [0,1)$, where $g(X) = 0.b_1b_2b_3b_4\ldots$ is the number for which $b_i = 1$ if $i \in X$ and $b_i = 0$ if $i \notin X$. For example, $g(\{1,3\}) = 0.10100\overline{0}$, and $g(\{2,4,6,8,\ldots\}) = 0.010101\overline{01}$. Also $g(\emptyset) = 0$ and $g(\mathbb{N}) = 0.111\overline{1}$. To see that g is injective, suppose $X \neq Y$. Then there is at least one integer i that belongs to one of X or Y, but not the other. Consequently $g(X) \neq g(Y)$ because they differ in the ith decimal place. This shows g is injective.

From the injections $f : [0,1) \to \mathscr{P}(\mathbb{N})$ and $g : \mathscr{P}(\mathbb{N}) \to [0,1)$, the Cantor-Bernstein-Schröeder theorem guarantees a bijection $h : [0,1) \to \mathscr{P}(\mathbb{N})$. Hence $|[0,1)| = |\mathscr{P}(\mathbb{N})|$. As $|\mathbb{R}| = |[0,1)|$, we conclude $|\mathbb{R}| = |\mathscr{P}(\mathbb{N})|$. ■

We know that $|\mathbb{R}| \neq |\mathbb{N}|$. But we just proved $|\mathbb{R}| = |\mathscr{P}(\mathbb{N})|$. This suggests that the cardinality of $\mathbb{R}$ is not "too far" from $|\mathbb{N}| = \aleph_0$. We close with a few informal remarks on this mysterious relationship between $\aleph_0$ and $|\mathbb{R}|$.

We established earlier in this chapter that $\aleph_0 < |\mathbb{R}|$. For nearly a century after Cantor formulated his theories on infinite sets, mathematicians struggled with the question of whether or not there exists a set A for which

$$\aleph_0 < |A| < |\mathbb{R}|.$$

It was commonly suspected that no such set exists, but no one was able to prove or disprove this. The assertion that no such A exists came to be called the **continuum hypothesis**.

Theorem 13.11 states that $|\mathbb{R}| = |\mathscr{P}(\mathbb{N})|$. Placing this in the context of the chain (13.2) on page 230, we have the following relationships.

$$\begin{array}{ccccccccc} \aleph_0 & & |\mathbb{R}| & & & & & & \\ \| & & \| & & & & & & \\ |\mathbb{N}| & < & |\mathscr{P}(\mathbb{N})| & < & |\mathscr{P}(\mathscr{P}(\mathbb{N}))| & < & |\mathscr{P}(\mathscr{P}(\mathscr{P}(\mathbb{N})))| & < & \cdots \end{array}$$

From this, we can see that the continuum hypothesis asserts that no set has a cardinality between that of $\mathbb{N}$ and its power set.

Although this may seem intuitively plausible, it eluded proof since Cantor first posed it in the 1880s. In fact, the real state of affairs is almost paradoxical. In 1931, the logician Kurt Gödel proved that for any sufficiently strong and consistent axiomatic system, there exist statements which can neither be proved nor disproved within the system.

Later he proved that the negation of the continuum hypothesis cannot be proved within the standard axioms of set theory (i.e., the Zermelo-Fraenkel axioms, mentioned in Section 1.10). This meant that either the continuum hypothesis is false and cannot be proven false, or it is true.

In 1964, Paul Cohen discovered another startling truth: Given the laws of logic and the axioms of set theory, no proof can deduce the continuum hypothesis. In essence he proved that the continuum hypothesis cannot be *proved*.

Taken together, Gödel and Cohens' results mean that the standard axioms of mathematics cannot "decide" whether the continuum hypothesis is true or false; that no logical conflict can arise from either asserting or denying the continuum hypothesis. We are free to either accept it as true or accept it as false, and the two choices lead to different—but equally consistent—versions of set theory.

On the face of it, this seems to undermine the foundation of logic, and everything we have done in this book. The continuum hypothesis should be a *statement* – it should be either true or false. How could it be both?

Here is an analogy that may help make sense of this. Consider the number systems $\mathbb{Z}_n$. What if we asked whether $[2]=[0]$ is true or false? Of course the answer depends on n. The expression $[2]=[0]$ is true in $\mathbb{Z}_2$ and false in $\mathbb{Z}_3$. Moreover, if we assert that $[2]=[0]$ is true, we are logically forced to the conclusion that this is taking place in the system $\mathbb{Z}_2$. If we assert that $[2]=[0]$ is false, then we are dealing with some other $\mathbb{Z}_n$. The fact that $[2]=[0]$ can be either true or false does not necessarily mean that there is some inherent inconsistency within the individual number systems $\mathbb{Z}_n$. The equation $[2]=[0]$ is a true statement in the "universe" of $\mathbb{Z}_2$ and a false statement in the universe of (say) $\mathbb{Z}_3$.

It is the same with the continuum hypothesis. Saying it's true leads to one system of set theory. Saying it's false leads to some other system of set theory. Gödel and Cohens' discoveries mean that these two types of set theory, although different, are equally consistent and valid mathematical universes.

So what should you believe? Fortunately, it does not make much difference, because most important mathematical results do not hinge on the continuum hypothesis. (They are true in both universes.) Unless you undertake a deep study of the foundations of mathematics, you will be fine accepting the continuum hypothesis as true. Most mathematicians are agnostics on this issue, but they tend to prefer the version of set theory in which the continuum hypothesis holds.

The situation with the continuum hypothesis is a testament to the immense complexity of mathematics. It is a reminder of the importance of rigor and careful, systematic methods of reasoning that begin with the ideas introduced in this book.

Exercises for Section 13.4

1. Show that if $A \subseteq B$ and there is an injection $g: B \to A$, then $|A| = |B|$.
2. Show that $|\mathbb{R}^2| = |\mathbb{R}|$. Suggestion: Begin by showing $|(0,1)\times(0,1)| = |(0,1)|$.
3. Let $\mathscr{F}$ be the set of all functions $\mathbb{N} \to \{0,1\}$. Show that $|\mathbb{R}| = |\mathscr{F}|$.
4. Let $\mathscr{F}$ be the set of all functions $\mathbb{R} \to \{0,1\}$. Show that $|\mathbb{R}| < |\mathscr{F}|$.
5. Consider the subset $B = \{(x,y) : x^2 + y^2 \leq 1\} \subseteq \mathbb{R}^2$. Show that $|B| = |\mathbb{R}^2|$.
6. Show that $|\mathscr{P}(\mathbb{N}\times\mathbb{N})| = |\mathscr{P}(\mathbb{N})|$.
7. Prove or disprove: If there is a injection $f: A \to B$ and a surjection $g: A \to B$, then there is a bijection $h: A \to B$.

Conclusion

If you have internalized the ideas in this book, then you have a set of rhetorical tools for deciphering and communicating mathematics. These tools are indispensable at any level. But of course it takes more than mere tools to build something. Planning, creativity, inspiration, skill, talent, intuition, passion and persistence are also vitally important. It is safe to say that if you have come this far, then you probably possess a sufficient measure of these traits.

The quest to understand mathematics has no end, but you are well equipped for the journey. It is my hope that the things you have learned from this book will lead you to a higher plane of understanding, creativity and expression.

Good luck and best wishes.

R.H.

Solutions

Chapter 1 Exercises

Section 1.1

1. $\{5x-1 : x \in \mathbb{Z}\} = \{\ldots -11, -6, -1, 4, 9, 14, 19, 24, 29, \ldots\}$
3. $\{x \in \mathbb{Z} : -2 \le x < 7\} = \{-2, -1, 0, 1, 2, 3, 4, 5, 6\}$
5. $\{x \in \mathbb{R} : x^2 = 3\} = \{-\sqrt{3}, \sqrt{3}\}$
7. $\{x \in \mathbb{R} : x^2 + 5x = -6\} = \{-2, -3\}$
9. $\{x \in \mathbb{R} : \sin \pi x = 0\} = \{\ldots, -2, -1, 0, 1, 2, 3, 4, \ldots\} = \mathbb{Z}$
11. $\{x \in \mathbb{Z} : |x| < 5\} = \{-4, -3, -2, -1, 0, 1, 2, 3, 4\}$
13. $\{x \in \mathbb{Z} : |6x| < 5\} = \{0\}$
15. $\{5a + 2b : a, b \in \mathbb{Z}\} = \{\ldots, -2, -1, 0, 1, 2, 3, \ldots\} = \mathbb{Z}$
17. $\{2, 4, 8, 16, 32, 64 \ldots\} = \{2^x : x \in \mathbb{N}\}$
19. $\{\ldots, -6, -3, 0, 3, 6, 9, 12, 15, \ldots\} = \{3x : x \in \mathbb{Z}\}$
21. $\{0, 1, 4, 9, 16, 25, 36, \ldots\} = \{x^2 : x \in \mathbb{Z}\}$
23. $\{3, 4, 5, 6, 7, 8\} = \{x \in \mathbb{Z} : 3 \le x \le 8\} = \{x \in \mathbb{N} : 3 \le x \le 8\}$
25. $\{\ldots, \frac{1}{8}, \frac{1}{4}, \frac{1}{2}, 1, 2, 4, 8, \ldots\} = \{2^n : n \in \mathbb{Z}\}$
27. $\{\ldots, -\pi, -\frac{\pi}{2}, 0, \frac{\pi}{2}, \pi, \frac{3\pi}{2}, 2\pi, \frac{5\pi}{2}, \ldots\} = \left\{\frac{k\pi}{2} : k \in \mathbb{Z}\right\}$

29. $|\{\{1\}, \{2, \{3, 4\}\}, \emptyset\}| = 3$
31. $|\{\{\{1\}, \{2, \{3, 4\}\}, \emptyset\}\}| = 1$
33. $|\{x \in \mathbb{Z} : |x| < 10\}| = 19$
35. $|\{x \in \mathbb{Z} : x^2 < 10\}| = 7$
37. $|\{x \in \mathbb{N} : x^2 < 0\}| = 0$

39. $\{(x, y) : x \in [1, 2], y \in [1, 2]\}$

41. $\{(x, y) : x \in [-1, 1], y = 1\}$

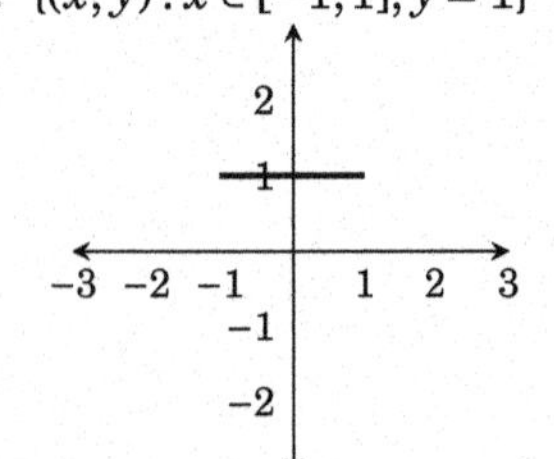

43. $\{(x, y) : |x| = 2, y \in [0, 1]\}$

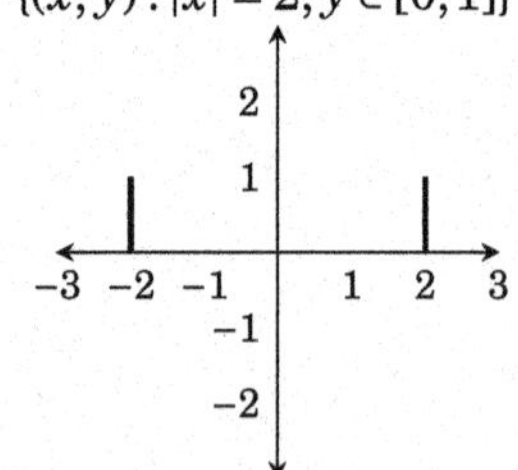

45. $\{(x, y) : x, y \in \mathbb{R}, x^2 + y^2 = 1\}$

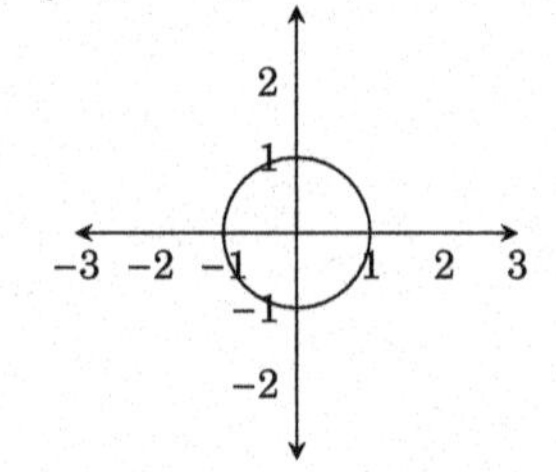

47. $\{(x,y): x,y \in \mathbb{R}, y \geq x^2 - 1\}$

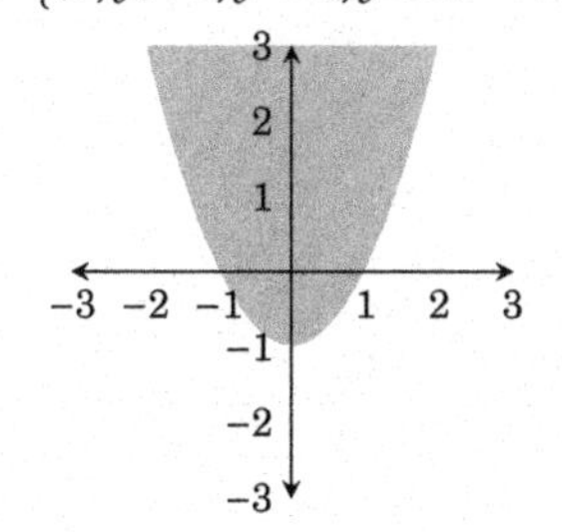

49. $\{(x, x+y): x \in \mathbb{R}, y \in \mathbb{Z}\}$

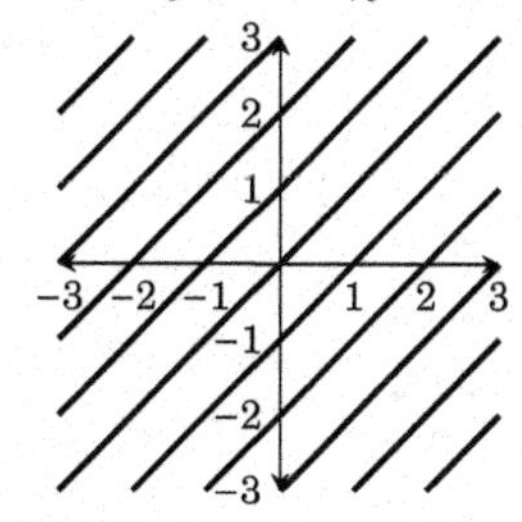

51. $\{(x,y) \in \mathbb{R}^2 : (y-x)(y+x) = 0\}$

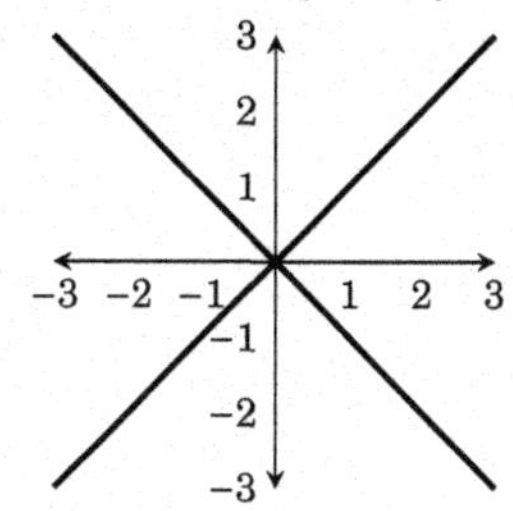

Section 1.2

1. Suppose $A = \{1,2,3,4\}$ and $B = \{a,c\}$.

(a) $A \times B = \{(1,a),(1,c),(2,a),(2,c),(3,a),(3,c),(4,a),(4,c)\}$

(b) $B \times A = \{(a,1),(a,2),(a,3),(a,4),(c,1),(c,2),(c,3),(c,4)\}$

(c) $A \times A = \{(1,1),(1,2),(1,3),(1,4),(2,1),(2,2),(2,3),(2,4),$
$(3,1),(3,2),(3,3),(3,4),(4,1),(4,2),(4,3),(4,4)\}$

(d) $B \times B = \{(a,a),(a,c),(c,a),(c,c)\}$

(e) $\emptyset \times B = \{(a,b): a \in \emptyset, b \in B\} = \emptyset$ (There are no ordered pairs (a,b) with $a \in \emptyset$.)

(f) $(A \times B) \times B =$
$\{((1,a),a),((1,c),a),((2,a),a),((2,c),a),((3,a),a),((3,c),a),((4,a),a),((4,c),a),$
$((1,a),c),((1,c),c),((2,a),c),((2,c),c),((3,a),c),((3,c),c),((4,a),c),((4,c),c)\}$

(g) $A \times (B \times B) =$
$\{(1,(a,a)),(1,(a,c)),(1,(c,a)),(1,(c,c)),$
$(2,(a,a)),(2,(a,c)),(2,(c,a)),(2,(c,c)),$
$(3,(a,a)),(3,(a,c)),(3,(c,a)),(3,(c,c)),$
$(4,(a,a)),(4,(a,c)),(4,(c,a)),(4,(c,c))\}$

(h) $B^3 = \{(a,a,a),(a,a,c),(a,c,a),(a,c,c),(c,a,a),(c,a,c),(c,c,a),(c,c,c)\}$

3. $\{x \in \mathbb{R} : x^2 = 2\} \times \{a,c,e\} = \{(-\sqrt{2},a),(\sqrt{2},a),(-\sqrt{2},c),(\sqrt{2},c),(-\sqrt{2},e),(\sqrt{2},e)\}$

5. $\{x \in \mathbb{R} : x^2 = 2\} \times \{x \in \mathbb{R} : |x| = 2\} = \{(-\sqrt{2},-2),(\sqrt{2},2),(-\sqrt{2},2),(\sqrt{2},-2)\}$

7. $\{\emptyset\} \times \{0,\emptyset\} \times \{0,1\} = \{(\emptyset,0,0),(\emptyset,0,1),(\emptyset,\emptyset,0),(\emptyset,\emptyset,1)\}$

Sketch the following Cartesian products on the x-y plane.

9. $\{1,2,3\} \times \{-1,0,1\}$

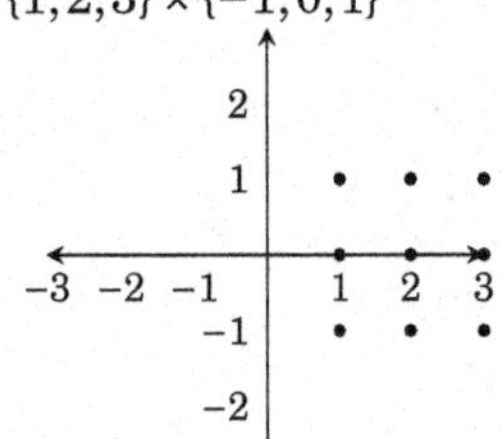

11. $[0,1] \times [0,1]$

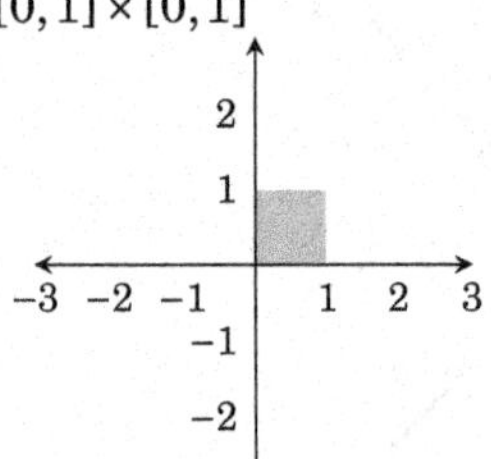

13. $\{1,1.5,2\} \times [1,2]$

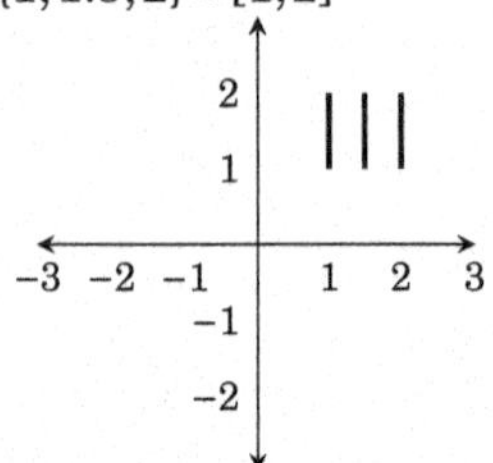

15. $\{1\} \times [0,1]$

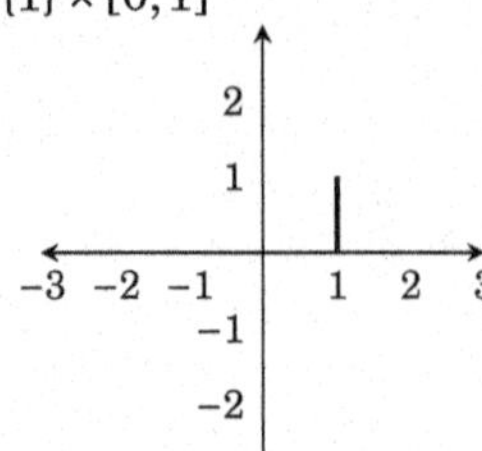

17. $\mathbb{N} \times \mathbb{Z}$

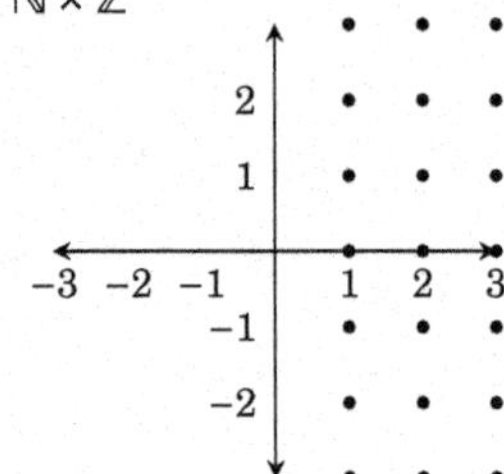

19. $[0,1] \times [0,1] \times [0,1]$

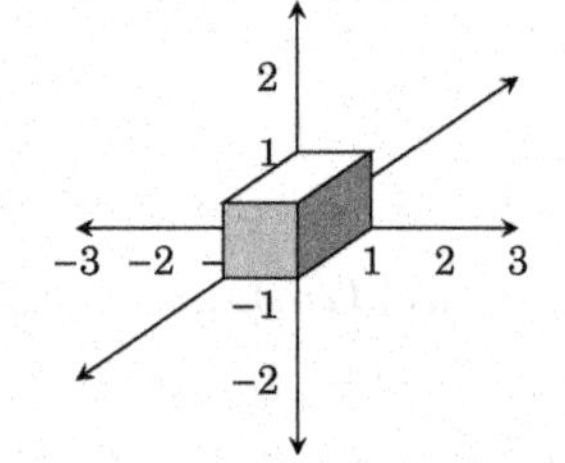

Section 1.3

A. List all the subsets of the following sets.

1. The subsets of $\{1,2,3,4\}$ are: $\{\}$, $\{1\}$, $\{2\}$, $\{3\}$, $\{4\}$, $\{1,2\}$, $\{1,3\}$, $\{1,4\}$, $\{2,3\}$, $\{2,4\}$, $\{3,4\}$, $\{1,2,3\}$, $\{1,2,4\}$, $\{1,3,4\}$, $\{2,3,4\}$, $\{1,2,3,4\}$.

3. The subsets of $\{\{\mathbb{R}\}\}$ are: $\{\}$ and $\{\{\mathbb{R}\}\}$.

5. The subsets of $\{\emptyset\}$ are $\{\}$ and $\{\emptyset\}$.

7. The subsets of $\{\mathbb{R}, \{\mathbb{Q}, \mathbb{N}\}\}$ are $\{\}$, $\{\mathbb{R}\}$,$\{\{\mathbb{Q}, \mathbb{N}\}\}$, $\{\mathbb{R}, \{\mathbb{Q}, \mathbb{N}\}\}$.

B. Write out the following sets by listing their elements between braces.

9. $\big\{X : X \subseteq \{3,2,a\} \text{ and } |X| = 2\big\} = \{\{3,2\},\{3,a\},\{2,a\}\}$

11. $\big\{X : X \subseteq \{3,2,a\} \text{ and } |X| = 4\big\} = \{\} = \emptyset$

C. Decide if the following statements are true or false.

13. $\mathbb{R}^3 \subseteq \mathbb{R}^3$ is **true** because any set is a subset of itself.

15. $\big\{(x,y) : x - 1 = 0\big\} \subseteq \big\{(x,y) : x^2 - x = 0\big\}$. This is true. (The even-numbered ones are both false. You have to explain why.)

Section 1.4

A. Find the indicated sets.

1. $\mathscr{P}(\{\{a,b\},\{c\}\}) = \{\emptyset,\{\{a,b\}\},\{\{c\}\},\{\{a,b\},\{c\}\}\}$

3. $\mathscr{P}(\{\{\emptyset\},5\}) = \{\emptyset,\{\{\emptyset\}\},\{5\},\{\{\emptyset\},5\}\}$

5. $\mathscr{P}(\mathscr{P}(\{2\})) = \{\emptyset,\{\emptyset\},\{\{2\}\},\{\emptyset,\{2\}\}\}$

7. $\mathscr{P}(\{a,b\}) \times \mathscr{P}(\{0,1\}) =$

$$\left\{ \begin{array}{cccc} (\emptyset,\emptyset), & (\emptyset,\{0\}), & (\emptyset,\{1\}), & (\emptyset,\{0,1\}), \\ (\{a\},\emptyset), & (\{a\},\{0\}), & (\{a\},\{1\}), & (\{a\},\{0,1\}), \\ (\{b\},\emptyset), & (\{b\},\{0\}), & (\{b\},\{1\}), & (\{b\},\{0,1\}), \\ (\{a,b\},\emptyset), & (\{a,b\},\{0\}), & (\{a,b\},\{1\}), & (\{a,b\},\{0,1\}) \end{array} \right\}$$

9. $\mathscr{P}(\{a,b\} \times \{0\}) = \{\emptyset,\{(a,0)\},\{(b,0)\},\{(a,0),(b,0)\}\}$

11. $\{X \subseteq \mathscr{P}(\{1,2,3\}) : |X| \le 1\} =$
$\{\emptyset,\{\emptyset\},\{\{1\}\},\{\{2\}\},\{\{3\}\},\{\{1,2\}\},\{\{1,3\}\},\{\{2,3\}\},\{\{1,2,3\}\}\}$

B. Suppose that $|A| = m$ and $|B| = n$. Find the following cardinalities.

13. $|\mathscr{P}(\mathscr{P}(\mathscr{P}(A)))| = 2^{\left(2^{(2^m)}\right)}$

15. $|\mathscr{P}(A \times B)| = 2^{mn}$

17. $|\{X \in \mathscr{P}(A) : |X| \le 1\}| = m+1$

19. $|\mathscr{P}(\mathscr{P}(\mathscr{P}(A \times \emptyset)))| = |\mathscr{P}(\mathscr{P}(\mathscr{P}(\emptyset)))| = 4$

Section 1.5

1. Suppose $A = \{4,3,6,7,1,9\}$, $B = \{5,6,8,4\}$ and $C = \{5,8,4\}$. Find:

(a) $A \cup B = \{1,3,4,5,6,7,8,9\}$

(b) $A \cap B = \{4,6\}$

(c) $A - B = \{3,7,1,9\}$

(d) $A - C = \{3,6,7,1,9\}$

(e) $B - A = \{5,8\}$

(f) $A \cap C = \{4\}$

(g) $B \cap C = \{5,8,4\}$

(h) $B \cup C = \{5,6,8,4\}$

(i) $C - B = \emptyset$

3. Suppose $A = \{0,1\}$ and $B = \{1,2\}$. Find:

(a) $(A \times B) \cap (B \times B) = \{(1,1),(1,2)\}$

(b) $(A \times B) \cup (B \times B) = \{(0,1),(0,2),(1,1),(1,2),(2,1),(2,2)\}$

(c) $(A \times B) - (B \times B) = \{(0,1),(0,2)\}$

(d) $(A \cap B) \times A = \{(1,0),(1,1)\}$

(e) $(A \times B) \cap B = \emptyset$

(f) $\mathscr{P}(A) \cap \mathscr{P}(B) = \{\emptyset,\{1\}\}$

(g) $\mathscr{P}(A) - \mathscr{P}(B) = \{\{0\},\{0,1\}\}$

(h) $\mathscr{P}(A \cap B) = \{\{\},\{1\}\}$

(i) $\{\emptyset,\{(0,1)\},\{(0,2)\},\{(1,1)\},\{(1,2)\},\{(0,1),(0,2)\},\{(0,1),(1,1)\},\{(0,1),(1,2)\},\{(0,2),(1,1)\},$
$\{(0,2),(1,2)\},\{(1,1),(1,2)\},\{(0,2),(1,1),(1,2)\},\{(0,1),(1,1),(1,2)\},\{(0,1),(0,2),(1,2)\},$
$\{(0,1),(0,2),(1,1)\},\{(0,1),(0,2),(1,1),(1,2)\}\}$

5. Sketch the sets $X = [1,3] \times [1,3]$ and $Y = [2,4] \times [2,4]$ on the plane $\mathbb{R}^2$. On separate drawings, shade in the sets $X \cup Y$, $X \cap Y$, $X - Y$ and $Y - X$. (Hint: X and Y are Cartesian products of intervals. You may wish to review how you drew sets like $[1,3] \times [1,3]$ in the Section 1.2.)

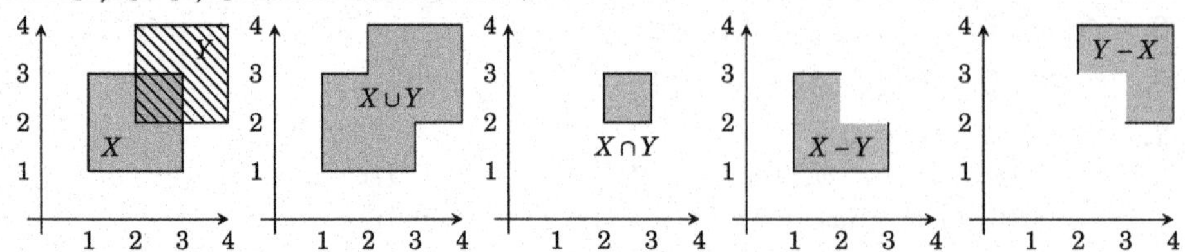

7. Sketch the sets $X = \{(x,y) \in \mathbb{R}^2 : x^2 + y^2 \le 1\}$ and $Y = \{(x,y) \in \mathbb{R}^2 : x \ge 0\}$ on $\mathbb{R}^2$. On separate drawings, shade in the sets $X \cup Y$, $X \cap Y$, $X - Y$ and $Y - X$.

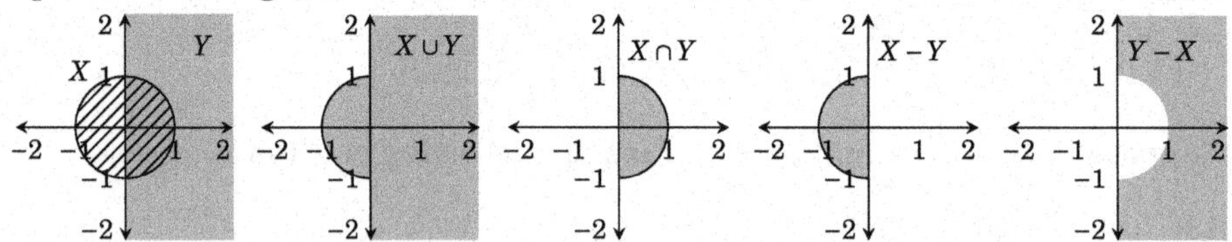

9. The first statement is true. (A picture should convince you; draw one if necessary.) The second statement is false: Notice for instance that $(0.5, 0.5)$ is in the right-hand set, but not the left-hand set.

Section 1.6

1. Suppose $A = \{4,3,6,7,1,9\}$ and $B = \{5,6,8,4\}$ have universal set $U = \{n \in \mathbb{Z} : 0 \le n \le 10\}$.

(a) $\overline{A} = \{0,2,5,8,10\}$

(b) $\overline{B} = \{0,1,2,3,7,9,10\}$

(c) $A \cap \overline{A} = \emptyset$

(d) $A \cup \overline{A} = \{0,1,2,3,4,5,6,7,8,9,10\} = U$

(e) $A - \overline{A} = A$

(f) $A - \overline{B} = \{4,6\}$

(g) $\overline{A} - \overline{B} = \{5,8\}$

(h) $\overline{A} \cap B = \{5,8\}$

(i) $\overline{\overline{A} \cap B} = \{0,1,2,3,4,6,7,9,10\}$

3. Sketch the set $X = [1,3] \times [1,2]$ on the plane $\mathbb{R}^2$. On separate drawings, shade in the sets $\overline{X}$, and $\overline{X} \cap ([0,2] \times [0,3])$.

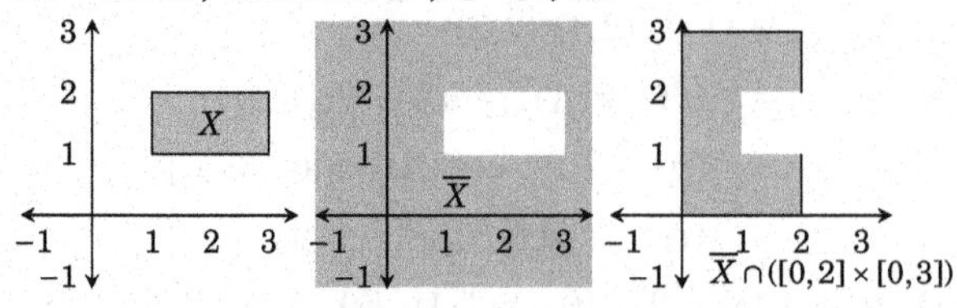

5. Sketch the set $X = \{(x,y) \in \mathbb{R}^2 : 1 \le x^2 + y^2 \le 4\}$ on the plane $\mathbb{R}^2$. On a separate drawing, shade in the set $\overline{X}$.

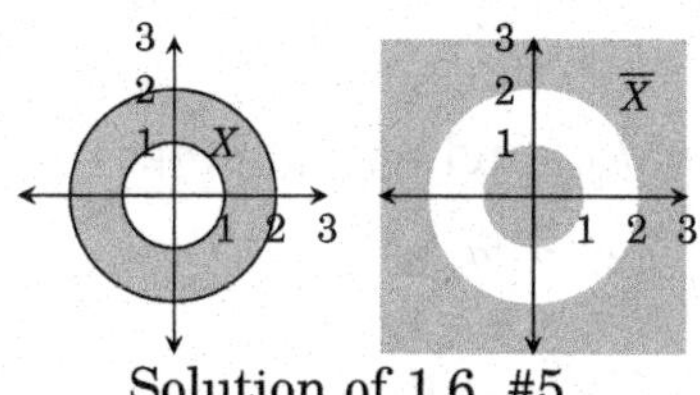

Solution of 1.6, #5.

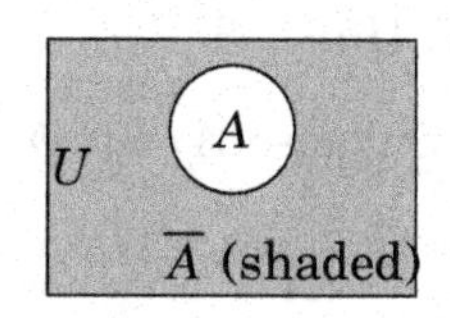

Solution of 1.7, #1.

Section 1.7

1. Draw a Venn diagram for $\overline{A}$. (Solution above right)

3. Draw a Venn diagram for $(A-B)\cap C$.

Scratch work is shown on the right. The set $A-B$ is indicated with vertical shading. The set C is indicated with horizontal shading. The intersection of $A-B$ and C is thus the overlapping region that is shaded with both vertical and horizontal lines. The final answer is drawn on the far right, where the set $(A-B)\cap C$ is shaded in gray.

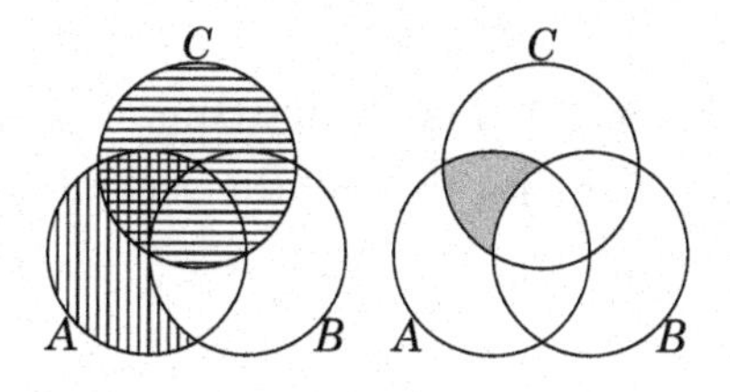

5. Draw Venn diagrams for $A\cup(B\cap C)$ and $(A\cup B)\cap(A\cup C)$. Based on your drawings, do you think $A\cup(B\cap C) = (A\cup B)\cap(A\cup C)$?

If you do the drawings carefully, you will find that your Venn diagrams are the same for both $A\cup(B\cap C)$ and $(A\cup B)\cap(A\cup C)$. Each looks as illustrated on the right. Based on this, we are inclined to say that the equation $A\cup(B\cap C) = (A\cup B)\cap(A\cup C)$ holds for all sets A, B and C.

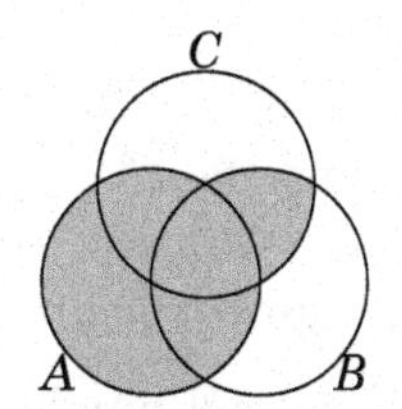

7. Suppose sets A and B are in a universal set U. Draw Venn diagrams for $\overline{A\cap B}$ and $\overline{A}\cup\overline{B}$. Based on your drawings, do you think it's true that $\overline{A\cap B} = \overline{A}\cup\overline{B}$?

The diagrams for $\overline{A\cap B}$ and $\overline{A}\cup\overline{B}$ look exactly alike. In either case the diagram is the shaded region illustrated on the right. Thus we would expect that the equation $\overline{A\cap B} = \overline{A}\cup\overline{B}$ is true for any sets A and B.

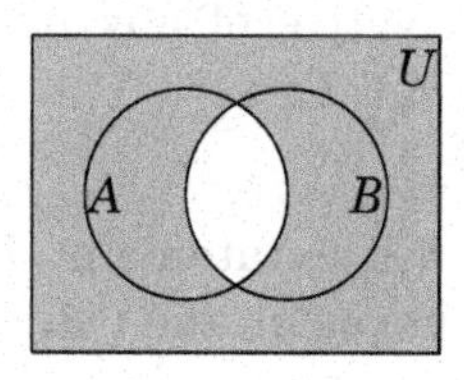

9. Draw a Venn diagram for $(A\cap B)-C$.

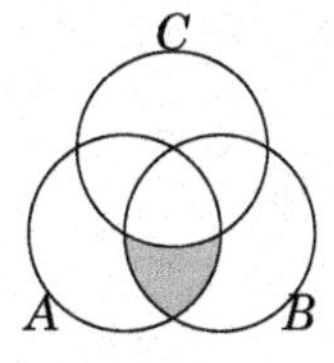

11. The simplest answer is $(B\cap C)-A$.

13. One answer is $(A\cup B\cup C)-(A\cap B\cap C)$.

Section 1.8

1. Suppose $A_1 = \{a,b,d,e,g,f\}$, $A_2 = \{a,b,c,d\}$, $A_3 = \{b,d,a\}$ and $A_4 = \{a,b,h\}$.
   (a) $\bigcup_{i=1}^{4} A_i = \{a,b,c,d,e,f,g,h\}$ **(b)** $\bigcap_{i=1}^{4} A_i = \{a,b\}$
3. For each $n \in \mathbb{N}$, let $A_n = \{0,1,2,3,\ldots,n\}$.
   (a) $\bigcup_{i\in\mathbb{N}} A_i = \{0\} \cup \mathbb{N}$ **(b)** $\bigcap_{i\in\mathbb{N}} A_i = \{0,1\}$
5. **(a)** $\bigcup_{i\in\mathbb{N}} [i,i+1] = [1,\infty)$ **(b)** $\bigcap_{i\in\mathbb{N}} [i,i+1] = \emptyset$
7. **(a)** $\bigcup_{i\in\mathbb{N}} \mathbb{R}\times[i,i+1] = \{(x,y) : x,y \in \mathbb{R}, y \geq 1\}$ **(b)** $\bigcap_{i\in\mathbb{N}} \mathbb{R}\times[i,i+1] = \emptyset$
9. **(a)** $\bigcup_{X\in\mathscr{P}(\mathbb{N})} X = \mathbb{N}$ **(b)** $\bigcap_{X\in\mathscr{P}(\mathbb{N})} X = \emptyset$
11. Yes, this is always true.
13. The first is true, the second is false.

Chapter 2 Exercises

Section 2.1

Decide whether or not the following are statements. In the case of a statement, say if it is true or false.

1. Every real number is an even integer. (Statement, False)
3. If x and y are real numbers and $5x = 5y$, then $x = y$. (Statement, True)
5. Sets $\mathbb{Z}$ and $\mathbb{N}$ are infinite. (Statement, True)
7. The derivative of any polynomial of degree 5 is a polynomial of degree 6. (Statement, False)
9. $\cos(x) = -1$
   This is not a statement. It is an open sentence because whether it's true or false depends on the value of x.
11. The integer x is a multiple of 7.
    This is an open sentence, and not a statement.
13. Either x is a multiple of 7, or it is not.
    This is a statement, for the sentence is true no matter what x is.
15. In the beginning God created the heaven and the earth.
    This is a statement, for it is either definitely true or definitely false. There is some controversy over whether it's true or false, but no one claims that it is neither true nor false.

Section 2.2

Express each statement as one of the forms $P \wedge Q$, $P \vee Q$, or $\sim P$. Be sure to also state exactly what statements P and Q stand for.

1. The number 8 is both even and a power of 2.
$P \wedge Q$
P: 8 is even
Q: 8 is a power of 2
Note: Do not say "Q: a power of 2," because that is not a statement.

3. $x \neq y$ $\quad\quad$ $\sim(x = y)$ $\quad\quad$ (Also $\sim P$ where $P : x = y$.)

5. $y \geq x$ $\quad\quad$ $\sim(y < x)$ $\quad\quad$ (Also $\sim P$ where $P : y < x$.)

7. The number x equals zero, but the number y does not.
$P \wedge \sim Q$
$P : x = 0$
$Q : y = 0$

9. $x \in A - B$
$(x \in A) \wedge \sim(x \in B)$

11. $A \in \{X \in \mathscr{P}(\mathbb{N}) : |\overline{X}| < \infty\}$
$(A \subseteq \mathbb{N}) \wedge (|\overline{A}| < \infty)$.

13. Human beings want to be good, but not too good, and not all the time.
$P \wedge \sim Q \wedge \sim R$
P: Human beings want to be good.
Q: Human beings want to be too good.
R: Human beings want to be good all the time.

Section 2.3

Without changing their meanings, convert each of the following sentences into a sentence having the form "*If P, then Q.*"

1. A matrix is invertible provided that its determinant is not zero.
Answer: If a matrix has a determinant not equal to zero, then it is invertible.

3. For a function to be integrable, it is necessary that it is continuous.
Answer: If function is integrable, then it is continuous.

5. An integer is divisible by 8 only if it is divisible by 4.
Answer: If an integer is divisible by 8, then it is divisible by 4.

7. A series converges whenever it converges absolutely.
Answer: If a series converges absolutely, then it converges.

9. A function is integrable provided the function is continuous.
Answer: If a function is continuous, then that function is integrable.

11. You fail only if you stop writing.
Answer: If you fail, then you have stopped writing.

13. Whenever people agree with me I feel I must be wrong.
Answer: If people agree with me, then I feel I must be wrong.

Section 2.4

Without changing their meanings, convert each of the following sentences into a sentence having the form "*P if and only if Q*."

1. For a matrix to be invertible, it is necessary and sufficient that its determinant is not zero.
   Answer: A matrix is invertible if and only if its determinant is not zero.
3. If $xy = 0$ then $x = 0$ or $y = 0$, and conversely.
   Answer: $xy = 0$ if and only if $x = 0$ or $y = 0$
5. For an occurrence to become an adventure, it is necessary and sufficient for one to recount it.
   Answer: An occurrence becomes an adventure if and only if one recounts it.

Section 2.5

1. Write a truth table for $P \vee (Q \Rightarrow R)$

P	Q	R	$Q \Rightarrow R$	$P \vee (Q \Rightarrow R)$
T	T	T	T	**T**
T	T	F	F	**T**
T	F	T	T	**T**
T	F	F	T	**T**
F	T	T	T	**T**
F	T	F	F	**F**
F	F	T	T	**T**
F	F	F	T	**T**

3. Write a truth table for $\sim (P \Rightarrow Q)$

P	Q	$P \Rightarrow Q$	$\sim (P \Rightarrow Q)$
T	T	T	**F**
T	F	F	**T**
F	T	T	**F**
F	F	T	**F**

5. Write a truth table for $(P \wedge \sim P) \vee Q$

P	Q	$(P \wedge \sim P)$	$(P \wedge \sim P) \vee Q$
T	T	F	**T**
T	F	F	**F**
F	T	F	**T**
F	F	F	**F**

7. Write a truth table for $(P \wedge \sim P) \Rightarrow Q$

P	Q	$(P \wedge \sim P)$	$(P \wedge \sim P) \Rightarrow Q$
T	T	F	**T**
T	F	F	**T**
F	T	F	**T**
F	F	F	**T**

9. Write a truth table for $\sim (\sim P \vee \sim Q)$.

P	Q	$\sim P$	$\sim Q$	$\sim P \vee \sim Q$	$\sim (\sim P \vee \sim Q)$
T	T	F	F	F	**T**
T	F	F	T	T	**F**
F	T	T	F	T	**F**
F	F	T	T	T	**F**

11. Suppose P is false and that the statement $(R \Rightarrow S) \Leftrightarrow (P \wedge Q)$ is true. Find the truth values of R and S. (This can be done without a truth table.)

Answer: Since P is false, it follows that $(P \wedge Q)$ is false also. But then in order for $(R \Rightarrow S) \Leftrightarrow (P \wedge Q)$ to be true, it must be that $(R \Rightarrow S)$ is false. The only way for $(R \Rightarrow S)$ to be false is if $\boxed{R \text{ is true and } S \text{ is false.}}$

Section 2.6

A. Use truth tables to show that the following statements are logically equivalent.

1. $P \wedge (Q \vee R) = (P \wedge Q) \vee (P \wedge R)$

P	Q	R	$Q \vee R$	$P \wedge Q$	$P \wedge R$	$P \wedge (Q \vee R)$	$(P \wedge Q) \vee (P \wedge R)$
T	*T*	*T*	*T*	*T*	*T*	**T**	**T**
T	*T*	*F*	*T*	*T*	*F*	**T**	**T**
T	*F*	*T*	*T*	*F*	*T*	**T**	**T**
T	*F*	*F*	*F*	*F*	*F*	**F**	**F**
F	*T*	*T*	*T*	*F*	*F*	**F**	**F**
F	*T*	*F*	*T*	*F*	*F*	**F**	**F**
F	*F*	*T*	*T*	*F*	*F*	**F**	**F**
F	*F*	*F*	*F*	*F*	*F*	**F**	**F**

Thus since their columns agree, the two statements are logically equivalent.

3. $P \Rightarrow Q = (\sim P) \vee Q$

P	Q	$\sim P$	$(\sim P) \vee Q$	$P \Rightarrow Q$
T	*T*	*F*	**T**	**T**
T	*F*	*F*	**F**	**F**
F	*T*	*T*	**T**	**T**
F	*F*	*T*	**T**	**T**

Thus since their columns agree, the two statements are logically equivalent.

5. $\sim (P \vee Q \vee R) = (\sim P) \wedge (\sim Q) \wedge (\sim R)$

P	Q	R	$P \vee Q \vee R$	$\sim P$	$\sim Q$	$\sim R$	$\sim (P \vee Q \vee R)$	$(\sim P) \wedge (\sim Q) \wedge (\sim R)$
T	*T*	*T*	*T*	*F*	*F*	*F*	**F**	**F**
T	*T*	*F*	*T*	*F*	*F*	*T*	**F**	**F**
T	*F*	*T*	*T*	*F*	*T*	*F*	**F**	**F**
T	*F*	*F*	*T*	*F*	*T*	*T*	**F**	**F**
F	*T*	*T*	*T*	*T*	*F*	*F*	**F**	**F**
F	*T*	*F*	*T*	*T*	*F*	*T*	**F**	**F**
F	*F*	*T*	*T*	*T*	*T*	*F*	**F**	**F**
F	*F*	*F*	*F*	*T*	*T*	*T*	**T**	**T**

Thus since their columns agree, the two statements are logically equivalent.

7. $P \Rightarrow Q \;=\; (P \wedge \sim Q) \Rightarrow (Q \wedge \sim Q)$

P	Q	$\sim Q$	$P \wedge \sim Q$	$Q \wedge \sim Q$	$(P \wedge \sim Q) \Rightarrow (Q \wedge \sim Q)$	$P \Rightarrow Q$
T	*T*	*F*	*F*	*F*	**T**	**T**
T	*F*	*T*	*T*	*F*	**F**	**F**
F	*T*	*F*	*F*	*F*	**T**	**T**
F	*F*	*T*	*F*	*F*	**T**	**T**

Thus since their columns agree, the two statements are logically equivalent.

B. Decide whether or not the following pairs of statements are logically equivalent.

9. By DeMorgan's law, we have $\sim(\sim P \vee \sim Q) = \sim\sim P \wedge \sim\sim Q = P \wedge Q$. Thus the two statements are logically equivalent.

11. $(\sim P) \wedge (P \Rightarrow Q)$ and $\sim(Q \Rightarrow P)$

P	Q	$\sim P$	$P \Rightarrow Q$	$Q \Rightarrow P$	$(\sim P) \wedge (P \Rightarrow Q)$	$\sim(Q \Rightarrow P)$
T	*T*	*F*	*T*	*T*	**F**	**F**
T	*F*	*F*	*F*	*T*	**F**	**F**
F	*T*	*T*	*T*	*F*	**T**	**T**
F	*F*	*T*	*T*	*T*	**T**	**F**

The columns for the two statements do not quite agree, thus the two statements are **not logically equivalent.**

Section 2.7

Write the following as English sentences. Say whether the statements are true or false.

1. $\forall x \in \mathbb{R}, x^2 > 0$
Answer: For every real number x, $x^2 > 0$.
Also: For every real number x, it follows that $x^2 > 0$.
Also: The square of any real number is positive. (etc.)
This statement is FALSE. Reason: 0 is a real number, but it's not true that $0^2 > 0$.

3. $\exists a \in \mathbb{R}, \forall x \in \mathbb{R}, ax = x$.
Answer: There exists a real number a for which $ax = x$ for every real number x.
This statement is TRUE. Reason: Consider $a = 1$.

5. $\forall n \in \mathbb{N}, \exists X \in \mathscr{P}(\mathbb{N}), |X| < n$
Answer: For every natural number n, there is a subset X of $\mathbb{N}$ with $|X| < n$.
This statement is TRUE. Reason: Suppose $n \in \mathbb{N}$. Let $X = \emptyset$. Then $|X| = 0 < n$.

7. $\forall X \subseteq \mathbb{N}, \exists n \in \mathbb{Z}, |X| = n$
Answer: For any subset X of $\mathbb{N}$, there exists an integer n for which $|X| = n$. This statement is FALSE. For example, the set $X = \{2,4,6,8,\ldots\}$ of all even natural numbers is infinite, so there does not exist any integer n for which $|X| = n$.

9. $\forall n \in \mathbb{Z}, \exists m \in \mathbb{Z}, m = n+5$
Answer: For every integer n there is another integer m such that $m = n+5$. This statement is TRUE.

Section 2.9

Translate each of the following sentences into symbolic logic.

1. If f is a polynomial and its degree is greater than 2, then f' is not constant.
Translation: $(P \wedge Q) \Rightarrow R$, where
$P: f$ is a polynomial,
$Q: f$ has degree greater than 2,
$R: f'$ is not constant.

3. If x is prime then $\sqrt{x}$ is not a rational number.
Translation: $P \Rightarrow \sim Q$, where
$P: x$ is prime,
$Q: \sqrt{x}$ is a rational number.

5. For every positive number ε, there is a positive number δ for which $|x-a| < \delta$ implies $|f(x) - f(a)| < \varepsilon$.
Translation: $\forall \varepsilon \in \mathbb{R}, \varepsilon > 0, \exists \delta \in \mathbb{R}, \delta > 0, (|x-a| < \delta) \Rightarrow (|f(x) - f(a)| < \varepsilon)$

7. There exists a real number a for which $a + x = x$ for every real number x.
Translation: $\exists a \in \mathbb{R}, \forall x \in \mathbb{R}, a + x = x$

9. If x is a rational number and $x \neq 0$, then $\tan(x)$ is not a rational number.
Translation: $((x \in \mathbb{Q}) \wedge (x \neq 0)) \Rightarrow (\tan(x) \notin \mathbb{Q})$

11. There is a Providence that protects idiots, drunkards, children and the United States of America.

One translation is as follows. Let R be union of the set of idiots, the set of drunkards, the set of children, and the set consisting of the USA. Let P be the open sentence $P(x)$: *x is a Providence*. Let S be the open sentence $S(x,y)$: *x protects y*. Then the translation is $\exists x, \forall y \in R, P(x) \wedge S(x,y)$.

(Notice that, although this is mathematically correct, some humor has been lost in the translation.)

13. Everything is funny as long as it is happening to somebody else.
Translation: $\forall x, (\sim M(x) \wedge S(x)) \Rightarrow F(x)$,
where $M(x)$: *x is happening to me*, $S(x)$: *x is happening to someone*, and $F(x)$: *x is funny*.

Section 2.10

Negate the following sentences.

1. The number x is positive, but the number y is not positive.
The "but" can be interpreted as "and." Using DeMorgan's law, the negation is:
The number x is not positive or the number y is positive.

3. For every prime number p there, is another prime number q with $q > p$.
Negation: *There is a prime number p such that for every prime number q, $q \le p$.*
Also: *There exists a prime number p for which $q \le p$ for every prime number q.* (etc.)

5. For every positive number ε there is a positive number M for which $|f(x)-b| < \varepsilon$ whenever $x > M$.
To negate this, it may be helpful to first write it in symbolic form. The statement is $\forall \varepsilon \in (0,\infty), \exists M \in (0,\infty), (x > M) \Rightarrow (|f(x)-b| < \varepsilon)$.
Working out the negation, we have

$$\begin{aligned} \sim \big(\forall \varepsilon \in (0,\infty), \exists M \in (0,\infty), (x > M) \Rightarrow (|f(x)-b| < \varepsilon)\big) &= \\ \exists \varepsilon \in (0,\infty), \sim \big(\exists M \in (0,\infty), (x > M) \Rightarrow (|f(x)-b| < \varepsilon)\big) &= \\ \exists \varepsilon \in (0,\infty), \forall M \in (0,\infty), \sim \big((x > M) \Rightarrow (|f(x)-b| < \varepsilon)\big). \end{aligned}$$

Finally, using the idea from Example 2.14, we can negate the conditional statement that appears here to get

$$\exists \varepsilon \in (0,\infty), \forall M \in (0,\infty), \exists x, (x > M) \wedge \sim (|f(x)-b| < \varepsilon)\big).$$

Negation: *There exists a positive number ε with the property that for every positive number M, there is a number x for which $x > M$ and $|f(x)-b| \ge \varepsilon$.*

7. I don't eat anything that has a face.
Negation: *I will eat some things that have a face.*
(Note. If your answer was *"I will eat anything that has a face."* then that is wrong, both morally and mathematically.)

9. If $\sin(x) < 0$, then it is not the case that $0 \le x \le \pi$.
Negation: *There exists a number x for which $\sin(x) < 0$ and $0 \le x \le \pi$.*

11. You can fool all of the people all of the time.

There are several ways to negate this, including:
There is a person that you can't fool all the time. or
There is a person x and a time y for which x is not fooled at time y.
(But Abraham Lincoln said it better.)

Chapter 3 Exercises

Section 3.1

1. Consider lists made from the letters *T, H, E, O, R, Y*, with repetition allowed.

(a) How many length-4 lists are there? Answer: $6 \cdot 6 \cdot 6 \cdot 6 = \mathbf{1296}$.

(b) How many length-4 lists are there that begin with T?
Answer: $1 \cdot 6 \cdot 6 \cdot 6 = \mathbf{216}$.

(c) How many length-4 lists are there that do not begin with T?
Answer: $5 \cdot 6 \cdot 6 \cdot 6 = \mathbf{1080}$.

3. How many ways can you make a list of length 3 from symbols A,B,C,D,E,F if...

(a) ... repetition is allowed. Answer: $6 \cdot 6 \cdot 6 = \mathbf{216}$.

(b) ... repetition is not allowed. Answer: $6 \cdot 5 \cdot 4 = \mathbf{120}$.

(c) ... repetition is not allowed and the list must contain the letter A.
Answer: $5 \cdot 4 + 5 \cdot 4 + 5 \cdot 4 = \mathbf{60}$.

(d) ... repetition is allowed and the list must contain the letter A.
Answer: $6 \cdot 6 \cdot 6 - 5 \cdot 5 \cdot 5 = \mathbf{91}$.

(Note: See Example 3.2 if a more detailed explanation is required.)

5. Five cards are dealt off of a standard 52-card deck and lined up in a row. How many such line-ups are there in which all five cards are of the same color? (i.e., all black or all red.)
There are $26 \cdot 25 \cdot 24 \cdot 23 \cdot 22 = 7,893,600$ possible black-card line-ups and $26 \cdot 25 \cdot 24 \cdot 23 \cdot 22 = 7,893,600$ possible red-card line-ups, so the answer is $7,893,600 + 7,893,600 = \mathbf{15,787,200}$.

7. This problems involves 8-digit binary strings such as 10011011 or 00001010. (i.e., 8-digit numbers composed of 0's and 1's.)

(a) How many such strings are there? Answer: $2 \cdot 2 \cdot 2 \cdot 2 \cdot 2 \cdot 2 \cdot 2 \cdot 2 = \mathbf{256}$.

(b) How many such strings end in 0? Answer: $2 \cdot 2 \cdot 2 \cdot 2 \cdot 2 \cdot 2 \cdot 2 \cdot 1 = \mathbf{128}$.

(c) How many such strings have the property that their second and fourth digits are 1's? Answer: $2 \cdot 1 \cdot 2 \cdot 1 \cdot 2 \cdot 2 \cdot 2 \cdot 2 = \mathbf{64}$.

(d) How many such strings are such that their second **or** fourth digits are 1's? Answer: These strings can be divided into three types. Type 1 consists of those strings of form $*1*0****$, Type 2 consist of strings of form $*0*1****$, and Type 3 consists of those of form $*1*1****$. By the multiplication principle there are $2^6 = 64$ strings of each type, so **there are $3 \cdot 64 = 192$ 8-digit binary strings whose second or fourth digits are 1's.**

9. This problem concerns 4-letter codes that can be made from the letters of the English Alphabet.

(a) How many such codes can be made? Answer: $26 \cdot 26 \cdot 26 \cdot 26 = \mathbf{456976}$

(b) How many such codes have no two consecutive letters the same?
We use the multiplication principle. There are 26 choices for the first letter. The second letter can't be the same as the first letter, so there are only 25 choices for it. The third letter can't be the same as the second letter, so there are only 25 choices for it. The fourth letter can't be the same as the third letter, so there are only 25 choices for it. **Thus there are $26 \cdot 25 \cdot 25 \cdot 25 = 406{,}250$ codes with no two consecutive letters the same.**

11. This problem concerns lists of length 6 made from the letters A,B,C,D,E,F,G,H. How many such lists are possible if repetition is not allowed and the list contains two consecutive vowels?
Answer: There are just two vowels A and E to choose from. The lists we want to make can be divided into five types. They have one of the forms $VV****$, or $*VV***$, or $**VV**$, or $***VV*$, or $****VV$, where V indicates a vowel and $*$ indicates a consonant. By the multiplication principle, there are $2\cdot1\cdot6\cdot5\cdot4\cdot3 = 720$ lists of form $VV****$. In fact, that for the same reason there are 720 lists of each form. Thus the answer to the question is $5\cdot720 =$ **3600**

Section 3.2

1. Answer $n =$ **14**.

3. Answer: $5! =$ **120**.

5. $\frac{120!}{118!} = \frac{120\cdot119\cdot118!}{118!} = 120\cdot119 =$ **14,280**.

7. Answer: $5!4! = 2880$.

9. The case $x = 1$ is straightforward. For $x = 2,3$ and 4, use integration by parts. For $x = \pi$, you are on your own.

Section 3.3

1. Suppose a set A has 37 elements. How many subsets of A have 10 elements? How many subsets have 30 elements? How many have 0 elements?
Answers: $\binom{37}{10} =$ **348,330,136**; $\binom{37}{30} =$ **10,295,472**; $\binom{37}{0} =$ **1**.

3. A set X has exactly 56 subsets with 3 elements. What is the cardinality of X?
The answer will be n, where $\binom{n}{3} = 56$. After some trial and error, you will discover $\binom{8}{3} = 56$, so $|X| = 8$.

5. How many 16-digit binary strings contain exactly seven 1's?
Answer: Make such a string as follows. Start with a list of 16 blank spots. Choose 7 of the blank spots for the 1's and put 0's in the other spots. There are $\binom{16}{7} =$ **114,40** ways to do this.

7. $|\{X \in \mathscr{P}(\{0,1,2,3,4,5,6,7,8,9\}) : |X| < 4\}| = \binom{10}{0} + \binom{10}{1} + \binom{10}{2} + \binom{10}{3} = 1+10+45+120 =$ **176**.

9. This problem concerns lists of length six made from the letters A,B,C,D,E,F, without repetition. How many such lists have the property that the D occurs before the A?
Answer: Make such a list as follows. Begin with six blank spaces and select two of these spaces. Put the D in the first selected space and the A in the second. There are $\binom{6}{2} = 15$ ways of doing this. For each of these 15 choices there are $4! = 24$ ways of filling in the remaining spaces. Thus the answer to the question is $15 \times 24 =$ **360** such lists.

11. How many 10-digit integers contain no 0's and exactly three 6's?
Answer: Make such a number as follows: Start with 10 blank spaces and choose three of these spaces for the 6's. There are $\binom{10}{3} = 120$ ways of doing this. For each of these 120 choices we can fill in the remaining seven blanks with choices from the digits 1,2,3,4,5,7,8,9, and there are 8^7 to do this. Thus the answer to the question is $\binom{10}{3} \cdot 8^7 = \mathbf{251{,}658{,}240}$.

13. Assume $n,k \in \mathbb{Z}$ with $0 \le k \le n$. Then $\binom{n}{k} = \frac{n!}{(n-k)!k!} = \frac{n!}{k!(n-k)!} = \frac{n!}{(n-(n-k))!(n-k)!} = \binom{n}{n-k}$.

Section 3.4

1. Write out Row 11 of Pascal's triangle.
Answer: 1 11 55 165 330 462 462 330 165 55 11 1

3. Use the binomial theorem to find the coefficient of x^8 in $(x+2)^{13}$.
Answer: According to the binomial theorem, the coefficient of x^8y^5 in $(x+y)^{13}$ is $\binom{13}{8}x^8y^5 = 1287x^8y^5$. Now plug in $y=2$ to get the final answer of $41184x^8$.

5. Use the binomial theorem to show $\sum_{k=0}^{n}\binom{n}{k} = 2^n$. Hint: Observe that $2^n = (1+1)^n$. Now use the binomial theorem to work out $(x+y)^n$ and plug in $x=1$ and $y=1$.

7. Use the binomial theorem to show $\sum_{k=0}^{n} 3^k\binom{n}{k} = 4^n$.
Hint: Observe that $4^n = (1+3)^n$. Now look at the hint for the previous problem.

9. Use the binomial theorem to show $\binom{n}{0}-\binom{n}{1}+\binom{n}{2}-\binom{n}{3}+\binom{n}{4}-\binom{n}{5}+\ldots\pm\binom{n}{n} = 0$.
Hint: Observe that $0 = 0^n = (1+(-1))^n$. Now use the binomial theorem.

11. Use the binomial theorem to show $9^n = \sum_{k=0}^{n}(-1)^k\binom{n}{k}10^{n-k}$.
Hint: Observe that $9^n = (10+(-1))^n$. Now use the binomial theorem.

13. Assume $n \ge 3$. Then $\binom{n}{3} = \binom{n-1}{3}+\binom{n-1}{2} = \binom{n-2}{3}+\binom{n-2}{2}+\binom{n-1}{2} = \cdots = \binom{2}{2}+\binom{3}{2}+\cdots+\binom{n-1}{2}$.

Section 3.5

1. At a certain university 523 of the seniors are history majors or math majors (or both). There are 100 senior math majors, and 33 seniors are majoring in both history and math. How many seniors are majoring in history?
Answer: Let A be the set of senior math majors and B be the set of senior history majors. From $|A\cup B| = |A|+|B|-|A\cap B|$ we get $523 = 100+|B|-33$, so $|B| = 523+33-100 = 456$. **There are 456 history majors.**

3. How many 4-digit positive integers are there that are even or contain no 0's?
Answer: Let A be the set of 4-digit even positive integers, and let B be the set of 4-digit positive integers that contain no 0's. We seek $|A\cup B|$. By the multiplication principle $|A| = 9\cdot 10\cdot 10\cdot 5 = 4500$. (Note the first digit cannot be 0 and the last digit must be even.) Also $|B| = 9\cdot 9\cdot 9\cdot 9 = 6561$. Further, $A\cap B$ consists of all even 4-digit integers that have no 0's. It follows that $|A\cap B| = 9\cdot 9\cdot 9\cdot 4 = 2916$. Then the answer to our question is $|A\cup B| = |A|+|B|-|A\cap B| = 4500+6561-2916 =$ **8145**.

5. How many 7-digit binary strings begin in 1 or end in 1 or have exactly four 1's?
Answer: Let A be the set of such strings that begin in 1. Let B be the set of such strings that end in 1. Let C be the set of such strings that have exactly four 1's. Then the answer to our question is $|A \cup B \cup C|$. Using Equation (3.4) to compute this number, we have $|A \cup B \cup C| = |A|+|B|+|C|-|A \cap B|-|A \cap C|-|B \cap C|+|A \cap B \cap C| = 2^6 + 2^6 + \binom{7}{4} - 2^5 - \binom{6}{3} - \binom{6}{3} + \binom{5}{2} = 64+64+35-32-20-20+10 = \mathbf{101}$.

7. This problem concerns 4-card hands dealt off of a standard 52-card deck. How many 4-card hands are there for which all four cards are of the same suit or all four cards are red?
Answer: Let A be the set of 4-card hands for which all four cards are of the same suit. Let B be the set of 4-card hands for which all four cards are red. Then $A \cap B$ is the set of 4-card hands for which the four cards are either all hearts or all diamonds. The answer to our question is $|A \cup B| = |A|+|B|-|A \cap B| = 4\binom{13}{4} + \binom{26}{4} - 2\binom{13}{4} = 2\binom{13}{4} + \binom{26}{4} = 1430 + 14950 = \mathbf{16380}$.

9. A 4-letter list is made from the letters *L,I,S,T,E,D* according to the following rule: Repetition is allowed, and the first two letters on the list are vowels or the list ends in *D*.
Answer: Let A be the set of such lists for which the first two letters are vowels, so $|A| = 2 \cdot 2 \cdot 6 \cdot 6 = 144$. Let B be the set of such lists that end in D, so $|B| = 6 \cdot 6 \cdot 6 \cdot 1 = 216$. Then $A \cap B$ is the set of such lists for which the first two entries are vowels and the list ends in D. Thus $|A \cap B| = 2 \cdot 2 \cdot 6 \cdot 1 = 24$. The answer to our question is $|A \cup B| = |A| + |B| - |A \cap B| = 144 + 216 - 24 = \mathbf{336}$.

Chapter 4 Exercises

1. If x is an even integer, then x^2 is even.

Proof. Suppose x is even. Thus $x = 2a$ for some $a \in \mathbb{Z}$.
Consequently $x^2 = (2a)^2 = 4a^2 = 2(2a^2)$.
Therefore $x^2 = 2b$, where b is the integer $2a^2$.
Thus x^2 is even by definition of an even number. ■

3. If a is an odd integer, then $a^2 + 3a + 5$ is odd.

Proof. Suppose a is odd.
Thus $a = 2c + 1$ for some integer c, by definition of an odd number.
Then $a^2 + 3a + 5 = (2c+1)^2 + 3(2c+1) + 5 = 4c^2 + 4c + 1 + 6c + 3 + 5 = 4c^2 + 10c + 9$
$= 4c^2 + 10c + 8 + 1 = 2(2c^2 + 5c + 4) + 1$.
This shows $a^2 + 3a + 5 = 2b + 1$, where $b = 2c^2 + 5c + 4 \in \mathbb{Z}$.
Therefore $a^2 + 3a + 5$ is odd. ■

5. Suppose $x, y \in \mathbb{Z}$. If x is even, then xy is even.

Proof. Suppose $x, y \in \mathbb{Z}$ and x is even.
Then $x = 2a$ for some integer a, by definition of an even number.
Thus $xy = (2a)(y) = 2(ay)$.
Therefore $xy = 2b$ where b is the integer ay, so xy is even. ■

7. Suppose $a,b \in \mathbb{Z}$. If $a \mid b$, then $a^2 \mid b^2$.

Proof. Suppose $a \mid b$.
By definition of divisibility, this means $b = ac$ for some integer c.
Squaring both sides of this equation produces $b^2 = a^2c^2$.
Then $b^2 = a^2d$, where $d = c^2 \in \mathbb{Z}$.
By definition of divisibility, this means $a^2 \mid b^2$. ■

9. Suppose a is an integer. If $7 \mid 4a$, then $7 \mid a$.

Proof. Suppose $7 \mid 4a$.
By definition of divisibility, this means $4a = 7c$ for some integer c.
Since $4a = 2(2a)$ it follows that $4a$ is even, and since $4a = 7c$, we know $7c$ is even.
But then c can't be odd, because that would make $7c$ odd, not even.
Thus c is even, so $c = 2d$ for some integer d.
Now go back to the equation $4a = 7c$ and plug in $c = 2d$. We get $4a = 14d$.
Dividing both sides by 2 gives $2a = 7d$.
Now, since $2a = 7d$, it follows that $7d$ is even, and thus d cannot be odd.
Then d is even, so $d = 2e$ for some integer e.
Plugging $d = 2e$ back into $2a = 7d$ gives $2a = 14e$.
Dividing both sides of $2a = 14e$ by 2 produces $a = 7e$.
Finally, the equation $a = 7e$ means that $7 \mid a$, by definition of divisibility. ■

11. Suppose $a,b,c,d \in \mathbb{Z}$. If $a \mid b$ and $c \mid d$, then $ac \mid bd$.

Proof. Suppose $a \mid b$ and $c \mid d$.
As $a \mid b$, the definition of divisibility means there is an integer x for which $b = ax$.
As $c \mid d$, the definition of divisibility means there is an integer y for which $d = cy$.
Since $b = ax$, we can multiply one side of $d = cy$ by b and the other by ax.
This gives $bd = axcy$, or $bd = (ac)(xy)$.
Since $xy \in \mathbb{Z}$, the definition of divisibility applied to $bd = (ac)(xy)$ gives $ac \mid bd$. ■

13. Suppose $x,y \in \mathbb{R}$. If $x^2 + 5y = y^2 + 5x$, then $x = y$ or $x + y = 5$.

Proof. Suppose $x^2 + 5y = y^2 + 5x$.
Then $x^2 - y^2 = 5x - 5y$, and factoring gives $(x-y)(x+y) = 5(x-y)$.
Now consider two cases.
Case 1. If $x - y \neq 0$ we can divide both sides of $(x-y)(x+y) = 5(x-y)$ by the non-zero quantity $x - y$ to get $x + y = 5$.
Case 2. If $x - y = 0$, then $x = y$. (By adding y to both sides.)
Thus $x = y$ or $x + y = 5$. ■

15. If $n \in \mathbb{Z}$, then n^2+3n+4 is even.

Proof. Suppose $n \in \mathbb{Z}$. We consider two cases.
Case 1. Suppose n is even. Then $n = 2a$ for some $a \in \mathbb{Z}$.
Therefore $n^2+3n+4 = (2a)^2+3(2a)+4 = 4a^2+6a+4 = 2(2a^2+3a+2)$.
So $n^2+3n+4 = 2b$ where $b = 2a^2+3a+2 \in \mathbb{Z}$, so n^2+3n+4 is even.
Case 2. Suppose n is odd. Then $n = 2a+1$ for some $a \in \mathbb{Z}$.
Therefore $n^2+3n+4 = (2a+1)^2+3(2a+1)+4 = 4a^2+4a+1+6a+3+4 = 4a^2+10a+8$ $= 2(2a^2+5a+4)$. So $n^2+3n+4 = 2b$ where $b = 2a^2+5a+4 \in \mathbb{Z}$, so n^2+3n+4 is even.

In either case n^2+3n+4 is even. ■

17. If two integers have opposite parity, then their product is even.

Proof. Suppose a and b are two integers with opposite parity. Thus one is even and the other is odd. Without loss of generality, suppose a is even and b is odd. Therefore there are integers c and d for which $a = 2c$ and $b = 2d+1$. Then the product of a and b is $ab = 2c(2d+1) = 2(2cd+c)$. Therefore $ab = 2k$ where $k = 2cd+c \in \mathbb{Z}$. Therefore the product ab is even. ■

19. Suppose $a,b,c \in \mathbb{Z}$. If $a^2 \mid b$ and $b^3 \mid c$ then $a^6 \mid c$.

Proof. Since $a^2 \mid b$ we have $b = ka^2$ for some $k \in \mathbb{Z}$. Since $b^3 \mid c$ we have $c = hb^3$ for some $h \in \mathbb{Z}$. Thus $c = h(ka^2)^3 = hk^3a^6$. Hence $a^6 \mid c$. ■

21. If p is prime and $0 < k < p$ then $p \mid \binom{p}{k}$.

Proof. From the formula $\binom{p}{k} = \frac{p!}{(p-k)!k!}$, we get $p! = \binom{p}{k}(p-k)!k!$. Now, since the prime number p is a factor of $p!$ on the left, it must also be a factor of $\binom{p}{k}(p-k)!k!$ on the right. Thus the prime number p appears in the prime factorization of $\binom{p}{k}(p-k)!k!$.
Now, $k!$ is a product of numbers smaller than p, so its prime factorization contains no p's. Similarly the prime factorization of $(p-k)!$ contains no p's. But we noted that the prime factorization of $\binom{p}{k}(p-k)!k!$ must contain a p, so it follows that the prime factorization of $\binom{p}{k}$ contains a p. Thus $\binom{p}{k}$ is a multiple of p, so p divides $\binom{p}{k}$. ■

23. If $n \in \mathbb{N}$ then $\binom{2n}{n}$ is even.

Proof. By definition, $\binom{2n}{n}$ is the number of n-element subsets of a set A with $2n$ elements. For each subset $X \subseteq A$ with $|X| = n$, the complement $\overline{X}$ is a different set, but it also has $2n - n = n$ elements. Imagine listing out all the n-elements subset of a set A. It could be done in such a way that the list has form

$$X_1, \overline{X_1},\ X_2, \overline{X_2},\ X_3, \overline{X_3},\ X_4, \overline{X_4},\ X_5, \overline{X_5} \ldots$$

This list has an even number of items, for they are grouped in pairs. Thus $\binom{2n}{n}$ is even. ■

25. If $a,b,c \in \mathbb{N}$ and $c \le b \le a$ then $\binom{a}{b}\binom{b}{c} = \binom{a}{b-c}\binom{a-b+c}{c}$.

Proof. Assume $a,b,c \in \mathbb{N}$ with $c \le b \le a$. Then we have $\binom{a}{b}\binom{b}{c} = \frac{a!}{(a-b)!b!}\frac{b!}{(b-c)!c!} = \frac{a!}{(a-b+c)!(a-b)!}\frac{(a-b+c)!}{(b-c)!c!} = \frac{a!}{(b-c)!(a-b+c)!}\frac{(a-b+c)!}{(a-b)!c!} = \binom{a}{b-c}\binom{a-b+c}{c}$. ∎

27. Suppose $a,b \in \mathbb{N}$. If $\gcd(a,b) > 1$, then $b \mid a$ or b is not prime.

Proof. Suppose $\gcd(a,b) > 1$. Let $c = \gcd(a,b) > 1$. Then since c is a divisor of both a and b, we have $a = cx$ and $b = cy$ for integers x and y. We divide into two cases according to whether or not b is prime.
Case I. Suppose b is prime. Then the above equation $b = cy$ with $c > 1$ forces $c = b$ and $y = 1$. Then $a = cx$ becomes $a = bx$, which means $b \mid a$. We conclude that the statement "$b \mid a$ *or* b *is not prime,*" is true.
Case II. Suppose b is not prime. Then the statement "$b \mid a$ *or* b *is not prime,*" is automatically true. ∎

Chapter 5 Exercises

1. Proposition Suppose $n \in \mathbb{Z}$. If n^2 is even, then n is even.

Proof. (Contrapositive) Suppose n is not even. Then n is odd, so $n = 2a+1$ for some integer a, by definition of an odd number. Thus $n^2 = (2a+1)^2 = 4a^2+4a+1 = 2(2a^2+2a)+1$. Consequently $n^2 = 2b+1$, where b is the integer $2a^2+2a$, so n^2 is odd. Therefore n^2 is not even. ∎

3. Proposition Suppose $a,b \in \mathbb{Z}$. If $a^2(b^2-2b)$ is odd, then a and b are odd.

Proof. (Contrapositive) Suppose it is not the case that a and b are odd. Then, by DeMorgan's law, at least one of a and b is even. Let us look at these cases separately.
Case 1. Suppose a is even. Then $a = 2c$ for some integer c. Thus $a^2(b^2-2b) = (2c)^2(b^2-2b) = 2(2c^2(b^2-2b))$, which is even.
Case 2. Suppose b is even. Then $b = 2c$ for some integer c. Thus $a^2(b^2-2b) = a^2((2c)^2-2(2c)) = 2(a^2(2c^2-2c))$, which is even.
(A third case involving a and b both even is unnecessary, for either of the two cases above cover this case.) Thus in either case $a^2(b^2-2b)$ is even, so it is not odd. ∎

5. Proposition Suppose $x \in \mathbb{R}$. If $x^2+5x < 0$ then $x < 0$.

Proof. (Contrapositive) Suppose it is not the case that $x < 0$, so $x \ge 0$. Then neither x^2 nor $5x$ is negative, so $x^2+5x \ge 0$. Thus it is not true that $x^2+5x < 0$. ∎

7. Proposition Suppose $a, b \in \mathbb{Z}$. If both ab and $a+b$ are even, then both a and b are even.

Proof. (Contrapositive) Suppose it is not the case that both a and b are even. Then at least one of them is odd. There are three cases to consider.
Case 1. Suppose a is even and b is odd. Then there are integers c and d for which $a = 2c$ and $b = 2d+1$. Then $ab = 2c(2d+1)$, which is even; and $a+b = 2c+2d+1 = 2(c+d)+1$, which is odd. Thus it is not the case that both ab and $a+b$ are even.
Case 2. Suppose a is odd and b is even. Then there are integers c and d for which $a = 2c+1$ and $b = 2d$. Then $ab = (2c+1)(2d) = 2(d(2c+1))$, which is even; and $a+b = 2c+1+2d = 2(c+d)+1$, which is odd. Thus it is not the case that both ab and $a+b$ are even.
Case 3. Suppose a is odd and b is odd. Then there are integers c and d for which $a = 2c+1$ and $b = 2d+1$. Then $ab = (2c+1)(2d+1) = 4cd+2c+2d+1 = 2(2cd+c+d)+1$, which is odd; and $a+b = 2c+1+2d+1 = 2(c+d+1)$, which is even. Thus it is not the case that both ab and $a+b$ are even.
These cases show that it is not the case that ab and $a+b$ are both even. (Note that unlike Exercise 3 above, we really did need all three cases here, for each case involved specific parities for **both** a and b.) ■

9. Proposition Suppose $n \in \mathbb{Z}$. If $3 \nmid n^2$, then $3 \nmid n$.

Proof. (Contrapositive) Suppose it is not the case that $3 \nmid n$, so $3 \mid n$. This means that $n = 3a$ for some integer a. Consequently $n^2 = 9a^2$, from which we get $n^2 = 3(3a^2)$. This shows that there in an integer $b = 3a^2$ for which $n^2 = 3b$, which means $3 \mid n^2$. Therefore it is not the case that $3 \nmid n^2$. ■

11. Proposition Suppose $x, y \in \mathbb{Z}$. If $x^2(y+3)$ is even, then x is even or y is odd.

Proof. (Contrapositive) Suppose it is not the case that x is even or y is odd. Using DeMorgan's law, this means x is not even and y is not odd, which is to say x is odd and y is even. Thus there are integers a and b for which $x = 2a+1$ and $y = 2b$. Consequently $x^2(y+3) = (2a+1)^2(2b+3) = (4a^2+4a+1)(2b+3) = 8a^2b+8ab+2b+12a^2+12a+3 = 8a^2b+8ab+2b+12a^2+12a+2+1 = 2(4a^2b+4ab+b+6a^2+6a+1)+1$. This shows $x^2(y+3) = 2c+1$ for $c = 4a^2b+4ab+b+6a^2+6a+1 \in \mathbb{Z}$. Consequently, $x^2(y+3)$ is not even. ■

13. Proposition Suppose $x \in \mathbb{R}$. If $x^5+7x^3+5x \geq x^4+x^2+8$, then $x \geq 0$.

Proof. (Contrapositive) Suppose it is not true that $x \geq 0$. Then $x < 0$, that is x is negative. Consequently, the expressions x^5, $7x^3$ and $5x$ are all negative (note the odd powers) so $x^5+7x^3+5x < 0$. Similarly the terms x^4, x^2, and 8 are all positive (note the even powers), so $0 < x^4+x^2+8$. From this we get $x^5+7x^3+5x < x^4+x^2+8$, so it is not true that $x^5+7x^3+5x \geq x^4+x^2+8$. ■

15. Proposition Suppose $x \in \mathbb{Z}$. If x^3-1 is even, then x is odd.

Proof. (Contrapositive) Suppose x is not odd. Thus x is even, so $x=2a$ for some integer a. Then $x^3-1=(2a)^3-1=8a^3-1=8a^3-2+1=2(4a^3-1)+1$. Therefore $x^3-1=2b+1$ where $b=4a^3-1\in\mathbb{Z}$, so x^3-1 is odd. Thus x^3-1 is not even. ■

17. Proposition If n is odd, then $8 \mid (n^2-1)$.

Proof. (Direct) Suppose n is odd, so $n=2a+1$ for some integer a. Then $n^2-1=(2a+1)^2-1=4a^2+4a=4(a^2+a)=4a(a+1)$. So far we have $n^2-1=4a(a+1)$, but we want a factor of 8, not 4. But notice that one of a or $a+1$ must be even, so $a(a+1)$ is even and hence $a(a+1)=2c$ for some integer c. Now we have $n^2-1=4a(a+1)=4(2c)=8c$. But $n^2-1=8c$ means $8 \mid (n^2-1)$. ■

19. Proposition Let $a,b\in\mathbb{Z}$ and $n\in\mathbb{N}$. If $a\equiv b \pmod n$ and $a\equiv c \pmod n$, then $c\equiv b \pmod n$.

Proof. (Direct) Suppose $a\equiv b \pmod n$ and $a\equiv c \pmod n$.
This means $n \mid (a-b)$ and $n \mid (a-c)$.
Thus there are integers d and e for which $a-b=nd$ and $a-c=ne$.
Subtracting the second equation from the first gives $c-b=nd-ne$.
Thus $c-b=n(d-e)$, so $n \mid (c-b)$ by definition of divisibility.
Therefore $c\equiv b \pmod n$ by definition of congruence modulo n. ■

21. Proposition Let $a,b\in\mathbb{Z}$ and $n\in\mathbb{N}$. If $a\equiv b \pmod n$, then $a^3\equiv b^3 \pmod n$.

Proof. (Direct) Suppose $a\equiv b \pmod n$. This means $n \mid (a-b)$, so there is an integer c for which $a-b=nc$. Then:

$$\begin{aligned}
a-b &= nc\\
(a-b)(a^2+ab+b^2) &= nc(a^2+ab+b^2)\\
a^3+a^2b+ab^2-ba^2-ab^2-b^3 &= nc(a^2+ab+b^2)\\
a^3-b^3 &= nc(a^2+ab+b^2).
\end{aligned}$$

Since $a^2+ab+b^2\in\mathbb{Z}$, the equation $a^3-b^3=nc(a^2+ab+b^2)$ implies $n \mid (a^3-b^3)$, and therefore $a^3\equiv b^3 \pmod n$. ■

23. Proposition Let $a,b,c\in\mathbb{Z}$ and $n\in\mathbb{N}$. If $a\equiv b \pmod n$, then $ca\equiv cb \pmod n$.

Proof. (Direct) Suppose $a\equiv b \pmod n$. This means $n \mid (a-b)$, so there is an integer d for which $a-b=nd$. Multiply both sides of this by c to get $ac-bc=ndc$. Consequently, there is an integer $e=dc$ for which $ac-bc=ne$, so $n \mid (ac-bc)$ and consequently $ac\equiv bc \pmod n$. ■

25. If $n\in\mathbb{N}$ and 2^n-1 is prime, then n is prime.

Proof. Assume n is not prime. Write $n=ab$ for some $a,b>1$. Then $2^n-1=2^{ab}-1=\left(2^b-1\right)\left(2^{ab-b}+2^{ab-2b}+2^{ab-3b}+\cdots+2^{ab-ab}\right)$. Hence 2^n-1 is composite. ■

27. If $a \equiv 0 \pmod 4$ or $a \equiv 1 \pmod 4$ then $\binom{a}{2}$ is even.

Proof. We prove this directly. Assume $a \equiv 0 \pmod 4$. Then $\binom{a}{2} = \frac{a(a-1)}{2}$. Since $a = 4k$ for some $k \in \mathbb{N}$, we have $\binom{a}{2} = \frac{4k(4k-1)}{2} = 2k(4k-1)$. Hence $\binom{a}{2}$ is even. Now assume $a \equiv 1 \pmod 4$. Then $a = 4k+1$ for some $k \in \mathbb{N}$. Hence $\binom{a}{2} = \frac{(4k+1)(4k)}{2} = 2k(4k+1)$. Hence, $\binom{a}{2}$ is even. This proves the result. ■

29. If integers a and b are not both zero, then $\gcd(a,b) = \gcd(a-b,b)$.

Proof. (Direct) Suppose integers a and b are not both zero. Let $d = \gcd(a,b)$. Because d is a divisor of both a and b, we have $a = dx$ and $b = dy$ for some integers x and y. Then $a-b = dx - dy = d(x-y)$, so it follows that d is also a common divisor of $a-b$ and b. Therefore it can't be greater than the greatest common divisor of $a-b$ and b, which is to say $\gcd(a,b) = d \le \gcd(a-b,b)$.

Now let $e = \gcd(a-b,b)$. Then e divides both $a-b$ and b, that is, $a-b = ex$ and $b = ey$ for integers x and y. Then $a = (a-b)+b = ex+ey = e(x+y)$, so now we see that e is a divisor of both a and b. Thus it is not more than their greatest common divisor, that is, $\gcd(a-b,b) = e \le \gcd(a,b)$.

The above two paragraphs have given $\gcd(a,b) \le \gcd(a-b,b)$ and $\gcd(a-b,b) \le \gcd(a,b)$. Thus $\gcd(a,b) = \gcd(a-b,b)$.

■

31. Suppose the division algorithm applied to a and b yields $a = qb + r$. Then $\gcd(a,b) = \gcd(r,b)$.

Proof. Suppose $a = qb+r$. Let $d = \gcd(a,b)$, so d is a common divisor of a and b; thus $a = dx$ and $b = dy$ for some integers x and y. Then $dx = a = qb + r = qdy + r$, hence $dx = qdy + r$, and so $r = dx - qdy = d(x - qy)$. Thus d is a divisor of r (and also of b), so $\gcd(a,b) = d \le \gcd(r,b)$.

On the other hand, let $e = \gcd(r,b)$, so $r = ex$ and $b = ey$ for some integers x and y. Then $a = qb + r = qey + ex = e(qy+x)$. Hence e is a divisor of a (and of course also of b) so $\gcd(r,b) = e \le \gcd(a,b)$.

We've now shown $\gcd(a,b) \le \gcd(r,b)$ and $\gcd(r,b) \le \gcd(a,b)$, so $\gcd(r,b) = \gcd(a,b)$.

■

Chapter 6 Exercises

1. Suppose n is an integer. If n is odd, then n^2 is odd.

Proof. Suppose for the sake of contradiction that n is odd and n^2 is not odd. Then n^2 is even. Now, since n is odd, we have $n = 2a+1$ for some integer a. Thus $n^2 = (2a+1)^2 = 4a^2 + 4a + 1 = 2(2a^2 + 2a) + 1$. This shows $n^2 = 2b + 1$, where b is the integer $b = 2a^2 + 2a$. Therefore we have n^2 is odd and n^2 is even, a contradiction. ■

3. Prove that $\sqrt[3]{2}$ is irrational.

Proof. Suppose for the sake of contradiction that $\sqrt[3]{2}$ is not irrational. Therefore it is rational, so there exist integers a and b for which $\sqrt[3]{2} = \frac{a}{b}$. Let us assume that this fraction is reduced, so a and b are not both even. Now we have $\sqrt[3]{2}^3 = \left(\frac{a}{b}\right)^3$, which gives $2 = \frac{a^3}{b^3}$, or $2b^3 = a^3$. From this we see that a^3 is even, from which we deduce that a is even. (For if a were odd, then $a^3 = (2c+1)^3 = 8c^3 + 12c^2 + 6c + 1 = 2(4c^3 + 6c^2 + 3c) + 1$ would be odd, not even.) Since a is even, it follows that $a = 2d$ for some integer d. The equation $2b^3 = a^3$ from above then becomes $2b^3 = (2d)^3$, or $2b^3 = 8d^3$. Dividing by 2, we get $b^3 = 4d^3$, and it follows that b^3 is even. Thus b is even also. (Using the same argument we used when a^3 was even.) At this point we have discovered that both a and b are even, contradicting the fact (observed above) that the a and b are not both even. ■

Here is an alternative proof.

Proof. Suppose for the sake of contradiction that $\sqrt[3]{2}$ is not irrational. Therefore there exist integers a and b for which $\sqrt[3]{2} = \frac{a}{b}$. Cubing both sides, we get $2 = \frac{a^3}{b^3}$. From this, $a^3 = b^3 + b^3$, which contradicts Fermat's last theorem. ■

5. Prove that $\sqrt{3}$ is irrational.

Proof. Suppose for the sake of contradiction that $\sqrt{3}$ is not irrational. Therefore it is rational, so there exist integers a and b for which $\sqrt{3} = \frac{a}{b}$. Let us assume that this fraction is reduced, so a and b have no common factor. Notice that $\sqrt{3}^2 = \left(\frac{a}{b}\right)^2$, so $3 = \frac{a^2}{b^2}$, or $3b^2 = a^2$. This means $3 \mid a^2$.

Now we are going to show that if $a \in \mathbb{Z}$ and $3 \mid a^2$, then $3 \mid a$. (This is a proof-within-a-proof.) We will use contrapositive proof to prove this conditional statement. Suppose $3 \nmid a$. Then there is a remainder of either 1 or 2 when 3 is divided into a.

Case 1. There is a remainder of 1 when 3 is divided into a. Then $a = 3m + 1$ for some integer m. Consequently, $a^2 = 9m^2 + 6m + 1 = 3(3m^2 + 2m) + 1$, and this means 3 divides into a^2 with a remainder of 1. Thus $3 \nmid a^2$.

Case 2. There is a remainder of 2 when 3 is divided into a. Then $a = 3m + 2$ for some integer m. Consequently, $a^2 = 9m^2 + 12m + 4 = 9m^2 + 12m + 3 + 1 = 3(3m^2 + 4m + 1) + 1$, and this means 3 divides into a^2 with a remainder of 1. Thus $3 \nmid a^2$.

In either case we have $3 \nmid a^2$, so we've shown $3 \nmid a$ implies $3 \nmid a^2$. Therefore, if $3 \mid a^2$, then $3 \mid a$.

Now go back to $3 \mid a^2$ in the first paragraph. This combined with the result of the second paragraph implies $3 \mid a$, so $a = 3d$ for some integer d. Now also in the first paragraph we had $3b^2 = a^2$, which now becomes $3b^2 = (3d)^2$ or $3b^2 = 9d^2$, so $b^2 = 3d^2$. But this means $3 \mid b^2$, and the second paragraph implies $3 \mid b$. Thus we have concluded that $3 \mid a$ and $3 \mid b$, but this contradicts the fact that the fraction $\frac{a}{b}$ is reduced. ■

7. If $a,b \in \mathbb{Z}$, then $a^2 - 4b - 3 \neq 0$.

Proof. Suppose for the sake of contradiction that $a,b \in \mathbb{Z}$ but $a^2 - 4b - 3 = 0$. Then we have $a^2 = 4b + 3 = 2(2b+1)+1$, which means a^2 is odd. Therefore a is odd also, so $a = 2c+1$ for some integer c. Plugging this back into $a^2 - 4b - 3 = 0$ gives us

$$\begin{aligned}
(2c+1)^2 - 4b - 3 &= 0 \\
4c^2 + 4c + 1 - 4b - 3 &= 0 \\
4c^2 + 4c - 4b &= 2 \\
2c^2 + 2c - 2b &= 1 \\
2(c^2 + c - b) &= 1.
\end{aligned}$$

From this last equation, we see that 1 is an even number, a contradiction. ■

9. Suppose $a,b \in \mathbb{R}$ and $a \neq 0$. If a is rational and ab is irrational, then b is irrational.

Proof. Suppose for the sake of contradiction that a is rational and ab is irrational and b is **not** irrational. Thus we have a and b rational, and ab irrational. Since a and b are rational, we know there are integers c,d,e,f for which $a = \frac{c}{d}$ and $b = \frac{e}{f}$. Then $ab = \frac{ce}{df}$, and since both ce and df are integers, it follows that ab is rational. But this is a contradiction because we started out with ab irrational. ■

11. There exist no integers a and b for which $18a + 6b = 1$.

Proof. Suppose for the sake of contradiction that there do exist integers a and b for which $18a + 6b = 1$. Then $1 = 2(9a + 3b)$, which means 1 is even, a contradiction. ■

13. For every $x \in [\pi/2, \pi]$, $\sin x - \cos x \geq 1$.

Proof. Suppose for the sake of contradiction that $x \in [\pi/2, \pi]$, but $\sin x - \cos x < 1$. Since $x \in [\pi/2, \pi]$, we know $\sin x \geq 0$ and $\cos x \leq 0$, so $\sin x - \cos x \geq 0$. Therefore we have $0 \leq \sin x - \cos x < 1$. Now the square of any number between 0 and 1 is still a number between 0 and 1, so we have $0 \leq (\sin x - \cos x)^2 < 1$, or $0 \leq \sin^2 x - 2\sin x \cos x + \cos^2 x < 1$. Using the fact that $\sin^2 x + \cos^2 x = 1$, this becomes $0 \leq -2\sin x \cos x + 1 < 1$. Subtracting 1, we obtain $-2\sin x \cos x < 0$. But above we remarked that $\sin x \geq 0$ and $\cos x \leq 0$, and hence $-2\sin x \cos x \geq 0$. We now have the contradiction $-2\sin x \cos x < 0$ and $-2\sin x \cos x \geq 0$. ■

15. If $b \in \mathbb{Z}$ and $b \nmid k$ for every $k \in \mathbb{N}$, then $b = 0$.

Proof. Suppose for the sake of contradiction that $b \in \mathbb{Z}$ and $b \nmid k$ for every $k \in \mathbb{N}$, but $b \neq 0$.
Case 1. Suppose $b > 0$. Then $b \in \mathbb{N}$, so $b|b$, contradicting $b \nmid k$ for every $k \in \mathbb{N}$.
Case 2. Suppose $b < 0$. Then $-b \in \mathbb{N}$, so $b|(-b)$, again a contradiction ■

17. For every $n \in \mathbb{Z}$, $4 \nmid (n^2+2)$.

Proof. Assume there exists $n \in \mathbb{Z}$ with $4 \mid (n^2+2)$. Then for some $k \in \mathbb{Z}$, $4k = n^2+2$ or $2k = n^2+2(1-k)$. If n is odd, this means $2k$ is odd, and we've reached a contradiction. If n is even then $n = 2j$ and we get $k = 2j^2+1-k$ for some $j \in \mathbb{Z}$. Hence $2(k-j^2) = 1$, so 1 is even, a contradiction. ■

Remark. It is fairly easy to see that two more than a perfect square is always either 2 (mod 4) or 3 (mod 4). This would end the proof immediately.

19. The product of 5 consecutive integers is a multiple of 120.

Proof. Given any collection of 5 consecutive integers, at least one must be a multiple of two, at least one must be a multiple of three, at least one must be a multiple of four and at least one must be a multiple of 5. Hence the product is a multiple of $5 \cdot 4 \cdot 3 \cdot 2 = 120$. In particular, the product is a multiple of 60. ■

21. Hints for Exercises 20–23. For Exercises 20, first show that the equation $a^2+b^2 = 3c^2$ has no solutions (other than the trivial solution $(a,b,c) = (0,0,0)$) in the integers. To do this, investigate the remainders of a sum of squares (mod 4). After you've done this, prove that the only solution is indeed the trivial solution.
Now, assume that the equation $x^2+y^2-3=0$ has a rational solution. Use the definition of rational numbers to yield a contradiction.

Chapter 7 Exercises

1. Suppose $x \in \mathbb{Z}$. Then x is even if and only if $3x+5$ is odd.

Proof. We first use direct proof to show that if x is even, then $3x+5$ is odd. Suppose x is even. Then $x = 2n$ for some integer n. Thus $3x+5 = 3(2n)+5 = 6n+5 = 6n+4+1 = 2(3n+2)+1$. Thus $3x+5$ is odd because it has form $2k+1$, where $k = 3n+2 \in \mathbb{Z}$.

Conversely, we need to show that if $3x+5$ is odd, then x is even. We will prove this using contrapositive proof. Suppose x is *not* even. Then x is odd, so $x = 2n+1$ for some integer n. Thus $3x+5 = 3(2n+1)+5 = 6n+8 = 2(3n+4)$. This means says $3x+5$ is twice the integer $3n+4$, so $3x+5$ is even, not odd. ■

3. Given an integer a, then a^3+a^2+a is even if and only if a is even.

Proof. First we will prove that if a^3+a^2+a is even then a is even. This is done with contrapositive proof. Suppose a is not even. Then a is odd, so there is an integer n for which $a = 2n+1$. Then

$$\begin{aligned} a^3+a^2+a &= (2n+1)^3+(2n+1)^2+(2n+1) \\ &= 8n^3+12n^2+6n+1+4n^2+4n+1+2n+1 \\ &= 8n^3+16n^2+12n+2+1 \\ &= 2(4n^3+8n^2+6n+1)+1. \end{aligned}$$

This expresses a^3+a^2+a as twice an integer plus 1, so a^3+a^2+a is odd, not even. We have now shown that if a^3+a^2+a is even then a is even.

Conversely, we need to show that if a is even, then a^3+a^2+a is even. We will use direct proof. Suppose a is even, so $a=2n$ for some integer n. Then $a^3+a^2+a = (2n)^3+(2n)^2+2n = 8n^3+4n^2+2n = 2(4n^3+2n^2+n)$. Therefore, a^3+a^2+a is even because it's twice an integer. ■

5. An integer a is odd if and only if a^3 is odd.

Proof. Suppose that a is odd. Then $a=2n+1$ for some integer n, and $a^3 = (2n+1)^3 = 8n^3+12n^2+6n+1 = 2(4n^3+6n^2+3n)+1$. This shows that a^3 is twice an integer, plus 1, so a^3 is odd. Thus we've proved that if a is odd then a^3 is odd.

Conversely we need to show that if a^3 is odd, then a is odd. For this we employ contrapositive proof. Suppose a is not odd. Thus a is even, so $a=2n$ for some integer n. Then $a^3 = (2n)^3 = 8n^3 = 2(4n^3)$ is even (not odd). ■

7. Suppose $x, y \in \mathbb{R}$. Then $(x+y)^2 = x^2+y^2$ if and only if $x=0$ or $y=0$.

Proof. First we prove with direct proof that if $(x+y)^2 = x^2+y^2$, then $x=0$ or $y=0$. Suppose $(x+y)^2 = x^2+y^2$. From this we get $x^2+2xy+y^2 = x^2+y^2$, so $2xy=0$, and hence $xy=0$. Thus $x=0$ or $y=0$.

Conversely, we need to show that if $x=0$ or $y=0$, then $(x+y)^2 = x^2+y^2$. This will be done with cases.
Case 1. If $x=0$ then $(x+y)^2 = (0+y)^2 = y^2 = 0^2+y^2 = x^2+y^2$.
Case 2. If $y=0$ then $(x+y)^2 = (x+0)^2 = x^2 = x^2+0^2 = x^2+y^2$.
Either way, we have $(x+y)^2 = x^2+y^2$. ■

9. Suppose $a \in \mathbb{Z}$. Prove that $14 \mid a$ if and only if $7 \mid a$ and $2 \mid a$.

Proof. First we prove that if $14 \mid a$, then $7 \mid a$ and $2 \mid a$. Direct proof is used. Suppose $14 \mid a$. This means $a = 14m$ for some integer m. Therefore $a = 7(2m)$, which means $7 \mid a$, and also $a = 2(7m)$, which means $2 \mid a$. Thus $7 \mid a$ and $2 \mid a$.

Conversely, we need to prove that if $7 \mid a$ and $2 \mid a$, then $14 \mid a$. Once again direct proof if used. Suppose $7 \mid a$ and $2 \mid a$. Since $2 \mid a$ it follows that $a = 2m$ for some integer m, and that in turn implies that a is even. Since $7 \mid a$ it follows that $a = 7n$ for some integer n. Now, since a is known to be even, and $a = 7n$, it follows that n is even (if it were odd, then $a = 7n$ would be odd). Thus $n = 2p$ for an appropriate integer p, and plugging $n = 2p$ back into $a = 7n$ gives $a = 7(2p)$, so $a = 14p$. Therefore $14 \mid a$. ■

11. Suppose $a, b \in \mathbb{Z}$. Prove that $(a-3)b^2$ is even if and only if a is odd or b is even.

Proof. First we will prove that if $(a-3)b^2$ is even, then a is odd or b is even. For this we use contrapositive proof. Suppose it is not the case that a is odd or b is even. Then by DeMorgan's law, a is even and b is odd. Thus there are integers m and n for which $a = 2m$ and $b = 2n+1$. Now observe $(a-3)b^2 = (2m-3)(2n+1)^2 = (2m-3)(4n^2+4n+1) = 8mn^2+8mn+2m-12n^2-12n-3 = 8mn^2+8mn+2m-12n^2-12n-4+1 = 2(4mn^2+4mn+m-6n^2-6n-2)+1$. This shows $(a-3)b^2$ is odd, so it's not even.

Conversely, we need to show that if a is odd or b is even, then $(a-3)b^2$ is even. For this we use direct proof, with cases.
Case 1. Suppose a is odd. Then $a = 2m+1$ for some integer m. Thus $(a-3)b^2 = (2m+1-3)b^2 = (2m-2)b^2 = 2(m-1)b^2$. Thus in this case $(a-3)b^2$ is even.
Case 2. Suppose b is even. Then $b = 2n$ for some integer n. Thus $(a-3)b^2 = (a-3)(2n)^2 = (a-3)4n^2 = 2(a-3)2n^2 =$. Thus in this case $(a-3)b^2$ is even.
Therefore, in any event, $(a-3)b^2$ is even. ■

13. Suppose $a, b \in \mathbb{Z}$. If $a+b$ is odd, then a^2+b^2 is odd.
Hint: Use direct proof. Suppose $a+b$ is odd. Argue that this means a and b have opposite parity. Then use cases.

15. Suppose $a, b \in \mathbb{Z}$. Prove that $a+b$ is even if and only if a and b have the same parity.

Proof. First we will show that if $a+b$ is even, then a and b have the same parity. For this we use contrapositive proof. Suppose it is not the case that a and b have the same parity. Then one of a and b is even and the other is odd. Without loss of generality, let's say that a is even and b is odd. Thus there are integers m and n for which $a = 2m$ and $b = 2n+1$. Then $a+b = 2m+2n+1 = 2(m+n)+1$, so $a+b$ is odd, not even.

Conversely, we need to show that if a and b have the same parity, then $a+b$ is even. For this, we use direct proof with cases. Suppose a and b have the same parity.
Case 1. Both a and b are even. Then there are integers m and n for which $a = 2m$ and $b = 2n$, so $a+b = 2m+2n = 2(m+n)$ is clearly even.
Case 2. Both a and b are odd. Then there are integers m and n for which $a = 2m+1$ and $b = 2n+1$, so $a+b = 2m+1+2n+1 = 2(m+n+1)$ is clearly even.
Either way, $a+b$ is even. This completes the proof. ■

17. There is a prime number between 90 and 100.

Proof. Simply observe that 97 is prime. ■

19. If $n \in \mathbb{N}$, then $2^0+2^1+2^2+2^3+2^4+\cdots+2^n = 2^{n+1}-1$.

Proof. We use direct proof. Suppose $n \in \mathbb{N}$. Let S be the number

$$S = 2^0+2^1+2^2+2^3+2^4+\cdots+2^{n-1}+2^n. \tag{1}$$

In what follows, we will solve for S and show $S = 2^{n+1}-1$. Multiplying both sides of (1) by 2 gives

$$2S = 2^1+2^2+2^3+2^4+2^5+\cdots+2^n+2^{n+1}. \tag{2}$$

Now subtract Equation (1) from Equation (2) to obtain $2S-S = -2^0+2^{n+1}$, which simplifies to $S = 2^{n+1}-1$. Combining this with Equation (1) produces $2^0+2^1+2^2+2^3+2^4+\cdots+2^n = 2^{n+1}-1$, so the proof is complete. ■

21. Every real solution of $x^3+x+3=0$ is irrational.

Proof. Suppose for the sake of contradiction that this polynomial has a rational solution $\frac{a}{b}$. We may assume that this fraction is fully reduced, so a and b are not both even. We have $\left(\frac{a}{b}\right)^3+\frac{a}{b}+3=0$. Clearing the denominator gives

$$a^3+ab^2+3b^3=0.$$

Consider two cases: First, if both a and b are odd, the left-hand side is a sum of three odds, which is odd, meaning 0 is odd, a contradiction. Second, if one of a and b is odd and the other is even, then the middle term of $a^3+ab^2+3b^3$ is even, while a^3 and $3b^2$ have opposite parity. Then $a^3+ab^2+3b^3$ is the sum of two evens and an odd, which is odd, again contradicting the fact that 0 is even. ■

23. Suppose a, b and c are integers. If $a \mid b$ and $a \mid (b^2-c)$, then $a \mid c$.

Proof. (Direct) Suppose $a \mid b$ and $a \mid (b^2-c)$. This means that $b = ad$ and $b^2-c = ae$ for some integers d and e. Squaring the first equation produces $b^2 = a^2d^2$. Subtracting $b^2-c = ae$ from $b^2 = a^2d^2$ gives $c = a^2d^2-ae = a(ad^2-e)$. As $ad^2-e \in \mathbb{Z}$, it follows that $a \mid c$. ■

25. If $p>1$ is an integer and $n \nmid p$ for each integer n for which $2 \le n \le \sqrt{p}$, then p is prime.

Proof. (Contrapositive) Suppose that p is not prime, so it factors as $p = mn$ for $1 < m, n < p$.

Observe that it is not the case that both $m > \sqrt{p}$ and $n > \sqrt{p}$, because if this were true the inequalities would multiply to give $mn > \sqrt{p}\sqrt{p} = p$, which contradicts $p = mn$.

Therefore $m \le \sqrt{p}$ or $n \le \sqrt{p}$. Without loss of generality, say $n \le \sqrt{p}$. Then the equation $p = mn$ gives $n \mid p$, with $1 < n \le \sqrt{p}$. Therefore it is not true that $n \nmid p$ for each integer n for which $2 \le n \le \sqrt{p}$. ■

27. Suppose $a, b \in \mathbb{Z}$. If a^2+b^2 is a perfect square, then a and b are not both odd.

Proof. (Contradiction) Suppose a^2+b^2 is a perfect square, and a and b are both odd. As a^2+b^2 is a perfect square, say c is the integer for which $c^2 = a^2+b^2$. As a and b are odd, we have $a = 2m+1$ and $b = 2n+1$ for integers m and n. Then

$$c^2 = a^2+b^2 = (2m+1)^2+(2n+1)^2 = 4(m^2+n^2+mn)+2.$$

This is even, so c is even also; let $c = 2k$. Now the above equation results in $(2k)^2 = 4(m^2+n^2+mn)+2$, which simplifies to $2k^2 = 2(m^2+n^2+mn)+1$. Thus $2k^2$ is both even and odd, a contradiction. ■

29. If $a \mid bc$ and $\gcd(a,b) = 1$, then $a \mid c$.

Proof. (Direct) Suppose $a \mid bc$ and $\gcd(a,b) = 1$. The fact that $a \mid bc$ means $bc = az$ for some integer z. The fact that $\gcd(a,b) = 1$ means that $ax+by = 1$ for some integers x and y (by Proposition 7.1 on page 126). From this we get $acx+bcy = c$; substituting $bc = az$ yields $acx+azy = c$, that is, $a(cx+zy) = c$. Therefore $a \mid c$. ■

31. If $n \in \mathbb{Z}$, then $\gcd(n, n+1) = 1$.

Proof. Suppose d is a positive integer that is a common divisor of n and $n+1$. Then $n = dx$ and $n+1 = dy$ for integers x and y. Then $1 = (n+1)-n = dy-dx = d(y-x)$. Now, $1 = d(y-x)$ is only possible if $d = \pm 1$ and $y-x = \pm 1$. Thus the greatest common divisor of n and $n+1$ can be no greater than 1. But 1 does divide both n and $n+1$, so $\gcd(n, n+1) = 1$. ■

33. If $n \in \mathbb{Z}$, then $\gcd(2n+1, 4n^2+1) = 1$.

Proof. Note that $4n^2+1 = (2n+1)(2n-1)+2$. Therefore, it suffices to show that $\gcd(2n+1, (2n+1)(2n-1)+2) = 1$. Let d be a common positive divisor of both $2n+1$ and $(2n+1)(2n-1)+2$, so $2n+1 = dx$ and $(2n+1)(2n-1)+2 = dy$ for integers x and y. Substituting the first equation into the second gives $dx(2n-1)+2 = dy$, so $2 = dy - dx(2n-1) = d(y-2nx-x)$. This means d divides 2, so d equals 1 or 2. But the equation $2n+1 = dx$ means d must be odd. Therefore $d = 1$, that is, $\gcd(2n+1, (2n+1)(2n-1)+2) = 1$. ■

35. Suppose $a, b \in \mathbb{N}$. Then $a = \gcd(a,b)$ if and only if $a \mid b$.

Proof. Suppose $a = \gcd(a,b)$. This means a is a divisor of both a and b. In particular $a \mid b$.

Conversely, suppose $a \mid b$. Then a divides both a and b, so $a \le \gcd(a,b)$. On the other hand, since $\gcd(a,b)$ divides a, we have $a = \gcd(a,b) \cdot x$ for some integer x. As all integers involved are positive, it follows that $a \ge gcd(a,b)$.

It has been established that $a \le \gcd(a,b)$ and $a \ge gcd(a,b)$. Thus $a = \gcd(a,b)$. ■

Chapter 8 Exercises

1. Prove that $\{12n : n \in \mathbb{Z}\} \subseteq \{2n : n \in \mathbb{Z}\} \cap \{3n : n \in \mathbb{Z}\}$.

Proof. Suppose $a \in \{12n : n \in \mathbb{Z}\}$. This means $a = 12n$ for some $n \in \mathbb{Z}$. Therefore $a = 2(6n)$ and $a = 3(4n)$. From $a = 2(6n)$, it follows that a is multiple of 2, so $a \in \{2n : n \in \mathbb{Z}\}$. From $a = 3(4n)$, it follows that a is multiple of 3, so $a \in \{3n : n \in \mathbb{Z}\}$. Thus by definition of the intersection of two sets, we have $a \in \{2n : n \in \mathbb{Z}\} \cap \{3n : n \in \mathbb{Z}\}$. Thus $\{12n : n \in \mathbb{Z}\} \subseteq \{2n : n \in \mathbb{Z}\} \cap \{3n : n \in \mathbb{Z}\}$. ■

3. If $k \in \mathbb{Z}$, then $\{n \in \mathbb{Z} : n \mid k\} \subseteq \{n \in \mathbb{Z} : n \mid k^2\}$.

Proof. Suppose $k \in \mathbb{Z}$. We now need to show $\{n \in \mathbb{Z} : n \mid k\} \subseteq \{n \in \mathbb{Z} : n \mid k^2\}$. Suppose $a \in \{n \in \mathbb{Z} : n \mid k\}$. Then it follows that $a \mid k$, so there is an integer c for which $k = ac$. Then $k^2 = a^2c^2$. Therefore $k^2 = a(ac^2)$, and from this the definition of divisibility gives $a \mid k^2$. But $a \mid k^2$ means that $a \in \{n \in \mathbb{Z} : n \mid k^2\}$. We have now shown $\{n \in \mathbb{Z} : n \mid k\} \subseteq \{n \in \mathbb{Z} : n \mid k^2\}$. ■

5. If p and q are integers, then $\{pn : n \in \mathbb{N}\} \cap \{qn : n \in \mathbb{N}\} \neq \emptyset$.

Proof. Suppose p and q are integers. Consider the integer pq. Observe that $pq \in \{pn : n \in \mathbb{N}\}$ and $pq \in \{qn : n \in \mathbb{N}\}$, so $pq \in \{pn : n \in \mathbb{N}\} \cap \{qn : n \in \mathbb{N}\}$. Therefore $\{pn : n \in \mathbb{N}\} \cap \{qn : n \in \mathbb{N}\} \neq \emptyset$. ■

7. Suppose A, B and C are sets. If $B \subseteq C$, then $A \times B \subseteq A \times C$.

Proof. This is a conditional statement, and we'll prove it with direct proof. Suppose $B \subseteq C$. (Now we need to prove $A \times B \subseteq A \times C$.)

Suppose $(a,b) \in A \times B$. Then by definition of the Cartesian product we have $a \in A$ and $b \in B$. But since $b \in B$ and $B \subseteq C$, we have $b \in C$. Since $a \in A$ and $b \in C$, it follows that $(a,b) \in A \times C$. Now we've shown $(a,b) \in A \times B$ implies $(a,b) \in A \times C$, so $A \times B \subseteq A \times C$.

In summary, we've shown that if $B \subseteq C$, then $A \times B \subseteq A \times C$. This completes the proof. ■

9. If A, B and C are sets then $A \cap (B \cup C) = (A \cap B) \cup (A \cap C)$.

Proof. We use the distributive law $P \wedge (Q \vee R) = (P \wedge Q) \vee (P \wedge R)$ from page 50.

$$
\begin{aligned}
A \cap (B \cup C) &= \{x : x \in A \;\wedge\; x \in B \cup C\} && \text{(def. of intersection)} \\
&= \{x : x \in A \;\wedge\; (x \in B \;\vee\; x \in C)\} && \text{(def. of union)} \\
&= \big\{x : \big(x \in A \;\wedge\; x \in B\big) \vee \big(x \in A \;\wedge\; x \in C\big)\big\} && \text{(distributive law)} \\
&= \{x : (x \in A \cap B) \vee (x \in A \cap C)\} && \text{(def. of intersection)} \\
&= (A \cap B) \cup (A \cap C) && \text{(def. of union)}
\end{aligned}
$$

The proof is complete. ■

11. If A and B are sets in a universal set U, then $\overline{A \cup B} = \overline{A} \cap \overline{B}$.

Proof. Just observe the following sequence of equalities.

$$
\begin{aligned}
\overline{A \cup B} &= U - (A \cup B) && \text{(def. of complement)}\\
&= \{x : (x \in U) \wedge (x \notin A \cup B)\} && \text{(def. of } -\text{)}\\
&= \{x : (x \in U) \wedge \sim (x \in A \cup B)\} &&\\
&= \{x : (x \in U) \wedge \sim ((x \in A) \vee (x \in B))\} && \text{(def. of } \cup\text{)}\\
&= \{x : (x \in U) \wedge (\sim (x \in A) \wedge \sim (x \in B))\} && \text{(DeMorgan)}\\
&= \{x : (x \in U) \wedge (x \notin A) \wedge (x \notin B)\} &&\\
&= \{x : (x \in U) \wedge (x \in U) \wedge (x \notin A) \wedge (x \notin B)\} && (x \in U) = (x \in U) \wedge (x \in U)\\
&= \{x : ((x \in U) \wedge (x \notin A)) \wedge ((x \in U) \wedge (x \notin B))\} && \text{(regroup)}\\
&= \{x : (x \in U) \wedge (x \notin A)\} \cap \{x : (x \in U) \wedge (x \notin B)\} && \text{(def. of } \cap\text{)}\\
&= (U - A) \cap (U - B) && \text{(def. of } -\text{)}\\
&= \overline{A} \cap \overline{B} && \text{(def. of complement)}
\end{aligned}
$$

The proof is complete. ■

13. If A, B and C are sets, then $A - (B \cup C) = (A - B) \cap (A - C)$.

Proof. Just observe the following sequence of equalities.

$$
\begin{aligned}
A - (B \cup C) &= \{x : (x \in A) \wedge (x \notin B \cup C)\} && \text{(def. of } -\text{)}\\
&= \{x : (x \in A) \wedge \sim (x \in B \cup C)\} &&\\
&= \{x : (x \in A) \wedge \sim ((x \in B) \vee (x \in C))\} && \text{(def. of } \cup\text{)}\\
&= \{x : (x \in A) \wedge (\sim (x \in B) \wedge \sim (x \in C))\} && \text{(DeMorgan)}\\
&= \{x : (x \in A) \wedge (x \notin B) \wedge (x \notin C)\} &&\\
&= \{x : (x \in A) \wedge (x \in A) \wedge (x \notin B) \wedge (x \notin C)\} && (x \in A) = (x \in A) \wedge (x \in A)\\
&= \{x : ((x \in A) \wedge (x \notin B)) \wedge ((x \in A) \wedge (x \notin C))\} && \text{(regroup)}\\
&= \{x : (x \in A) \wedge (x \notin B)\} \cap \{x : (x \in A) \wedge (x \notin C)\} && \text{(def. of } \cap\text{)}\\
&= (A - B) \cap (A - C) && \text{(def. of } -\text{)}
\end{aligned}
$$

The proof is complete. ■

15. If A, B and C are sets, then $(A \cap B) - C = (A - C) \cap (B - C)$.

Proof. Just observe the following sequence of equalities.

$$
\begin{aligned}
(A \cap B) - C &= \{x : (x \in A \cap B) \wedge (x \notin C)\} && \text{(def. of } -\text{)}\\
&= \{x : (x \in A) \wedge (x \in B) \wedge (x \notin C)\} && \text{(def. of } \cap\text{)}\\
&= \{x : (x \in A) \wedge (x \notin C) \wedge (x \in B) \wedge (x \notin C)\} && \text{(regroup)}\\
&= \{x : ((x \in A) \wedge (x \notin C)) \wedge ((x \in B) \wedge (x \notin C))\} && \text{(regroup)}\\
&= \{x : (x \in A) \wedge (x \notin C)\} \cap \{x : (x \in B) \wedge (x \notin C)\} && \text{(def. of } \cap\text{)}\\
&= (A - C) \cap (B - C) && \text{(def. of } \cap\text{)}
\end{aligned}
$$

The proof is complete. ■

17. If A, B and C are sets, then $A \times (B \cap C) = (A \times B) \cap (A \times C)$.

Proof. See Example 8.12. ■

19. Prove that $\{9^n : n \in \mathbb{Z}\} \subseteq \{3^n : n \in \mathbb{Z}\}$, but $\{9^n : n \in \mathbb{Z}\} \neq \{3^n : n \in \mathbb{Z}\}$.

Proof. Suppose $a \in \{9^n : n \in \mathbb{Z}\}$. This means $a = 9^n$ for some integer $n \in \mathbb{Z}$. Thus $a = 9^n = (3^2)^n = 3^{2n}$. This shows a is an integer power of 3, so $a \in \{3^n : n \in \mathbb{Z}\}$. Therefore $a \in \{9^n : n \in \mathbb{Z}\}$ implies $a \in \{3^n : n \in \mathbb{Z}\}$, so $\{9^n : n \in \mathbb{Z}\} \subseteq \{3^n : n \in \mathbb{Z}\}$.

But notice $\{9^n : n \in \mathbb{Z}\} \neq \{3^n : n \in \mathbb{Z}\}$ as $3 \in \{3^n : n \in \mathbb{Z}\}$, but $3 \notin \{9^n : n \in \mathbb{Z}\}$. ■

21. Suppose A and B are sets. Prove $A \subseteq B$ if and only if $A - B = \emptyset$.

Proof. First we will prove that if $A \subseteq B$, then $A - B = \emptyset$. Contrapositive proof is used. Suppose that $A - B \neq \emptyset$. Thus there is an element $a \in A - B$, which means $a \in A$ but $a \notin B$. Since not every element of A is in B, we have $A \not\subseteq B$.

Conversely, we will prove that if $A - B = \emptyset$, then $A \subseteq B$. Again, contrapositive proof is used. Suppose $A \not\subseteq B$. This means that it is not the case that every element of A is an element of B, so there is an element $a \in A$ with $a \notin B$. Therefore we have $a \in A - B$, so $A - B \neq \emptyset$. ■

23. For each $a \in \mathbb{R}$, let $A_a = \{(x, a(x^2-1)) \in \mathbb{R}^2 : x \in \mathbb{R}\}$. Prove that $\bigcap_{a \in \mathbb{R}} A_a = \{(-1,0),(1,0)\}$.

Proof. First we will show that $\{(-1,0),(1,0)\} \subseteq \bigcap_{a \in \mathbb{R}} A_a$. Notice that for any $a \in \mathbb{R}$, we have $(-1,0) \in A_a$ because A_a contains the ordered pair $(-1, a((-1)^2-1) = (-1,0)$. Similarly $(1,0) \in A_a$. Thus each element of $\{(-1,0),(1,0)\}$ belongs to every set A_a, so every element of $\bigcap_{a \in \mathbb{R}} A_a$, so $\{(-1,0),(1,0)\} \subseteq \bigcap_{a \in \mathbb{R}} A_a$.

Now we will show $\bigcap_{a \in \mathbb{R}} A_a \subseteq \{(-1,0),(1,0)\}$. Suppose $(c,d) \in \bigcap_{a \in \mathbb{R}} A_a$. This means (c,d) is in every set A_a. In particular $(c,d) \in A_0 = \{(x, 0(x^2-1)) : x \in \mathbb{R}\} = \{(x,0) : x \in \mathbb{R}\}$. It follows that $d = 0$. Then also we have $(c,d) = (c,0) \in A_1 = \{(x, 1(x^2-1)) : x \in \mathbb{R}\} = \{(x, x^2-1) : x \in \mathbb{R}\}$. Therefore $(c,0)$ has the form (c, c^2-1), that is $(c,0) = (c, c^2-1)$. From this we get $c^2 - 1 = 0$, so $c = \pm 1$. Therefore $(c,d) = (1,0)$ or $(c,d) = (-1,0)$, so $(c,d) \in \{(-1,0),(1,0)\}$. This completes the demonstration that $(c,d) \in \bigcap_{a \in \mathbb{R}} A_a$ implies $(c,d) \in \{(-1,0),(1,0)\}$, so it follows that $\bigcap_{a \in \mathbb{R}} A_a \subseteq \{(-1,0),(1,0)\}$.

Now it's been shown that $\{(-1,0),(1,0)\} \subseteq \bigcap_{a \in \mathbb{R}} A_a$ and $\bigcap_{a \in \mathbb{R}} A_a \subseteq \{(-1,0),(1,0)\}$, so it follows that $\bigcap_{a \in \mathbb{R}} A_a = \{(-1,0),(1,0)\}$. ■

25. Suppose A, B, C and D are sets. Prove that $(A \times B) \cup (C \times D) \subseteq (A \cup C) \times (B \cup D)$.

Proof. Suppose $(a,b) \in (A \times B) \cup (C \times D)$.
By definition of union, this means $(a,b) \in (A \times B)$ **or** $(a,b) \in (C \times D)$.
We examine these two cases individually.
Case 1. Suppose $(a,b) \in (A \times B)$. By definition of $\times$, it follows that $a \in A$ and $b \in B$. From this, it follows from the definition of $\cup$ that $a \in A \cup C$ and $b \in B \cup D$. Again from the definition of $\times$, we get $(a,b) \in (A \cup C) \times (B \cup D)$.

Case 2. Suppose $(a,b) \in (C \times D)$. By definition of $\times$, it follows that $a \in C$ and $b \in D$. From this, it follows from the definition of $\cup$ that $a \in A \cup C$ and $b \in B \cup D$. Again from the definition of $\times$, we get $(a,b) \in (A \cup C) \times (B \cup D)$.
In either case, we obtained $(a,b) \in (A \cup C) \times (B \cup D)$,
so we've proved that $(a,b) \in (A \times B) \cup (C \times D)$ implies $(a,b) \in (A \cup C) \times (B \cup D)$.
Therefore $(A \times B) \cup (C \times D) \subseteq (A \cup C) \times (B \cup D)$. ■

27. Prove $\{12a + 4b : a,b \in \mathbb{Z}\} = \{4c : c \in \mathbb{Z}\}$.

Proof. First we show $\{12a + 4b : a,b \in \mathbb{Z}\} \subseteq \{4c : c \in \mathbb{Z}\}$. Suppose $x \in \{12a + 4b : a,b \in \mathbb{Z}\}$. Then $x = 12a + 4b$ for some integers a and b. From this we get $x = 4(3a + b)$, so $x = 4c$ where c is the integer $3a + b$. Consequently $x \in \{4c : c \in \mathbb{Z}\}$. This establishes that $\{12a + 4b : a,b \in \mathbb{Z}\} \subseteq \{4c : c \in \mathbb{Z}\}$.
Next we show $\{4c : c \in \mathbb{Z}\} \subseteq \{12a + 4b : a,b \in \mathbb{Z}\}$. Suppose $x \in \{4c : c \in \mathbb{Z}\}$. Then $x = 4c$ for some $c \in \mathbb{Z}$. Thus $x = (12 + 4(-2))c = 12c + 4(-2c)$, and since c and $-2c$ are integers we have $x \in \{12a + 4b : a,b \in \mathbb{Z}\}$.
This proves that $\{12a + 4b : a,b \in \mathbb{Z}\} = \{4c : c \in \mathbb{Z}\}$. ■

29. Suppose $A \neq \emptyset$. Prove that $A \times B \subseteq A \times C$, if and only if $B \subseteq C$.

Proof. First we will prove that if $A \times B \subseteq A \times C$, then $B \subseteq C$. Using contrapositive, suppose that $B \not\subseteq C$. This means there is an element $b \in B$ with $b \notin C$. Since $A \neq \emptyset$, there exists an element $a \in A$. Now consider the ordered pair (a,b). Note that $(a,b) \in A \times B$, but $(a,b) \notin A \times C$. This means $A \times B \not\subseteq A \times C$.

Conversely, we will now show that if $B \subseteq C$, then $A \times B \subseteq A \times C$. We use direct proof. Suppose $B \subseteq C$. Assume that $(a,b) \in A \times B$. This means $a \in A$ and $b \in B$. But, as $B \subseteq C$, we also have $b \in C$. From $a \in A$ and $b \in C$, we get $(a,b) \in A \times C$. We've now shown $(a,b) \in A \times B$ implies $(a,b) \in A \times C$, so $A \times B \subseteq A \times C$. ■

31. Suppose $B \neq \emptyset$ and $A \times B \subseteq B \times C$. Prove $A \subseteq C$.

Proof. Suppose $B \neq \emptyset$ and $A \times B \subseteq B \times C$. In what follows, we show that $A \subseteq C$. Let $x \in A$. Because B is not empty, it contains some element b. Observe that $(x,b) \in A \times B$. But as $A \times B \subseteq B \times C$, we also have $(x,b) \in B \times C$, so, in particular, $x \in B$. As $x \in A$ and $x \in B$, we have $(x,x) \in A \times B$. But as $A \times B \subseteq B \times C$, it follows that $(x,x) \in B \times C$. This implies $x \in C$.

Now we've shown $x \in A$ implies $x \in C$, so $A \subseteq C$. ■

Chapter 9 Exercises

1. If $x,y \in \mathbb{R}$, then $|x + y| = |x| + |y|$.
This is **false.**
Disproof: Here is a counterexample: Let $x = 1$ and $y = -1$. Then $|x + y| = 0$ and $|x| + |y| = 2$, so it's not true that $|x + y| = |x| + |y|$.

3. If $n \in \mathbb{Z}$ and $n^5 - n$ is even, then n is even.
This is **false.**
Disproof: Here is a counterexample: Let $n = 3$. Then $n^5 - n = 3^5 - 3 = 240$, but n is not even.

5. If A, B, C and D are sets, then $(A \times B) \cup (C \times D) = (A \cup C) \times (B \cup D)$.
This is **false.**
Disproof: Here is a counterexample: Let $A = \{1,2\}$, $B = \{1,2\}$, $C = \{2,3\}$ and $D = \{2,3\}$. Then $(A \times B) \cup (C \times D) = \{(1,1),(1,2),(2,1),(2,2)\} \cup \{(2,2),(2,3),(3,2),(3,3)\} = \{(1,1),(1,2),(2,1),(2,2),(2,3),(3,2),(3,3)\}$. Also $(A \cup C) \times (B \cup D) = \{1,2,3\} \times \{1,2,3\} = \{(1,1),(1,2),(1,3),(2,1),(2,2),(2,3),(3,1),(3,2),(3,3)\}$, so you can see that $(A \times B) \cup (C \times D) \neq (A \cup C) \times (B \cup D)$.

7. If A, B and C are sets, and $A \times C = B \times C$, then $A = B$.
This is **false.**
Disproof: Here is a counterexample: Let $A = \{1\}$, $B = \{2\}$ and $C = \emptyset$. Then $A \times C = B \times C = \emptyset$, but $A \neq B$.

9. If A and B are sets, then $\mathscr{P}(A) - \mathscr{P}(B) \subseteq \mathscr{P}(A - B)$.
This is **false.**
Disproof: Here is a counterexample: Let $A = \{1,2\}$ and $B = \{1\}$. Then $\mathscr{P}(A) - \mathscr{P}(B) = \{\emptyset,\{1\},\{2\},\{1,2\}\} - \{\emptyset,\{1\}\} = \{\{2\},\{1,2\}\}$. Also $\mathscr{P}(A - B) = \mathscr{P}(\{2\}) = \{\emptyset,\{2\}\}$. In this example we have $\mathscr{P}(A) - \mathscr{P}(B) \not\subseteq \mathscr{P}(A - B)$.

11. If $a, b \in \mathbb{N}$, then $a + b < ab$.
This is **false.**
Disproof: Here is a counterexample: Let $a = 1$ and $b = 1$. Then $a + b = 2$ and $ab = 1$, so it's not true that $a + b < ab$.

13. There exists a set X for which $\mathbb{R} \subseteq X$ and $\emptyset \in X$. This is **true.**

Proof. Simply let $X = \mathbb{R} \cup \{\emptyset\}$. If $x \in \mathbb{R}$, then $x \in \mathbb{R} \cup \{\emptyset\} = X$, so $\mathbb{R} \subseteq X$. Likewise, $\emptyset \in \mathbb{R} \cup \{\emptyset\} = X$ because $\emptyset \in \{\emptyset\}$. ■

15. Every odd integer is the sum of three odd integers. This is **true.**

Proof. Suppose n is odd. Then $n = n + 1 + (-1)$, and therefore n is the sum of three odd integers. ■

17. For all sets A and B, if $A - B = \emptyset$, then $B \neq \emptyset$.
This is **false.**
Disproof: Here is a counterexample: Just let $A = \emptyset$ and $B = \emptyset$. Then $A - B = \emptyset$, but it's not true that $B \neq \emptyset$.

19. For every $r, s \in \mathbb{Q}$ with $r < s$, there is an irrational number u for which $r < u < s$.
This is **true.**

Proof. (Direct) Suppose $r, s \in \mathbb{Q}$ with $r < s$. Consider the number $u = r + \sqrt{2}\frac{s-r}{2}$. In what follows we will show that u is irrational and $r < u < s$. Certainly since

$s-r$ is positive, it follows that $r < r+\sqrt{2}\frac{s-r}{2} = u$. Also, since $\sqrt{2} < 2$ we have

$$u = r+\sqrt{2}\frac{s-r}{2} < r+2\frac{s-r}{2} = s,$$

and therefore $u < s$. Thus we can conclude $r < u < s$.

Now we just need to show that u is irrational. Suppose for the sake of contradiction that u is rational. Then $u = \frac{a}{b}$ for some integers a and b. Since r and s are rational, we have $r = \frac{c}{d}$ and $s = \frac{e}{f}$ for some $c,d,e,f \in \mathbb{Z}$. Now we have

$$\begin{aligned} u &= r+\sqrt{2}\frac{s-r}{2} \\ \frac{a}{b} &= \frac{c}{d}+\sqrt{2}\frac{\frac{e}{f}-\frac{c}{d}}{2} \\ \frac{ad-bc}{bd} &= \sqrt{2}\frac{ed-cf}{2df} \\ \frac{(ad-bc)2df}{bd(ed-cf)} &= \sqrt{2} \end{aligned}$$

This expresses $\sqrt{2}$ as a quotient of two integers, so $\sqrt{2}$ is rational, a contradiction. Thus u is irrational.

In summary, we have produced an irrational number u with $r < u < s$, so the proof is complete. ■

21. There exist two prime numbers p and q for which $p-q = 97$.
This statement is **false.**
Disproof: Suppose for the sake of contradiction that this is true. Let p and q be prime numbers for which $p-q = 97$. Now, since their difference is odd, p and q must have opposite parity, so one of p and q is even and the other is odd. But there exists only one even prime number (namely 2), so either $p = 2$ or $q = 2$. If $p = 2$, then $p-q = 97$ implies $q = 2-97 = -95$, which is not prime. On the other hand if $q = 2$, then $p-q = 97$ implies $p = 99$, but that's not prime either. Thus one of p or q is not prime, a contradiction.

23. If $x,y \in \mathbb{R}$ and $x^3 < y^3$, then $x < y$. This is **true.**

Proof. (Contrapositive) Suppose $x \geq y$. We need to show $x^3 \geq y^3$.
Case 1. Suppose x and y have opposite signs, that is one of x and y is positive and the other is negative. Then since $x \geq y$, x is positive and y is negative. Then, since the powers are odd, x^3 is positive and y^3 is negative, so $x^3 \geq y^3$.
Case 2. Suppose x and y do not have opposite signs. Then $x^2+xy+y^2 \geq 0$ and also $x-y \geq 0$ because $x \geq y$. Thus we have $x^3-y^3 = (x-y)(x^2+xy+y^2) \geq 0$. From this we get $x^3-y^3 \geq 0$, so $x^3 \geq y^3$.
In either case we have $x^3 \geq y^3$. ■

25. For all $a,b,c \in \mathbb{Z}$, if $a \mid bc$, then $a \mid b$ or $a \mid c$.
This is **false.**
Disproof: Let $a=6$, $b=3$ and $c=4$. Note that $a \mid bc$, but $a \nmid b$ and $a \nmid c$.

27. The equation $x^2 = 2^x$ has three real solutions.

Proof. By inspection, the numbers $x=2$ and $x=4$ are two solutions of this equation. But there is a third solution. Let m be the real number for which $m2^m = \frac{1}{2}$. Then negative number $x=-2m$ is a solution, as follows.

$$x^2 = (-2m)^2 = 4m^2 = 4\left(\frac{m2^m}{2^m}\right)^2 = 4\left(\frac{\frac{1}{2}}{2^m}\right)^2 = \frac{1}{2^{2m}} = 2^{-2m} = 2^x.$$

Therefore we have three solutions 2, 4 and m. ■

29. If $x,y \in \mathbb{R}$ and $|x+y| = |x-y|$, then $y=0$.
This is **false.**
Disproof: Let $x=0$ and $y=1$. Then $|x+y| = |x-y|$, but $y=1$.

31. No number appears in Pascal's triangle more than four times.
Disproof: The number 120 appears six times. Check that $\binom{10}{3} = \binom{10}{7} = \binom{16}{2} = \binom{16}{14} = \binom{120}{1} = \binom{120}{119} = 120$.

33. Suppose $f(x) = a_0 + a_1x + a_2x^2 + \cdots + a_nx^n$ is a polynomial of degree 1 or greater, and for which each coefficient a_i is in $\mathbb{N}$. Then there is an $n \in \mathbb{N}$ for which the integer $f(n)$ is not prime.

Proof. (Outline) Note that, because the coefficients are all positive and the degree is greater than 1, we have $f(1) > 1$. Let $b = f(1) > 1$. Now, the polynomial $f(x)-b$ has a root 1, so $f(x)-b = (x-1)g(x)$ for some polynomial g. Then $f(x) = (x-1)g(x)+b$. Now note that $f(b+1) = bg(b)+b = b(g(b)+1)$. If we can now show that $g(b)+1$ is an integer, then we have a nontrivial factoring $f(b+1) = b(g(b)+1)$, and $f(b+1)$ is not prime. To complete the proof, use the fact that $f(x)-b = (x-1)g(x)$ has integer coefficients, and deduce that $g(x)$ must also have integer coefficients. ■

Chapter 10 Exercises

1. For every integer $n \in \mathbb{N}$, it follows that $1+2+3+4+\cdots+n = \dfrac{n^2+n}{2}$.

Proof. We will prove this with mathematical induction.

(1) Observe that if $n=1$, this statement is $1 = \dfrac{1^2+1}{2}$, which is obviously true.

(2) Consider any integer $k \geq 1$. We must show that S_k implies S_{k+1}. In other words, we must show that if $1+2+3+4+\cdots+k = \frac{k^2+k}{2}$ is true, then

$$1+2+3+4+\cdots+k+(k+1) = \frac{(k+1)^2+(k+1)}{2}$$

is also true. We use direct proof.
Suppose $k \geq 1$ and $1+2+3+4+\cdots+k = \frac{k^2+k}{2}$. Observe that

$$\begin{aligned}
1+2+3+4+\cdots+k+(k+1) &= \\
(1+2+3+4+\cdots+k)+(k+1) &= \\
\frac{k^2+k}{2}+(k+1) &= \frac{k^2+k+2(k+1)}{2} \\
&= \frac{k^2+2k+1\ +\ k+1}{2} \\
&= \frac{(k+1)^2+(k+1)}{2}.
\end{aligned}$$

Therefore we have shown that $1+2+3+4+\cdots+k+(k+1) = \frac{(k+1)^2+(k+1)}{2}$. ■

3. For every integer $n \in \mathbb{N}$, it follows that $1^3+2^3+3^3+4^3+\cdots+n^3 = \frac{n^2(n+1)^2}{4}$.

Proof. We will prove this with mathematical induction.
(1) When $n=1$ the statement is $1^3 = \frac{1^2(1+1)^2}{4} = \frac{4}{4} = 1$, which is true.
(2) Now assume the statement is true for some integer $n = k \geq 1$, that is assume $1^3+2^3+3^3+4^3+\cdots+k^3 = \frac{k^2(k+1)^2}{4}$. Observe that this implies the statement is true for $n = k+1$.

$$\begin{aligned}
1^3+2^3+3^3+4^3+\cdots+k^3+(k+1)^3 &= \\
(1^3+2^3+3^3+4^3+\cdots+k^3)+(k+1)^3 &= \\
\frac{k^2(k+1)^2}{4}+(k+1)^3 &= \frac{k^2(k+1)^2}{4}+\frac{4(k+1)^3}{4} \\
&= \frac{k^2(k+1)^2+4(k+1)^3}{4} \\
&= \frac{(k+1)^2(k^2+4(k+1)^1)}{4} \\
&= \frac{(k+1)^2(k^2+4k+4)}{4} \\
&= \frac{(k+1)^2(k+2)^2}{4} \\
&= \frac{(k+1)^2((k+1)+1)^2}{4}
\end{aligned}$$

Therefore $1^3+2^3+3^3+4^3+\cdots+k^3+(k+1)^3 = \frac{(k+1)^2((k+1)+1)^2}{4}$, which means the statement is true for $n = k+1$. ■

5. If $n \in \mathbb{N}$, then $2^1 + 2^2 + 2^3 + \cdots + 2^n = 2^{n+1} - 2$.

Proof. The proof is by mathematical induction.
(1) When $n = 1$, this statement is $2^1 = 2^{1+1} - 2$, or $2 = 4 - 2$, which is true.
(2) Now assume the statement is true for some integer $n = k \geq 1$, that is assume $2^1 + 2^2 + 2^3 + \cdots + 2^k = 2^{k+1} - 2$. Observe this implies that the statement is true for $n = k + 1$, as follows:

$$\begin{aligned} 2^1 + 2^2 + 2^3 + \cdots + 2^k + 2^{k+1} &= \\ (2^1 + 2^2 + 2^3 + \cdots + 2^k) + 2^{k+1} &= \\ 2^{k+1} - 2 + 2^{k+1} &= 2 \cdot 2^{k+1} - 2 \\ &= 2^{k+2} - 2 \\ &= 2^{(k+1)+1} - 2 \end{aligned}$$

Thus we have $2^1 + 2^2 + 2^3 + \cdots + 2^k + 2^{k+1} = 2^{(k+1)+1} - 2$, so the statement is true for $n = k + 1$.
Thus the result follows by mathematical induction. ■

7. If $n \in \mathbb{N}$, then $1 \cdot 3 + 2 \cdot 4 + 3 \cdot 5 + 4 \cdot 6 + \cdots + n(n+2) = \dfrac{n(n+1)(2n+7)}{6}$.

Proof. The proof is by mathematical induction.
(1) When $n = 1$, we have $1 \cdot 3 = \frac{1(1+1)(2+7)}{6}$, which is the true statement $3 = \frac{18}{6}$.
(2) Now assume the statement is true for some integer $n = k \geq 1$, that is assume $1 \cdot 3 + 2 \cdot 4 + 3 \cdot 5 + 4 \cdot 6 + \cdots + k(k+2) = \frac{k(k+1)(2k+7)}{6}$. Now observe that

$$\begin{aligned} 1 \cdot 3 + 2 \cdot 4 + 3 \cdot 5 + 4 \cdot 6 + \cdots + k(k+2) + (k+1)((k+1)+2) &= \\ (1 \cdot 3 + 2 \cdot 4 + 3 \cdot 5 + 4 \cdot 6 + \cdots + k(k+2)) + (k+1)((k+1)+2) &= \\ \frac{k(k+1)(2k+7)}{6} + (k+1)((k+1)+2) &= \\ \frac{k(k+1)(2k+7)}{6} + \frac{6(k+1)(k+3)}{6} &= \\ \frac{k(k+1)(2k+7) + 6(k+1)(k+3)}{6} &= \\ \frac{(k+1)(k(2k+7) + 6(k+3))}{6} &= \\ \frac{(k+1)(2k^2 + 13k + 18)}{6} &= \\ \frac{(k+1)(k+2)(2k+9)}{6} &= \\ \frac{(k+1)((k+1)+1)(2(k+1)+7)}{6} & \end{aligned}$$

Thus we have $1 \cdot 3 + 2 \cdot 4 + 3 \cdot 5 + 4 \cdot 6 + \cdots + k(k+2) + (k+1)((k+1)+2) = \frac{(k+1)((k+1)+1)(2(k+1)+7)}{6}$, and this means the statement is true for $n = k + 1$.
Thus the result follows by mathematical induction. ■

9. For any integer $n \ge 0$, it follows that $24 \mid (5^{2n} - 1)$.

Proof. The proof is by mathematical induction.

(1) For $n = 0$, the statement is $24 \mid (5^{2\cdot 0} - 1)$. This is $24 \mid 0$, which is true.

(2) Now assume the statement is true for some integer $n = k \ge 1$, that is assume $24 \mid (5^{2k} - 1)$. This means $5^{2k} - 1 = 24a$ for some integer a, and from this we get $5^{2k} = 24a + 1$. Now observe that

$$\begin{aligned} 5^{2(k+1)} - 1 &= \\ 5^{2k+2} - 1 &= \\ 5^2 5^{2k} - 1 &= \\ 5^2(24a+1) - 1 &= \\ 25(24a+1) - 1 &= \\ 25 \cdot 24a + 25 - 1 &= 24(25a+1). \end{aligned}$$

This shows $5^{2(k+1)} - 1 = 24(25a+1)$, which means $24 \mid 5^{2(k+1)} - 1$.

This completes the proof by mathematical induction. ■

11. For any integer $n \ge 0$, it follows that $3 \mid (n^3 + 5n + 6)$.

Proof. The proof is by mathematical induction.

(1) When $n = 0$, the statement is $3 \mid (0^3 + 5 \cdot 0 + 6)$, or $3 \mid 6$, which is true.

(2) Now assume the statement is true for some integer $n = k \ge 0$, that is assume $3 \mid (k^3 + 5k + 6)$. This means $k^3 + 5k + 6 = 3a$ for some integer a. We need to show that $3 \mid ((k+1)^3 + 5(k+1) + 6)$. Observe that

$$\begin{aligned} (k+1)^3 + 5(k+1) + 6 &= k^3 + 3k^2 + 3k + 1 + 5k + 5 + 6 \\ &= (k^3 + 5k + 6) + 3k^2 + 3k + 6 \\ &= 3a + 3k^2 + 3k + 6 \\ &= 3(a + k^2 + k + 2). \end{aligned}$$

Thus we have deduced $(k+1)^3 - (k+1) = 3(a + k^2 + k + 2)$. Since $a + k^2 + k + 2$ is an integer, it follows that $3 \mid ((k+1)^3 + 5(k+1) + 6)$.

It follows by mathematical induction that $3 \mid (n^3 + 5n + 6)$ for every $n \ge 0$. ■

13. For any integer $n \ge 0$, it follows that $6 \mid (n^3 - n)$.

Proof. The proof is by mathematical induction.

(1) When $n = 0$, the statement is $6 \mid (0^3 - 0)$, or $6 \mid 0$, which is true.

(2) Now assume the statement is true for some integer $n = k \geq 0$, that is, assume $6 \mid (k^3 - k)$. This means $k^3 - k = 6a$ for some integer a. We need to show that $6 \mid ((k+1)^3 - (k+1))$. Observe that

$$\begin{aligned} (k+1)^3 - (k+1) &= k^3 + 3k^2 + 3k + 1 - k - 1 \\ &= (k^3 - k) + 3k^2 + 3k \\ &= 6a + 3k^2 + 3k \\ &= 6a + 3k(k+1). \end{aligned}$$

Thus we have deduced $(k+1)^3 - (k+1) = 6a + 3k(k+1)$. Since one of k or $(k+1)$ must be even, it follows that $k(k+1)$ is even, so $k(k+1) = 2b$ for some integer b. Consequently $(k+1)^3 - (k+1) = 6a + 3k(k+1) = 6a + 3(2b) = 6(a+b)$. Since $(k+1)^3 - (k+1) = 6(a+b)$ it follows that $6 \mid ((k+1)^3 - (k+1))$.

Thus the result follows by mathematical induction. ■

15. If $n \in \mathbb{N}$, then $\frac{1}{1\cdot 2} + \frac{1}{2\cdot 3} + \frac{1}{3\cdot 4} + \frac{1}{4\cdot 5} + \cdots + \frac{1}{n(n+1)} = 1 - \frac{1}{n+1}$.

Proof. The proof is by mathematical induction.

(1) When $n = 1$, the statement is $\frac{1}{1(1+1)} = 1 - \frac{1}{1+1}$, which simplifies to $\frac{1}{2} = \frac{1}{2}$.

(2) Now assume the statement is true for some integer $n = k \geq 1$, that is assume $\frac{1}{1\cdot 2} + \frac{1}{2\cdot 3} + \frac{1}{3\cdot 4} + \frac{1}{4\cdot 5} + \cdots + \frac{1}{k(k+1)} = 1 - \frac{1}{k+1}$. Next we show that the statement for $n = k+1$ is true. Observe that

$$\begin{aligned} &\frac{1}{1\cdot 2} + \frac{1}{2\cdot 3} + \frac{1}{3\cdot 4} + \frac{1}{4\cdot 5} + \cdots + \frac{1}{k(k+1)} + \frac{1}{(k+1)((k+1)+1)} = \\ &\left(\frac{1}{1\cdot 2} + \frac{1}{2\cdot 3} + \frac{1}{3\cdot 4} + \frac{1}{4\cdot 5} + \cdots + \frac{1}{k(k+1)}\right) + \frac{1}{(k+1)(k+2)} = \\ &\left(1 - \frac{1}{k+1}\right) + \frac{1}{(k+1)(k+2)} = \\ &1 - \frac{1}{k+1} + \frac{1}{(k+1)(k+2)} = \\ &1 - \frac{k+2}{(k+1)(k+2)} + \frac{1}{(k+1)(k+2)} = \\ &1 - \frac{k+1}{(k+1)(k+2)} = \\ &1 - \frac{1}{k+2} = \\ &1 - \frac{1}{(k+1)+1}. \end{aligned}$$

This establishes $\frac{1}{1\cdot 2} + \frac{1}{2\cdot 3} + \frac{1}{3\cdot 4} + \frac{1}{4\cdot 5} + \cdots + \frac{1}{(k+1)((k+1)+1} = 1 - \frac{1}{(k+1)+1}$, which is to say that the statement is true for $n = k+1$.

This completes the proof by mathematical induction. ■

17. Suppose $A_1, A_2, \ldots A_n$ are sets in some universal set U, and $n \geq 2$. Prove that $\overline{A_1 \cap A_2 \cap \cdots \cap A_n} = \overline{A_1} \cup \overline{A_2} \cup \cdots \cup \overline{A_n}$.

Proof. The proof is by strong induction.
(1) When $n = 2$ the statement is $\overline{A_1 \cap A_2} = \overline{A_1} \cup \overline{A_2}$. This is not an entirely obvious statement, so we have to prove it. Observe that

$$\begin{aligned}
\overline{A_1 \cap A_2} &= \{x : (x \in U) \wedge (x \notin A_1 \cap A_2)\} \text{ (definition of complement)} \\
&= \{x : (x \in U) \wedge \sim (x \in A_1 \cap A_2)\} \\
&= \{x : (x \in U) \wedge \sim ((x \in A_1) \wedge (x \in A_2))\} \text{ (definition of } \cap) \\
&= \{x : (x \in U) \wedge (\sim (x \in A_1) \vee \sim (x \in A_2))\} \text{ (DeMorgan)} \\
&= \{x : (x \in U) \wedge ((x \notin A_1) \vee (x \notin A_2))\} \\
&= \{x : (x \in U) \wedge (x \notin A_1) \vee (x \in U) \wedge (x \notin A_2)\} \text{ (distributive prop.)} \\
&= \{x : ((x \in U) \wedge (x \notin A_1))\} \cup \{x : ((x \in U) \wedge (x \notin A_2))\} \text{ (def. of } \cup) \\
&= \overline{A_1} \cup \overline{A_2} \text{ (definition of complement)}
\end{aligned}$$

(2) Let $k \geq 2$. Assume the statement is true if it involves k or fewer sets. Then

$$\begin{aligned}
\overline{A_1 \cap A_2 \cap \cdots \cap A_{k-1} \cap A_k \cap A_{k+1}} &= \\
\overline{A_1 \cap A_2 \cap \cdots \cap A_{k-1} \cap (A_k \cap A_{k+1})} &= \overline{A_1} \cup \overline{A_2} \cup \cdots \cup \overline{A_{k-1}} \cup \overline{A_k \cap A_{k+1}} \\
&= \overline{A_1} \cup \overline{A_2} \cup \cdots \cup \overline{A_{k-1}} \cup \overline{A_k} \cup \overline{A_{k+1}}
\end{aligned}$$

Thus the statement is true when it involves $k+1$ sets.
This completes the proof by strong induction. ■

19. Prove $\sum_{k=1}^{n} 1/k^2 \leq 2 - 1/n$ for every n.

Proof. This clearly holds for $n = 1$. Assume it holds for some $n \geq 1$. Then $\sum_{k=1}^{n+1} 1/k^2 \leq 2 - 1/n + 1/(n+1)^2 = 2 - \frac{(n+1)^2 - n}{n(n+1)^2} \leq 2 - 1/(n+1)$. The proof is complete. ■

21. If $n \in \mathbb{N}$, then $\frac{1}{1} + \frac{1}{2} + \frac{1}{3} + \cdots + \frac{1}{2^n} \geq 1 + \frac{n}{2}$.

Proof. If $n = 1$, the result is obvious.
Assume the proposition holds for some $n > 1$. Then

$$\begin{aligned}
\frac{1}{1} + \frac{1}{2} + \frac{1}{3} + \cdots + \frac{1}{2^{n+1}} &= \left(\frac{1}{1} + \frac{1}{2} + \frac{1}{3} + \cdots + \frac{1}{2^n}\right) + \left(\frac{1}{2^n+1} + \frac{1}{2^n+2} + \frac{1}{2^n+3} + \cdots + \frac{1}{2^{n+1}}\right) \\
&\geq \left(1 + \frac{n}{2}\right) + \left(\frac{1}{2^n+1} + \frac{1}{2^n+2} + \frac{1}{2^n+3} + \cdots + \frac{1}{2^{n+1}}\right).
\end{aligned}$$

Now, the sum $\left(\frac{1}{2^n+1} + \frac{1}{2^n+2} + \frac{1}{2^n+3} + \cdots + \frac{1}{2^{n+1}}\right)$ on the right has $2^{n+1} - 2^n = 2^n$ terms, all greater than or equal to $\frac{1}{2^{n+1}}$, so the sum is greater than $2^n \frac{1}{2^{n+1}} = \frac{1}{2}$. Therefore we get $\frac{1}{1} + \frac{1}{2} + \frac{1}{3} + \cdots + \frac{1}{2^{n+1}} \geq \left(1 + \frac{n}{2}\right) + \left(\frac{1}{2^n+1} + \frac{1}{2^n+2} + \frac{1}{2^n+3} + \cdots + \frac{1}{2^{n+1}}\right) \geq \left(1 + \frac{n}{2}\right) + \frac{1}{2} = 1 + \frac{n+1}{2}$. This means the result is true for $n+1$, so the theorem is proved. ■

23. Use induction to prove the binomial theorem $(x+y)^n = \sum_{i=0}^{n} \binom{n}{i} x^{n-i} y^i$.

Proof. Notice that when $n=1$, the formula is $(x+y)^1 = \binom{1}{0}x^1y^0 + \binom{1}{1}x^0y^1 = x+y$, which is true.

Now assume the theorem is true for some $n>1$. We will show that this implies that it is true for the power $n+1$. Just observe that

$$\begin{aligned}
(x+y)^{n+1} &= (x+y)(x+y)^n \\
&= (x+y)\sum_{i=0}^{n}\binom{n}{i}x^{n-i}y^i \\
&= \sum_{i=0}^{n}\binom{n}{i}x^{(n+1)-i}y^i + \sum_{i=0}^{n}\binom{n}{i}x^{n-i}y^{i+1} \\
&= \sum_{i=0}^{n}\left[\binom{n}{i}+\binom{n}{i-1}\right]x^{(n+1)-i}y^i + y^{n+1} \\
&= \sum_{i=0}^{n}\binom{n+1}{i}x^{(n+1)-i}y^i + \binom{n+1}{n+1}y^{n+1} \\
&= \sum_{i=0}^{n+1}\binom{n+1}{i}x^{(n+1)-i}y^i.
\end{aligned}$$

This shows that the formula is true for $(x+y)^{n+1}$, so the theorem is proved. ■

25. Concerning the Fibonacci sequence, prove that $F_1+F_2+F_3+F_4+\ldots+F_n = F_{n+2}-1$.

Proof. The proof is by induction.

(1) When $n=1$ the statement is $F_1 = F_{1+2}-1 = F_3-1 = 2-1 = 1$, which is true. Also when $n=2$ the statement is $F_1+F_2 = F_{2+2}-1 = F_4-1 = 3-1 = 2$, which is true, as $F_1+F_2 = 1+1 = 2$.

(2) Now assume $k \geq 1$ and $F_1+F_2+F_3+F_4+\ldots+F_k = F_{k+2}-1$. We need to show $F_1+F_2+F_3+F_4+\ldots+F_k+F_{k+1} = F_{k+3}-1$. Observe that

$$\begin{aligned}
F_1+F_2+F_3+F_4+\ldots+F_k+F_{k+1} &= \\
(F_1+F_2+F_3+F_4+\ldots+F_k)+F_{k+1} &= \\
F_{k+2}-1++F_{k+1} &= (F_{k+1}+F_{k+2})-1 \\
&= F_{k+3}-1.
\end{aligned}$$

This completes the proof by induction. ■

27. Concerning the Fibonacci sequence, prove that $F_1+F_3+\cdots+F_{2n-1} = F_{2n}$.

Proof. If $n=1$, the result is immediate. Assume for some $n>1$ we have $\sum_{i=1}^{n} F_{2i-1} = F_{2n}$. Then $\sum_{i=1}^{n+1} F_{2i-1} = F_{2n+1}+\sum_{i=1}^{n} F_{2i-1} = F_{2n+1}+F_{2n} = F_{2n+2} = F_{2(n+1)}$ as desired. ■

29. Prove that $\binom{n}{0}+\binom{n-1}{1}+\binom{n-2}{2}+\binom{n-3}{3}+\cdots+\binom{1}{n-1}+\binom{0}{n}=F_{n+1}$.

Proof. (Strong Induction) For $n=1$ this is $\binom{1}{0}+\binom{0}{1}=1+0=1=F_2=F_{1+1}$. Thus the assertion is true when $n=1$.

Now fix n and assume that $\binom{k}{0}+\binom{k-1}{1}+\binom{k-2}{2}+\binom{k-3}{3}+\cdots+\binom{1}{k-1}+\binom{0}{k}=F_{k+1}$ whenever $k<n$. In what follows we use the identity $\binom{n}{k}=\binom{n-1}{k-1}+\binom{n-1}{k}$. We also often use $\binom{a}{b}=0$ whenever it is untrue that $0\le b\le a$.

$$\begin{aligned}
&\binom{n}{0}+\binom{n-1}{1}+\binom{n-2}{2}+\cdots+\binom{1}{n-1}+\binom{0}{n}\\
=\;&\binom{n}{0}+\binom{n-1}{1}+\binom{n-2}{2}+\cdots+\binom{1}{n-1}\\
=\;&\binom{n-1}{-1}+\binom{n-1}{0}+\binom{n-2}{0}+\binom{n-2}{1}+\binom{n-3}{1}+\binom{n-3}{2}+\cdots+\binom{0}{n-1}+\binom{0}{n}\\
=\;&\binom{n-1}{0}+\binom{n-2}{0}+\binom{n-2}{1}+\binom{n-3}{1}+\binom{n-3}{2}+\cdots+\binom{0}{n-1}+\binom{0}{n}\\
=\;&\left[\binom{n-1}{0}+\binom{n-2}{1}+\cdots+\binom{0}{n-1}\right]+\left[\binom{n-2}{0}+\binom{n-3}{1}+\cdots+\binom{0}{n-2}\right]\\
=\;&F_n+F_{n-1}=F_n
\end{aligned}$$

This completes the proof. ■

31. Prove that $\sum_{k=0}^{n}\binom{k}{r}=\binom{n+1}{r+1}$, where $r\in\mathbb{N}$.
Hint: Use induction on the integer n. After doing the basis step, break up the expression $\binom{k}{r}$ as $\binom{k}{r}=\binom{k-1}{r-1}+\binom{k-1}{r}$. Then regroup, use the induction hypothesis, and recombine using the above identity.

33. Suppose that n infinitely long straight lines lie on the plane in such a way that no two are parallel, and no three intersect at a single point. Show that this arrangement divides the plane into $\frac{n^2+n+2}{2}$ regions.

Proof. The proof is by induction. For the basis step, suppose $n=1$. Then there is one line, and it clearly divides the plane into 2 regions, one on either side of the line. As $2=\frac{1^2+1+2}{2}=\frac{n^2+n+2}{2}$, the formula is correct when $n=1$.
Now suppose there are $n+1$ lines on the plane, and that the formula is correct for when there are n lines on the plane. Single out one of the $n+1$ lines on the plane, and call it ℓ. Remove line ℓ, so that there are now n lines on the plane.
By the induction hypothesis, these n lines divide the plane into $\frac{n^2+n+2}{2}$ regions. Now add line ℓ back. Doing this adds an additional $n+1$ regions. (The diagram illustrates the case where $n+1=5$. Without ℓ, there are $n=4$ lines. Adding ℓ back produces $n+1=5$ new regions.)

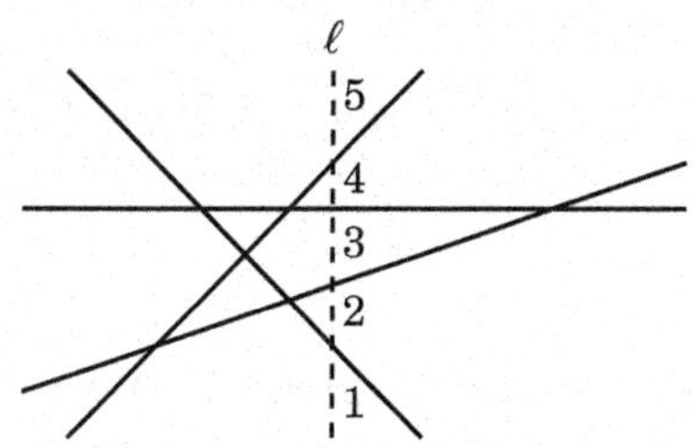

Thus, with $n+1$ lines there are all together $(n+1)+\frac{n^2+n+2}{2}$ regions. Observe

$$(n+1)+\frac{n^2+n+2}{2}=\frac{2n+2+n^2+n+2}{2}=\frac{(n+1)^2+(n+1)+2}{2}.$$

Thus, with $n+1$ lines, we have $\frac{(n+1)^2+(n+1)+2}{2}$ regions, which means that the formula is true for when there are $n+1$ lines. We have shown that if the formula is true for n lines, it is also true for $n+1$ lines. This completes the proof by induction. ■

35. If $n,k \in \mathbb{N}$, and n is even and k is odd, then $\binom{n}{k}$ is even.

Proof. Notice that if k is not a value between 0 and n, then $\binom{n}{k}=0$ is even; thus from here on we can assume that $0<k<n$. We will use strong induction.

For the basis case, notice that the assertion is true for the even values $n=2$ and $n=4$: $\binom{2}{1}=2$; $\binom{4}{1}=4$; $\binom{4}{3}=4$ (even in each case).
Now fix and even n assume that $\binom{m}{k}$ is even whenever m is even, k is odd, and $m<n$. Using the identity $\binom{n}{k}=\binom{n-1}{k-1}+\binom{n-1}{k}$ three times, we get

$$\begin{aligned}\binom{n}{k} &= \binom{n-1}{k-1}+\binom{n-1}{k}\\ &= \binom{n-2}{k-2}+\binom{n-2}{k-1}+\binom{n-2}{k-1}+\binom{n-2}{k}\\ &= \binom{n-2}{k-2}+2\binom{n-2}{k-1}+\binom{n-2}{k}.\end{aligned}$$

Now, $n-2$ is even, and k and $k-2$ are odd. By the inductive hypothesis, the outer terms of the above expression are even, and the middle is clearly even; thus we have expressed $\binom{n}{k}$ as the sum of three even integers, so it is even. ■

Chapter 11 Exercises

Section 11.0 Exercises

1. Let $A=\{0,1,2,3,4,5\}$. Write out the relation R that expresses $>$ on A. Then illustrate it with a diagram.

$R=\{(5,4),(5,3),(5,3),(5,3),(5,1),(5,0),(4,3),(4,2),(4,1),$
$(4,0),(3,2),(3,1),(3,0),(2,1),(2,0),(1,0)\}$

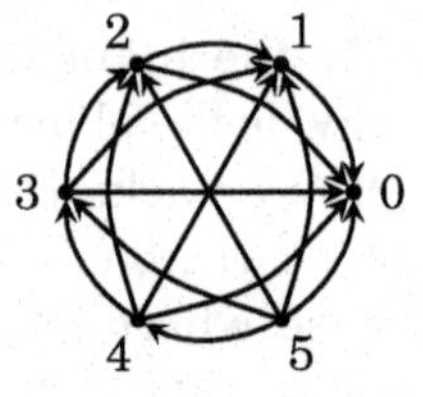

3. Let $A=\{0,1,2,3,4,5\}$. Write out the relation R that expresses $\geq$ on A. Then illustrate it with a diagram.

$$R = \{(5,5),(5,4),(5,3),(5,2),(5,1),(5,0),(4,4),(4,3),(4,2),(4,1),(4,0),(3,3),(3,2),(3,1),(3,0),(2,2),(2,1),(2,0),(1,1),(1,0),(0,0)\}$$

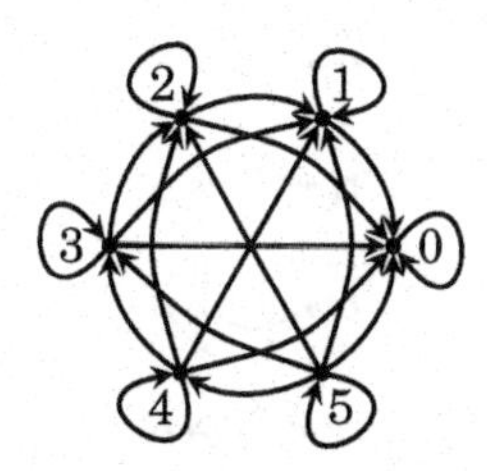

5. The following diagram represents a relation R on a set A. Write the sets A and R. Answer: $A = \{0,1,2,3,4,5\}$; $R = \{(3,3),(4,3),(4,2),(1,2),(2,5),(5,0)\}$

7. Write the relation $<$ on the set $A = \mathbb{Z}$ as a subset R of $\mathbb{Z}\times\mathbb{Z}$. This is an infinite set, so you will have to use set-builder notation.
Answer: $R = \{(x,y) \in \mathbb{Z}\times\mathbb{Z} : y - x \in \mathbb{N}\}$

9. How many different relations are there on the set $A = \{1,2,3,4,5,6\}$?
Consider forming a relation $R \subseteq A\times A$ on A. For each ordered pair $(x,y) \in A\times A$, we have two choices: we can either include (x,y) in R or not include it. There are $6\cdot 6 = 36$ ordered pairs in $A\times A$. By the multiplication principle, there are thus 2^{36} different subsets R and hence also this many relations on A.

11. Answer: $2^{(|A|^2)}$ **13.** Answer: $\neq$ **15.** Answer: $\equiv \pmod 3$

Section 11.1 Exercises

1. Consider the relation $R = \{(a,a),(b,b),(c,c),(d,d),(a,b),(b,a)\}$ on the set $A = \{a,b,c,d\}$. Which of the properties reflexive, symmetric and transitive does R possess and why? If a property does not hold, say why.
This **is reflexive** because $(x,x) \in R$ (i.e., xRx)for every $x \in A$.
It **is symmetric** because it is impossible to find an $(x,y) \in R$ for which $(y,x) \notin R$.
It **is transitive** because $(xRy \wedge yRz) \Rightarrow xRz$ always holds.

3. Consider the relation $R = \{(a,b),(a,c),(c,b),(b,c)\}$ on the set $A = \{a,b,c\}$. Which of the properties reflexive, symmetric and transitive does R possess and why? If a property does not hold, say why.
This **is not reflexive** because $(a,a) \notin R$ (for example).
It **is not symmetric** because $(a,b) \in R$ but $(b,a) \notin R$.
It **is not transitive** because cRb and bRc are true, but cRc is false.

5. Consider the relation $R = \{(0,0),(\sqrt{2},0),(0,\sqrt{2}),(\sqrt{2},\sqrt{2})\}$ on $\mathbb{R}$. Say whether this relation is reflexive, symmetric and transitive. If a property does not hold, say why.
This **is not reflexive** because $(1,1) \notin R$ (for example).
It **is symmetric** because it is impossible to find an $(x,y) \in R$ for which $(y,x) \notin R$.
It **is transitive** because $(xRy \wedge yRz) \Rightarrow xRz$ always holds.

7. There are 16 possible different relations R on the set $A = \{a,b\}$. Describe all of them. (A picture for each one will suffice, but don't forget to label the nodes.) Which ones are reflexive? Symmetric? Transitive?

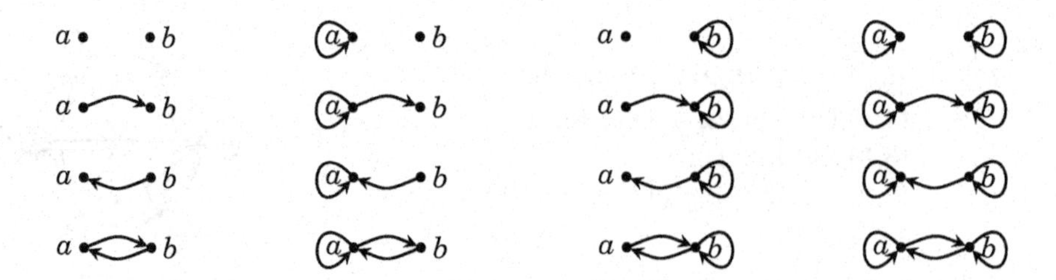

Only the four in the right column are reflexive. Only the eight in the first and fourth rows are symmetric. All of them are transitive **except** the first three on the fourth row.

9. Define a relation on $\mathbb{Z}$ by declaring xRy if and only if x and y have the same parity. Say whether this relation is reflexive, symmetric and transitive. If a property does not hold, say why. What familiar relation is this?
This **is reflexive** because xRx since x always has the same parity as x.
It **is symmetric** because if x and y have the same parity, then y and x must have the same parity (that is, $xRy \Rightarrow yRx$).
It **is transitive** because if x and y have the same parity and y and z have the same parity, then x and z must have the same parity. (That is $(xRy \wedge yRz) \Rightarrow xRz$ always holds.)
The relation is congruence modulo 2.

11. Suppose $A = \{a,b,c,d\}$ and $R = \{(a,a),(b,b),(c,c),(d,d)\}$. Say whether this relation is reflexive, symmetric and transitive. If a property does not hold, say why.
This **is reflexive** because $(x,x) \in R$ for every $x \in A$.
It **is symmetric** because it is impossible to find an $(x,y) \in R$ for which $(y,x) \notin R$.
It **is transitive** because $(xRy \wedge yRz) \Rightarrow xRz$ always holds.
(For example $(aRa \wedge aRa) \Rightarrow aRa$ is true, etc.)

13. Consider the relation $R = \{(x,y) \in \mathbb{R} \times \mathbb{R} : x - y \in \mathbb{Z}\}$ on $\mathbb{R}$. Prove that this relation is reflexive and symmetric, and transitive.

Proof. In this relation, xRy means $x - y \in \mathbb{Z}$.

To see that R is reflexive, take any $x \in \mathbb{R}$ and observe that $x - x = 0 \in \mathbb{Z}$, so xRx. Therefore R is reflexive.

To see that R is symmetric, we need to prove $xRy \Rightarrow yRx$ for all $x, y \in \mathbb{R}$. We use direct proof. Suppose xRy. This means $x - y \in \mathbb{Z}$. Then it follows that $-(x-y) = y - x$ is also in $\mathbb{Z}$. But $y - x \in \mathbb{Z}$ means yRx. We've shown xRy implies yRx, so R is symmetric.

To see that R is transitive, we need to prove $(xRy \wedge yRz) \Rightarrow xRz$ is always true. We prove this conditional statement with direct proof. Suppose xRy and yRz. Since xRy, we know $x - y \in \mathbb{Z}$. Since yRz, we know $y - z \in \mathbb{Z}$. Thus $x - y$ and $y - z$ are both integers; by adding these integers we get another integer $(x-y)+(y-z) = x - z$. Thus $x - z \in \mathbb{Z}$, and this means xRz. We've now shown that if xRy and yRz, then xRz. Therefore R is transitive. ■

15. Prove or disprove: If a relation is symmetric and transitive, then it is also reflexive.
This is **false**. For a counterexample, consider the relation $R = \{(a,a),(a,b),(b,a),(b,b)\}$ on the set $A = \{a,b,c\}$. This is symmetric and transitive but it is not reflexive.

17. Define a relation $\sim$ on $\mathbb{Z}$ as $x \sim y$ if and only if $|x-y| \leq 1$. Say whether $\sim$ is reflexive, symmetric and transitive.
This is reflexive because $|x-x| = 0 \leq 1$ for all integers x. It is symmetric because $x \sim y$ if and only if $|x-y| \leq 1$, if and only if $|y-x| \leq 1$, if and only if $y \sim x$. It is not transitive because, for example, $0 \sim 1$ and $1 \sim 2$, but is not the case that $0 \sim 2$.

Section 11.2 Exercises

1. Let $A = \{1,2,3,4,5,6\}$, and consider the following equivalence relation on A: $R = \{(1,1),(2,2),(3,3),(4,4),(5,5),(6,6),(2,3),(3,2),(4,5),(5,4),(4,6),(6,4),(5,6),(6,5)\}$. List the equivalence classes of R.

The equivalence classes are: $[1] = \{1\}$; $\quad [2] = [3] = \{2,3\}$; $\quad [4] = [5] = [6] = \{4,5,6\}$.

3. Let $A = \{a,b,c,d,e\}$. Suppose R is an equivalence relation on A. Suppose R has three equivalence classes. Also aRd and bRc. Write out R as a set.
Answer: $R = \{(a,a),(b,b),(c,c),(d,d),(e,e),(a,d),(d,a),(b,c),(c,b)\}$.

5. There are two different equivalence relations on the set $A = \{a,b\}$. Describe them all. Diagrams will suffice.
Answer: $R = \{(a,a),(b,b)\}$ and $R = \{(a,a),(b,b),(a,b),(b,a)\}$

7. Define a relation R on $\mathbb{Z}$ as xRy if and only if $3x-5y$ is even. Prove R is an equivalence relation. Describe its equivalence classes.

To prove that R is an equivalence relation, we must show it's reflexive, symmetric and transitive.
The relation R is reflexive for the following reason. If $x \in \mathbb{Z}$, then $3x-5x = -2x$ is even. But then since $3x-5x$ is even, we have xRx. Thus R is reflexive.

To see that R is symmetric, suppose xRy. We must show yRx. Since xRy, we know $3x-5y$ is even, so $3x-5y = 2a$ for some integer a. Now reason as follows:

$$\begin{aligned} 3x-5y &= 2a \\ 3x-5y+8y-8x &= 2a+8y-8x \\ 3y-5x &= 2(a+4y-4x). \end{aligned}$$

From this it follows that $3y-5x$ is even, so yRx. We've now shown xRy implies yRx, so R is symmetric.

To prove that R is transitive, assume that xRy and yRz. (We will show that this implies xRz.) Since xRy and yRz, it follows that $3x-5y$ and $3y-5z$ are both even, so $3x-5y = 2a$ and $3y-5z = 2b$ for some integers a and b. Adding these equations, we get $(3x-5y)+(3y-5z) = 2a+2b$, and this simplifies to $3x-5z = 2(a+b+y)$.

Therefore $3x-5z$ is even, so xRz. We've now shown that if xRy and yRz, then xRz, so R is transitive.

We've now shown that R is reflexive, symmetric and transitive, so it is an equivalence relation.

The completes the first part of the problem. Now we move on the second part. To find the equivalence classes, first note that

$$[0] = \{x \in \mathbb{Z} : xR0\} = \{x \in \mathbb{Z} : 3x - 5 \cdot 0 \text{ is even}\} = \{x \in \mathbb{Z} : 3x \text{ is even}\} = \{x \in \mathbb{Z} : x \text{ is even}\}.$$

Thus the equivalence class [0] consists of all even integers. Next, note that

$$[1] = \{x \in \mathbb{Z} : xR1\} = \{x \in \mathbb{Z} : 3x - 5 \cdot 1 \text{ is even}\} = \{x \in \mathbb{Z} : 3x - 5 \text{ is even}\} = \{x \in \mathbb{Z} : x \text{ is odd}\}.$$

Thus the equivalence class [1] consists of all odd integers.

Consequently there are just two equivalence classes $\{\ldots,-4,-2,0,2,4,\ldots\}$ and $\{\ldots,-3,-1,1,3,5,\ldots\}$.

9. Define a relation R on $\mathbb{Z}$ as xRy if and only if $4 \mid (x+3y)$. Prove R is an equivalence relation. Describe its equivalence classes.

This is reflexive, because for any $x \in \mathbb{Z}$ we have $4 \mid (x+3x)$, so xRx.

To prove that R is symmetric, suppose xRy. Then $4 \mid (x+3y)$, so $x+3y = 4a$ for some integer a. Multiplying by 3, we get $3x+9y = 12a$, which becomes $y+3x = 12a-8y$. Then $y+3x = 4(3a-2y)$, so $4 \mid (y+3x)$, hence yRx. Thus we've shown xRy implies yRx, so R is symmetric.

To prove transitivity, suppose xRy and yRz. Then $4|(x+3y)$ and $4|(y+3z)$, so $x+3y = 4a$ and $y+3z = 4b$ for some integers a and b. Adding these two equations produces $x+4y+3z = 4a+4b$, or $x+3z = 4a+4b-4y = 4(a+b-y)$. Consequently $4|(x+3z)$, so xRz, and R is transitive.

As R is reflexive, symmetric and transitive, it is an equivalence relation.

Now let's compute its equivalence classes.

$[0] = \{x \in \mathbb{Z} : xR0\} = \{x \in \mathbb{Z} : 4 \mid (x+3\cdot 0)\} = \{x \in \mathbb{Z} : 4 \mid x\} = \{\ldots -4,0,4,8,12,16\ldots\}$
$[1] = \{x \in \mathbb{Z} : xR1\} = \{x \in \mathbb{Z} : 4 \mid (x+3\cdot 1)\} = \{x \in \mathbb{Z} : 4 \mid (x+3)\} = \{\ldots -3,1,5,9,13,17\ldots\}$
$[2] = \{x \in \mathbb{Z} : xR2\} = \{x \in \mathbb{Z} : 4 \mid (x+3\cdot 2)\} = \{x \in \mathbb{Z} : 4 \mid (x+6)\} = \{\ldots -2,2,6,10,14,18\ldots\}$
$[3] = \{x \in \mathbb{Z} : xR3\} = \{x \in \mathbb{Z} : 4 \mid (x+3\cdot 3)\} = \{x \in \mathbb{Z} : 4 \mid (x+9)\} = \{\ldots -1,3,7,11,15,19\ldots\}$

11. Prove or disprove: If R is an equivalence relation on an infinite set A, then R has infinitely many equivalence classes.
This is **False**. Counterexample: consider the relation of congruence modulo 2. It is a relation on the infinite set $\mathbb{Z}$, but it has only two equivalence classes.

13. Answer: $m|A|$

15. Answer: 15

Section 11.3 Exercises

1. List all the partitions of the set $A = \{a, b\}$. Compare your answer to the answer to Exercise 5 of Section 11.2.
There are just two partitions $\{\{a\},\{b\}\}$ and $\{\{a,b\}\}$. These correspond to the two equivalence relations $R_1 = \{(a,a),(b,b)\}$ and $R_2 = \{(a,a),(a,b),(b,a),(b,b)\}$, respectively, on A.

3. Describe the partition of $\mathbb{Z}$ resulting from the equivalence relation $\equiv \pmod{4}$.
Answer: The partition is $\{[0],[1],[2],[3]\}$ =
$\big\{\{\ldots,-4,0,4,8,12,\ldots\},\{\ldots,-3,1,5,9,13,\ldots\},\ \{\ldots,-2,2,4,6,10,14,\ldots\},\ \{\ldots,-1,3,7,11,15,\ldots\}\big\}$

5. Answer: Congruence modulo 2, or "same parity."

Section 11.4 Exercises

1. Write the addition and multiplication tables for $\mathbb{Z}_2$.

+	[0]	[1]
[0]	[0]	[1]
[1]	[1]	[0]

·	[0]	[1]
[0]	[0]	[0]
[1]	[0]	[1]

3. Write the addition and multiplication tables for $\mathbb{Z}_4$.

+	[0]	[1]	[2]	[3]
[0]	[0]	[1]	[2]	[3]
[1]	[1]	[2]	[3]	[0]
[2]	[2]	[3]	[0]	[1]
[3]	[3]	[0]	[1]	[2]

·	[0]	[1]	[2]	[3]
[0]	[0]	[0]	[0]	[0]
[1]	[0]	[1]	[2]	[3]
[2]	[0]	[2]	[0]	[2]
[3]	[0]	[3]	[2]	[1]

5. Suppose $[a],[b] \in \mathbb{Z}_5$ and $[a]\cdot[b] = [0]$. Is it necessarily true that either $[a] = [0]$ or $[b] = [0]$?

The multiplication table for $\mathbb{Z}_5$ is shown in Section 11.4. In the body of that table, the only place that [0] occurs is in the first row or the first column. That row and column are both headed by [0]. It follows that if $[a]\cdot[b] = [0]$, then either $[a]$ or $[b]$ must be [0].

7. Do the following calculations in $\mathbb{Z}_9$, in each case expressing your answer as $[a]$ with $0 \le a \le 8$.
(a) $[8]+[8] = [7]$ **(b)** $[24]+[11] = [8]$ **(c)** $[21]\cdot[15] = [0]$ **(d)** $[8]\cdot[8] = [1]$

Chapter 12 Exercises

Section 12.1 Exercises

1. Suppose $A = \{0,1,2,3,4\}$, $B = \{2,3,4,5\}$ and $f = \{(0,3),(1,3),(2,4),(3,2),(4,2)\}$. State the domain and range of f. Find $f(2)$ and $f(1)$.
Domain is A; Range is $\{2,3,4\}$; $f(2) = 4$; $f(1) = 3$.

3. There are four different functions $f : \{a,b\} \to \{0,1\}$. List them all. Diagrams will suffice.
$f_1 = \{(a,0),(b,0)\}$ $f_2 = \{(a,1),(b,0)\}$, $f_3 = \{(a,0),(b,1)\}$ $f_4 = \{(a,1),(b,1)\}$

5. Give an example of a relation from $\{a,b,c,d\}$ to $\{d,e\}$ that is not a function.
One example is $\{(a,d),(a,e),(b,d),(c,d),(d,d)\}$.

7. Consider the set $f = \{(x,y) \in \mathbb{Z} \times \mathbb{Z} : 3x + y = 4\}$. Is this a function from $\mathbb{Z}$ to $\mathbb{Z}$? Explain.
Yes, since $3x + y = 4$ if and only if $y = 4 - 3x$, this is the function $f : \mathbb{Z} \to \mathbb{Z}$ defined as $f(x) = 4 - 3x$.

9. Consider the set $f = \{(x^2, x) : x \in \mathbb{R}\}$. Is this a function from $\mathbb{R}$ to $\mathbb{R}$? Explain.
No. This is not a function. Observe that f contains the ordered pairs $(4,2)$ and $(4,-2)$. Thus the real number 4 occurs as the first coordinate of more than one element of f.

11. Is the set $\theta = \{(X, |X|) : X \subseteq \mathbb{Z}_5\}$ a function? If so, what is its domain and range?
Yes, this is a function. The domain is $\mathscr{P}(\mathbb{Z}_5)$. The range is $\{0,1,2,3,4,5\}$.

Section 12.2 Exercises

1. Let $A = \{1,2,3,4\}$ and $B = \{a,b,c\}$. Give an example of a function $f : A \to B$ that is neither injective nor surjective.
Consider $f = \{(1,a),(2,a),(3,a),(4,a)\}$.
Then f is not injective because $f(1) = f(2)$.
Also f is not surjective because it sends no element of A to the element $c \in B$.

3. Consider the cosine function $\cos : \mathbb{R} \to \mathbb{R}$. Decide whether this function is injective and whether it is surjective. What if it had been defined as $\cos : \mathbb{R} \to [-1,1]$?
The function $\cos : \mathbb{R} \to \mathbb{R}$ is **not injective** because, for example, $\cos(0) = \cos(2\pi)$. It is **not surjective** because if $b = 5 \in \mathbb{R}$ (for example), there is no real number for which $\cos(x) = b$. The function $\cos : \mathbb{R} \to [-1,1]$ **is surjective.** but not injective.

5. A function $f : \mathbb{Z} \to \mathbb{Z}$ is defined as $f(n) = 2n + 1$. Verify whether this function is injective and whether it is surjective.
This function is injective. To see this, suppose $m,n \in \mathbb{Z}$ and $f(m) = f(n)$.
This means $2m + 1 = 2n + 1$, from which we get $2m = 2n$, and then $m = n$.
Thus f is injective.
This function is not surjective. To see this notice that $f(n)$ is odd for all $n \in \mathbb{Z}$. So given the (even) number 2 in the codomain $\mathbb{Z}$, there is no n with $f(n) = 2$.

7. A function $f : \mathbb{Z} \times \mathbb{Z} \to \mathbb{Z}$ is defined as $f((m,n)) = 2n - 4m$. Verify whether this function is injective and whether it is surjective.
This is **not injective** because $(0,2) \neq (-1,0)$, yet $f((0,2)) = f((-1,0)) = 4$. This is **not surjective** because $f((m,n)) = 2n - 4m = 2(n - 2m)$ is always even. If $b \in \mathbb{Z}$ is odd, then $f((m,n)) \neq b$, for all $(m,n) \in \mathbb{Z} \times \mathbb{Z}$.

9. Prove that the function $f : \mathbb{R} - \{2\} \to \mathbb{R} - \{5\}$ defined by $f(x) = \frac{5x+1}{x-2}$ is bijective.

Proof. First, let's check that f is injective. Suppose $f(x) = f(y)$. Then

$$\begin{aligned}
\frac{5x+1}{x-2} &= \frac{5y+1}{y-2} \\
(5x+1)(y-2) &= (5y+1)(x-2) \\
5xy - 10x + y - 2 &= 5yx - 10y + x - 2 \\
-10x + y &= -10y + x \\
11y &= 11x \\
y &= x.
\end{aligned}$$

Since $f(x) = f(y)$ implies $x = y$, it follows that f is injective.
Next, let's check that f is surjective. For this, take an arbitrary element $b \in \mathbb{R} - \{5\}$. We want to see if there is an $x \in \mathbb{R} - \{2\}$ for which $f(x) = b$, or $\frac{5x+1}{x-2} = b$. Solving this for x, we get:

$$\begin{aligned}
5x + 1 &= b(x-2) \\
5x + 1 &= bx - 2b \\
5x - xb &= -2b - 1 \\
x(5-b) &= -2b - 1.
\end{aligned}$$

Since we have assumed $b \in \mathbb{R} - \{5\}$, the term $(5 - b)$ is not zero, and we can divide with impunity to get $x = \dfrac{-2b-1}{5-b}$. This is an x for which $f(x) = b$, so f is surjective.
Since f is both injective and surjective, it is bijective. ■

11. Consider the function $\theta : \{0,1\} \times \mathbb{N} \to \mathbb{Z}$ defined as $\theta(a,b) = (-1)^a b$. Is θ injective? Is it surjective? Explain.
First we show that θ is injective. Suppose $\theta(a,b) = \theta(c,d)$. Then $(-1)^a b = (-1)^c d$. As b and d are both in $\mathbb{N}$, they are both positive. Then because $(-1)^a b = (-1)^c d$, it follows that $(-1)^a$ and $(-1)^c$ have the same sign. Since each of $(-1)^a$ and $(-1)^c$ equals ± 1, we have $(-1)^a = (-1)^c$, so then $(-1)^a b = (-1)^c d$ implies $b = d$. But also $(-1)^a = (-1)^c$ means a and c have the same parity, and because $a, c \in \{0,1\}$, it follows $a = c$. Thus $(a,b) = (c,d)$, so θ is injective.
Next note that θ **is not surjective** because $\theta(a,b) = (-1)^a b$ is either positive or negative, but never zero. Therefore there exist no element $(a,b) \in \{0,1\} \times \mathbb{N}$ for which $\theta(a,b) = 0 \in \mathbb{Z}$.

13. Consider the function $f : \mathbb{R}^2 \to \mathbb{R}^2$ defined by the formula $f(x,y) = (xy, x^3)$. Is f injective? Is it surjective?
Notice that $f(0,1) = (0,0)$ and $f(0,0) = (0,0)$, so f is **not injective**. To show that f is also **not surjective**, we will show that it's impossible to find an ordered pair (x,y) with $f(x,y) = (1,0)$. If there were such a pair, then $f(x,y) = (xy,x^3) = (1,0)$, which yields $xy = 1$ and $x^3 = 0$. From $x^3 = 0$ we get $x = 0$, so $xy = 0$, a contradiction.

15. This question concerns functions $f : \{A,B,C,D,E,F,G\} \to \{1,2,3,4,5,6,7\}$. How many such functions are there? How many of these functions are injective? How many are surjective? How many are bijective?
Function f can described as a list $(f(A), f(B), f(C), f(D), f(E), f(F), f(G))$, where there are seven choices for each entry. By the multiplication principle, the total number of functions f is $7^7 = 823543$.
If f is injective, then this list can't have any repetition, so there are $7! = 5040$ injective functions. Since any injective function sends the seven elements of the domain to seven distinct elements of the codomain, all of the injective functions are surjective, and vice versa. Thus there are 5040 surjective functions and 5040 bijective functions.

17. This question concerns functions $f : \{A,B,C,D,E,F,G\} \to \{1,2\}$. How many such functions are there? How many of these functions are injective? How many are surjective? How many are bijective?
Function f can described as a list $(f(A), f(B), f(C), f(D), f(E), f(F), f(G))$, where there are two choices for each entry. Therefore the total number of functions is $2^7 = 128$. It is impossible for any function to send all seven elements of $\{A,B,C,D,E,F,G\}$ to seven distinct elements of $\{1,2\}$, so none of these 128 functions is injective, hence none are bijective.
How many are surjective? Only two of the 128 functions are not surjective, and they are the "constant" functions $\{(A,1),(B,1),(C,1),(D,1),(E,1),(F,1),(G,1)\}$ and $\{(A,2),(B,2),(C,2),(D,2),(E,2),(F,2),(G,2)\}$. So there are 126 surjective functions.

Section 12.3 Exercises

1. If 6 integers are chosen at random, at least two will have the same remainder when divided by 5.

Proof. Write $\mathbb{Z}$ as follows: $\mathbb{Z} = \bigcup_{j=0}^{4} \{5k + j : k \in \mathbb{Z}\}$. This is a partition of $\mathbb{Z}$ into 5 sets. If six integers are picked at random, by the pigeonhole principle, at least two will be in the same set. However, each set corresponds to the remainder of a number after being divided by 5 (for example, $\{5k+1 : k \in \mathbb{Z}\}$ are all those integers that leave a remainder of 1 after being divided by 5). ■

3. Given any six positive integers, there are two for which their sum or difference is divisible by 9.

Proof. If for two of the integers n,m we had $n \equiv m \pmod 9$, then $n - m \equiv 0 \pmod 9$, and we would be done. Thus assume this is not the case. Observe that the only two element subsets of positive integers that sum to 9 are $\{1,8\},\{2,7\},\{3,6\}$, and $\{4,5\}$. However, since at least five of the six integers must have distinct remainders from 1, 2, ..., 8 it follows from the pigeonhole principle that two integers n,m are in the same set. Hence $n+m \equiv 0 \pmod 9$ as desired. ■

5. Prove that any set of 7 distinct natural numbers contains a pair of numbers whose sum or difference is divisible by 10.

Proof. Let $S = \{a_1,a_2,a_3,a_4,a_5,a_6,a_7\}$ be any set of 7 natural numbers. Let's say that $a_1 < a_2 < a_3 < \cdots < a_7$. Consider the set

$$\begin{aligned} A &= \{a_1-a_2,\ a_1-a_3,\ a_1-a_4,\ a_1-a_5,\ a_1-a_6,\ a_1-a_7, \\ &\quad\ a_1+a_2,\ a_1+a_3,\ a_1+a_4,\ a_1+a_5,\ a_1+a_6,\ a_1+a_7\} \end{aligned}$$

Thus $|A| = 12$. Now let $B = \{0,1,2,3,4,5,6,7,8,9\}$, so $|B| = 10$. Let $f : A \to B$ be the function for which $f(n)$ equals the last digit of n. (That is $f(97) = 7$, $f(12) = 2$, $f(230) = 0$, etc.) Then, since $|A| > |B|$, the pigeonhole principle guarantees that f is not injective. Thus A contains elements $a_1 \pm a_i$ and $a_1 \pm a_j$ for which $f(a_1 \pm a_i) = f(a_1 \pm a_j)$. This means the last digit of $a_1 \pm a_i$ is the same as the last digit of $a_1 \pm a_j$. Thus the last digit of the difference $(a_1 \pm a_i) - (a_1 \pm a_j) = \pm a_i \pm a_j$ is 0. Hence $\pm a_i \pm a_j$ is a sum or difference of elements of S that is divisible by 10. ■

Section 12.4 Exercises

1. Suppose $A = \{5,6,8\}$, $B = \{0,1\}$, $C = \{1,2,3\}$. Let $f : A \to B$ be the function $f = \{(5,1),(6,0),(8,1)\}$, and $g : B \to C$ be $g = \{(0,1),(1,1)\}$. Find $g \circ f$.
$g \circ f = \{(5,1),(6,1),(8,1)\}$

3. Suppose $A = \{1,2,3\}$. Let $f : A \to A$ be the function $f = \{(1,2),(2,2),(3,1)\}$, and let $g : A \to A$ be the function $g = \{(1,3),(2,1),(3,2)\}$. Find $g \circ f$ and $f \circ g$.
$g \circ f = \{(1,1),(2,1),(3,3)\}$; $f \circ g = \{(1,1),(2,2),(3,2)\}$.

5. Consider the functions $f,g : \mathbb{R} \to \mathbb{R}$ defined as $f(x) = \sqrt[3]{x+1}$ and $g(x) = x^3$. Find the formulas for $g \circ f$ and $f \circ g$.
$g \circ f(x) = x+1$; $f \circ g(x) = \sqrt[3]{x^3+1}$

7. Consider the functions $f,g : \mathbb{Z} \times \mathbb{Z} \to \mathbb{Z} \times \mathbb{Z}$ defined as $f(m,n) = (mn, m^2)$ and $g(m,n) = (m+1, m+n)$. Find the formulas for $g \circ f$ and $f \circ g$.
Note $g \circ f(m,n) = g(f(m,n)) = g(mn,m^2) = (mn+1, mn+m^2)$.
Thus $\boxed{g \circ f(m,n) = (mn+1, mn+m^2).}$
Note $f \circ g(m,n) = f(g(m,n)) = f(m+1, m+n) = ((m+1)(m+n), (m+1)^2)$.
Thus $\boxed{f \circ g(m,n) = (m^2+mn+m+n, m^2+2m+1).}$

9. Consider the functions $f : \mathbb{Z} \times \mathbb{Z} \to \mathbb{Z}$ defined as $f(m,n) = m+n$ and $g : \mathbb{Z} \to \mathbb{Z} \times \mathbb{Z}$ defined as $g(m) = (m,m)$. Find the formulas for $g \circ f$ and $f \circ g$.
$g \circ f(m,n) = (m+n, m+n)$
$f \circ g(m) = 2m$

Section 12.5 Exercises

1. Check that the function $f : \mathbb{Z} \to \mathbb{Z}$ defined by $f(n) = 6 - n$ is bijective. Then compute f^{-1}.
It is injective as follows. Suppose $f(m) = f(n)$. Then $6 - m = 6 - n$, which reduces to $m = n$.
It is surjective as follows. If $b \in \mathbb{Z}$, then $f(6-b) = 6-(6-b) = b$.
Inverse: $f^{-1}(n) = 6 - n$.

3. Let $B = \{2^n : n \in \mathbb{Z}\} = \{\ldots, \frac{1}{4}, \frac{1}{2}, 1, 2, 4, 8, \ldots\}$. Show that the function $f : \mathbb{Z} \to B$ defined as $f(n) = 2^n$ is bijective. Then find f^{-1}.
It is injective as follows. Suppose $f(m) = f(n)$, which means $2^m = 2^n$. Taking $\log_2$ of both sides gives $\log_2(2^m) = \log_2(2^n)$, which simplifies to $m = n$.
The function f is surjective as follows. Suppose $b \in B$. By definition of B this means $b = 2^n$ for some $n \in \mathbb{Z}$. Then $f(n) = 2^n = b$.
Inverse: $f^{-1}(n) = \log_2(n)$.

5. The function $f : \mathbb{R} \to \mathbb{R}$ defined as $f(x) = \pi x - e$ is bijective. Find its inverse.
Inverse: $f^{-1}(x) = \dfrac{x+e}{\pi}$.

7. Show that the function $f : \mathbb{R}^2 \to \mathbb{R}^2$ defined by the formula $f((x,y) = ((x^2+1)y, x^3)$ is bijective. Then find its inverse.
First we prove the function is injective. Assume $f(x_1, y_1) = f(x_2, y_2)$. Then $(x_1^2+1)y_1 = (x_2^2+1)y_2$ and $x_1^3 = x_2^3$. Since the real-valued function $f(x) = x^3$ is one-to-one, it follows that $x_1 = x_2$. Since $x_1 = x_2$, and $x_1^2 + 1 > 0$ we may divide both sides of $(x_1^2+1)y_1 = (x_1^2+1)y_2$ by (x_1^2+1) to get $y_1 = y_2$. Hence $(x_1, y_1) = (x_2, y_2)$.
Now we prove the function is surjective. Let $(a,b) \in \mathbb{R}^2$. Set $x = b^{1/3}$ and $y = a/(b^{2/3}+1)$. Then $f(x,y) = ((b^{2/3}+1)\frac{a}{b^{2/3}+1}, (b^{1/3})^3) = (a,b)$. It now follows that f is bijective.
Finally, we compute the inverse. Write $f(x,y) = (u,v)$. Interchange variables to get $(x,y) = f(u,v) = ((u^2+1)v, u^3)$. Thus $x = (u^2+1)v$ and $y = u^3$. Hence $u = y^{1/3}$ and $v = \frac{x}{y^{2/3}+1}$. Therefore $f^{-1}(x,y) = (u,v) = \left(y^{1/3}, \frac{x}{y^{2/3}+1}\right)$.

9. Consider the function $f : \mathbb{R} \times \mathbb{N} \to \mathbb{N} \times \mathbb{R}$ defined as $f(x,y) = (y, 3xy)$. Check that this is bijective; find its inverse.

To see that this is injective, suppose $f(a,b) = f(c,d)$. This means $(b, 3ab) = (d, 3cd)$. Since the first coordinates must be equal, we get $b = d$. As the second coordinates are equal, we get $3ab = 3dc$, which becomes $3ab = 3bc$. Note that, from the definition of f, $b \in \mathbb{N}$, so $b \neq 0$. Thus we can divide both sides of $3ab = 3bc$ by the non-zero quantity $3b$ to get $a = c$. Now we have $a = c$ and $b = d$, so $(a,b) = (c,d)$. It follows that f is injective.

Next we check that f is surjective. Given any (b,c) in the codomain $\mathbb{N} \times \mathbb{R}$, notice that $(\frac{c}{3b}, b)$ belongs to the domain $\mathbb{R} \times \mathbb{N}$, and $f(\frac{c}{3b}, b) = (b,c)$. Thus f is surjective. As it is both injective and surjective, it is bijective; thus the inverse exists.

To find the inverse, recall that we obtained $f(\frac{c}{3b}, b) = (b,c)$. Then $f^{-1}f(\frac{c}{3b}, b) = f^{-1}(b,c)$, which reduces to $(\frac{c}{3b}, b) = f^{-1}(b,c)$. Replacing b and c with x and y, respectively, we get $f^{-1}(x,y) = (\frac{y}{3x}, x)$.

Section 12.6 Exercises

1. Consider the function $f : \mathbb{R} \to \mathbb{R}$ defined as $f(x) = x^2 + 3$. Find $f([-3,5])$ and $f^{-1}([12,19])$. Answers: $f([-3,5]) = [3,28]$; $f^{-1}([12,19]) = [-4,-3] \cup [3,4]$.

3. This problem concerns functions $f : \{1,2,3,4,5,6,7\} \to \{0,1,2,3,4\}$. How many such functions have the property that $|f^{-1}(\{3\})| = 3$? Answer: $4^4 \binom{7}{3}$.

5. Consider a function $f : A \to B$ and a subset $X \subseteq A$. We observed in Section 12.6 that $f^{-1}(f(X)) \neq X$ in general. However $X \subseteq f^{-1}(f(X))$ is always true. Prove this.

Proof. Suppose $a \in X$. Thus $f(a) \in \{f(x) : x \in X\} = f(X)$, that is $f(a) \in f(X)$. Now, by definition of preimage, we have $f^{-1}(f(X)) = \{x \in A : f(x) \in f(X)\}$. Since $a \in A$ and $f(a) \in f(X)$, it follows that $a \in f^{-1}(f(X))$. This proves $X \subseteq f^{-1}(f(X))$. ■

7. Given a function $f : A \to B$ and subsets $W, X \subseteq A$, prove $f(W \cap X) \subseteq f(W) \cap f(X)$.

Proof. Suppose $b \in f(W \cap X)$. This means $b \in \{f(x) : x \in W \cap X\}$, that is $b = f(a)$ for some $a \in W \cap X$. Since $a \in W$ we have $b = f(a) \in \{f(x) : x \in W\} = f(W)$. Since $a \in X$ we have $b = f(a) \in \{f(x) : x \in X\} = f(X)$. Thus b is in both $f(W)$ and $f(X)$, so $b \in f(W) \cap f(X)$. This completes the proof that $f(W \cap X) \subseteq f(W) \cap f(X)$. ■

9. Given a function $f : A \to B$ and subsets $W, X \subseteq A$, prove $f(W \cup X) = f(W) \cup f(X)$.

Proof. First we will show $f(W \cup X) \subseteq f(W) \cup f(X)$. Suppose $b \in f(W \cup X)$. This means $b \in \{f(x) : x \in W \cup X\}$, that is, $b = f(a)$ for some $a \in W \cup X$. Thus $a \in W$ or $a \in X$. If $a \in W$, then $b = f(a) \in \{f(x) : x \in W\} = f(W)$. If $a \in X$, then $b = f(a) \in \{f(x) : x \in X\} = f(X)$. Thus b is in $f(W)$ or $f(X)$, so $b \in f(W) \cup f(X)$. This completes the proof that $f(W \cup X) \subseteq f(W) \cup f(X)$.
Next we will show $f(W) \cup f(X) \subseteq f(W \cup X)$. Suppose $b \in f(W) \cup f(X)$. This means $b \in f(W)$ or $b \in f(X)$. If $b \in f(W)$, then $b = f(a)$ for some $a \in W$. If $b \in f(X)$, then $b = f(a)$ for some $a \in X$. Either way, $b = f(a)$ for some a that is in W or X. That is, $b = f(a)$ for some $a \in W \cup X$. But this means $b \in f(W \cup X)$. This completes the proof that $f(W) \cup f(X) \subseteq f(W \cup X)$.
The previous two paragraphs show $f(W \cup X) = f(W) \cup f(X)$. ■

11. Given $f : A \to B$ and subsets $Y, Z \subseteq B$, prove $f^{-1}(Y \cup Z) = f^{-1}(Y) \cup f^{-1}(Z)$.

Proof. First we will show $f^{-1}(Y \cup Z) \subseteq f^{-1}(Y) \cup f^{-1}(Z)$. Suppose $a \in f^{-1}(Y \cup Z)$. By Definition 12.9, this means $f(a) \in Y \cup Z$. Thus, $f(a) \in Y$ or $f(a) \in Z$. If $f(a) \in Y$, then $a \in f^{-1}(Y)$, by Definition 12.9. Similarly, if $f(a) \in Z$, then $a \in f^{-1}(Z)$. Hence $a \in f^{-1}(Y)$ or $a \in f^{-1}(Z)$, so $a \in f^{-1}(Y) \cup f^{-1}(Z)$. Consequently $f^{-1}(Y \cup Z) \subseteq f^{-1}(Y) \cup f^{-1}(Z)$.
Next we show $f^{-1}(Y) \cup f^{-1}(Z) \subseteq f^{-1}(Y \cup Z)$. Suppose $a \in f^{-1}(Y) \cup f^{-1}(Z)$. This means $a \in f^{-1}(Y)$ or $a \in f^{-1}(Z)$. Hence, by Definition 12.9, $f(a) \in Y$ or $f(a) \in Z$, which means $f(a) \in Y \cup Z$. But by Definition 12.9, $f(a) \in Y \cup Z$ means $a \in f^{-1}(Y \cup Z)$. Consequently $f^{-1}(Y) \cup f^{-1}(Z) \subseteq f^{-1}(Y \cup Z)$.
The previous two paragraphs show $f^{-1}(Y \cup Z) = f^{-1}(Y) \cup f^{-1}(Z)$. ■

13. Let $f : A \to B$ be a function, and $X \subseteq A$. Prove or disprove: $f\big(f^{-1}(f(X))\big) = f(X)$.

Proof. First we will show $f\big(f^{-1}(f(X))\big) \subseteq f(X)$. Suppose $y \in f\big(f^{-1}(f(X))\big)$. By definition of image, this means $y = f(x)$ for some $x \in f^{-1}(f(X))$. But by definition of preimage, $x \in f^{-1}(f(X))$ means $f(x) \in f(X)$. Thus we have $y = f(x) \in f(X)$, as desired.

Next we show $f(X) \subseteq f\big(f^{-1}(f(X))\big)$. Suppose $y \in f(X)$. This means $y = f(x)$ for some $x \in X$. Then $f(x) = y \in f(X)$, which means $x \in f^{-1}(f(X))$. Then by definition of image, $f(x) \in f(f^{-1}(f(X)))$. Now we have $y = f(x) \in f(f^{-1}(f(X)))$, as desired.

The previous two paragraphs show $f\big(f^{-1}(f(X))\big) = f(X)$. ■

Chapter 13 Exercises

Section 13.1 Exercises

1. $\mathbb{R}$ and $(0,\infty)$
Observe that the function $f(x) = e^x$ sends $\mathbb{R}$ to $(0,\infty)$. It is injective because $f(x) = f(y)$ implies $e^x = e^y$, and taking ln of both sides gives $x = y$. It is surjective because if $b \in (0,\infty)$, then $f(\ln(b)) = b$. Therefore, because of the bijection $f : \mathbb{R} \to (0,\infty)$, it follows that $|\mathbb{R}| = |(0,\infty)|$.

3. $\mathbb{R}$ and $(0,1)$
Observe that the function $\frac{1}{\pi} f(x) = \cot^{-1}(x)$ sends $\mathbb{R}$ to $(0,1)$. It is injective and surjective by elementary trigonometry. Therefore, because of the bijection $f : \mathbb{R} \to (0,1)$, it follows that $|\mathbb{R}| = |(0,1)|$.

5. $A = \{3k : k \in \mathbb{Z}\}$ and $B = \{7k : k \in \mathbb{Z}\}$
Observe that the function $f(x) = \frac{7}{3}x$ sends A to B. It is injective because $f(x) = f(y)$ implies $\frac{7}{3}x = \frac{7}{3}y$, and multiplying both sides by $\frac{3}{7}$ gives $x = y$. It is surjective because if $b \in B$, then $b = 7k$ for some integer k. Then $3k \in A$, and $f(3k) = 7k = b$. Therefore, because of the bijection $f : A \to B$, it follows that $|A| = |B|$.

7. $\mathbb{Z}$ and $S = \left\{\ldots, \frac{1}{8}, \frac{1}{4}, \frac{1}{2}, 1, 2, 4, 8, 16, \ldots\right\}$
Observe that the function $f : \mathbb{Z} \to S$ defined as $f(n) = 2^n$ is bijective: It is injective because $f(m) = f(n)$ implies $2^m = 2^n$, and taking $\log_2$ of both sides produces $m = n$. It is surjective because any element b of S has form $b = 2^n$ for some integer n, and therefore $f(n) = 2^n = b$. Because of the bijection $f : \mathbb{Z} \to S$, it follows that $|\mathbb{Z}| = |S|$.

9. $\{0,1\} \times \mathbb{N}$ and $\mathbb{N}$
Consider the function $f : \{0,1\} \times \mathbb{N} \to \mathbb{N}$ defined as $f(a,n) = 2n - a$. This is injective because if $f(a,n) = f(b,m)$, then $2n - a = 2m - b$. Now if a were unequal to b, one of a or b would be 0 and the other would be 1, and one side of $2n - a = 2m - b$ would be odd and the other even, a contradiction. Therefore $a = b$. Then $2n - a = 2m - b$ becomes $2n - a = 2m - a$; add a to both sides and divide by 2 to get $m = n$. Thus we have $a = b$ and $m = n$, so $(a,n) = (b,m)$, so f is injective.

To see that f is surjective, take any $b \in \mathbb{N}$. If b is even, then $b = 2n$ for some integer n, and $f(0,n) = 2n - 0 = b$. If b is odd, then $b = 2n+1$ for some integer n. Then $f(1,n+1) = 2(n+1)-1 = 2n+1 = b$. Therefore f is surjective. Then f is a bijection, so $|\{0,1\} \times \mathbb{N}| = |\mathbb{N}|$.

11. $[0,1]$ and $(0,1)$

Proof. Consider the subset $X = \{\frac{1}{n} : n \in \mathbb{N}\} \subseteq [0,1]$. Let $f : [0,1] \to [0,1)$ be defined as $f(x) = x$ if $x \in [0,1] - X$ and $f(\frac{1}{n}) = \frac{1}{n+1}$ for any $\frac{1}{n} \in X$. It is easy to check that f is a bijection. Next let $Y = \{1 - \frac{1}{n} : n \in \mathbb{N}\} \subseteq [0,1)$, and define $g : [0,1) \to (0,1)$ as $g(x) = x$ if $x \in [0,1) - Y$ and $g(1-\frac{1}{n}) = 1 - \frac{1}{n+1}$ for any $1 - \frac{1}{n} \in Y$. As in the case of f, it is easy to check that g is a bijection. Therefore the composition $g \circ f : [0,1] \to (0,1)$ is a bijection. (See Theorem 12.2.) We conclude that $|[0,1]| = |(0,1)|$. ■

13. $\mathscr{P}(\mathbb{N})$ and $\mathscr{P}(\mathbb{Z})$
Outline: By Exercise 18 of Section 12.2, we have a bijection $f : \mathbb{N} \to \mathbb{Z}$ defined as $f(n) = \dfrac{(-1)^n(2n-1)+1}{4}$. Now define a function $\Phi : \mathscr{P}(\mathbb{N}) \to \mathscr{P}(\mathbb{Z})$ as $\Phi(X) = \{f(x) : x \in X\}$. Check that Φ is a bijection.

15. Find a formula for the bijection f in Example 13.2.
Hint: Consider the function f from Exercise 18 of Section 12.2.

Section 13.2 Exercises

1. Prove that the set $A = \{\ln(n) : n \in \mathbb{N}\} \subseteq \mathbb{R}$ is countably infinite.
Just note that its elements can be written in infinite list form as $\ln(1), \ln(2), \ln(3), \cdots$. Thus A is countably infinite.

3. Prove that the set $A = \{(5n, -3n) : n \in \mathbb{Z}\}$ is countably infinite.
Consider the function $f : \mathbb{Z} \to A$ defined as $f(n) = (5n, -3n)$. This is clearly surjective, and it is injective because $f(n) = f(m)$ gives $(5n, -3n) = (5m, -3m)$, so $5n = 5m$, hence $m = n$. Thus, because f is surjective, $|\mathbb{Z}| = |A|$, and $|A| = |\mathbb{Z}| = \aleph_0$. Therefore A is countably infinite.

5. Prove or disprove: There exists a countably infinite subset of the set of irrational numbers.
This is true. Just consider the set consisting of the irrational numbers $\frac{\pi}{1}, \frac{\pi}{2}, \frac{\pi}{3}, \frac{\pi}{4}, \cdots$.

7. Prove or disprove: The set $\mathbb{Q}^{100}$ is countably infinite.
This is true. Note $\mathbb{Q}^{100} = \mathbb{Q} \times \mathbb{Q} \times \cdots \times \mathbb{Q}$ (100 times), and since $\mathbb{Q}$ is countably infinite, it follows from the corollary of Theorem 13.5 that this product is countably infinite.

9. Prove or disprove: The set $\{0,1\} \times \mathbb{N}$ is countably infinite.
This is true. Note that $\{0,1\} \times \mathbb{N}$ can be written in infinite list form as $(0,1),(1,1),(0,2),(1,2),(0,3),(1,3),(0,4),(1,4),\cdots$. Thus the set is countably infinite.

11. Partition $\mathbb{N}$ into 8 countably infinite sets.

For each $i \in \{1,2,3,4,5,6,7,8\}$, let X_i be those natural numbers that are congruent to i modulo 8, that is,

$$
\begin{aligned}
X_1 &= \{1,9,17,25,33,\ldots\} \\
X_2 &= \{2,10,18,26,34,\ldots\} \\
X_3 &= \{3,11,19,27,35,\ldots\} \\
X_4 &= \{4,12,20,28,36,\ldots\} \\
X_5 &= \{5,13,21,29,37,\ldots\} \\
X_6 &= \{6,14,22,30,38,\ldots\} \\
X_7 &= \{7,15,13,31,39,\ldots\} \\
X_8 &= \{8,16,24,32,40,\ldots\}
\end{aligned}
$$

13. If $A = \{X \subset \mathbb{N} : X \text{ is finite}\}$, then $|A| = \aleph_0$.

Proof. This is **true.** To show this we will describe how to arrange the items of A in an infinite list $X_1, X_2, X_3, X_4, \ldots$.

For each natural number n, let p_n be the nth prime number. Thus $p_1 = 2$, $p_2 = 3$, $p_3 = 5$, $p_4 = 7$, $p_5 = 11$, and so on. Now consider any element $X \in A$. If $X \neq \emptyset$, then $X = \{n_1, n_2, n_3, \ldots, n_k\}$, where $k = |X|$ and $n_i \in \mathbb{N}$ for each $1 \leq i \leq k$. Define a function $f : A \to \mathbb{N} \cup \{0\}$ as follows: $f(\{n_1, n_2, n_3, \ldots, n_k\}) = p_{n_1} p_{n_2} \cdots p_{n_k}$. For example, $f(\{1,2,3\}) = p_1 p_2 p_3 = 2 \cdot 3 \cdot 5 = 30$, and $f(\{3,5\}) = p_3 p_5 = 5 \cdot 11 = 55$, etc. Also, we should not forget that $\emptyset \in A$, and we define $f(\emptyset) = 0$.

Note that $f : A \to \mathbb{N} \cup \{0\}$ is an injection: Let $X = \{n_1, n_2, n_3, \ldots, n_k\}$ and $Y = \{m_1, m_2, m_3, \ldots, m_\ell\}$, and $X \neq Y$. Then there is an integer a that belongs to one of X or Y but not the other. Then the prime factorization of one of the numbers $f(X)$ and $f(Y)$ uses the prime number p_a but the prime factorization of the other does not use p_a. It follows that $f(X) \neq f(Y)$ by the fundamental theorem of arithmetic. Thus f is injective.

So each set $X \in A$ is associated with an integer $f(X) \geq 0$, and no two different sets are associated with the same number. Thus we can list the elements in $X \in A$ in increasing order of the numbers $f(X)$. The list begins as

$$\emptyset,\ \{1\},\ \{2\},\ \{3\},\ \{1,2\},\ \{4\},\ \{1,3\},\ \{5\},\ \{6\},\ \{1,4\},\ \{2,3\},\ \{7\}, \ldots$$

It follows that A is countably infinite. ■

15. Hint: Use the fundamental theorem of arithmetic.

Section 13.3 Exercises

1. Suppose B is an uncountable set and A is a set. Given that there is a surjective function $f : A \to B$, what can be said about the cardinality of A?

The set A must be uncountable, as follows. For each $b \in B$, let a_b be an element of A for which $f(a_b) = b$. (Such an element must exist because f is surjective.) Now form the set $U = \{a_b : b \in B\}$. Then the function $f : U \to B$ is bijective, by construction. Then since B is uncountable, so is U. Therefore U is an uncountable subset of A, so A is uncountable by Theorem 13.9.

3. Prove or disprove: If A is uncountable, then $|A| = |\mathbb{R}|$.

This is false. Let $A = \mathscr{P}(\mathbb{R})$. Then A is uncountable, and by Theorem 13.7, $|\mathbb{R}| < |\mathscr{P}(\mathbb{R})| = |A|$.

5. Prove or disprove: The set $\{0,1\} \times \mathbb{R}$ is uncountable.

This is true. To see why, first note that the function $f : \mathbb{R} \to \{0\} \times \mathbb{R}$ defined as $f(x) = (0,x)$ is a bijection. Thus $|\mathbb{R}| = |\{0\} \times \mathbb{R}|$, and since $\mathbb{R}$ is uncountable, so is $\{0\} \times \mathbb{R}$. Then $\{0\} \times \mathbb{R}$ is an uncountable subset of the set $\{0,1\} \times \mathbb{R}$, so $\{0,1\} \times \mathbb{R}$ is uncountable by Theorem 13.9.

7. Prove or disprove: If $A \subseteq B$ and A is countably infinite and B is uncountable, then $B - A$ is uncountable.

This is true. To see why, suppose to the contrary that $B - A$ is countably infinite. Then $B = A \cup (B - A)$ is a union of countably infinite sets, and thus countable, by Theorem 13.6. This contradicts the fact that B is uncountable.

Exercises for Section 13.4

1. Show that if $A \subseteq B$ and there is an injection $g : B \to A$, then $|A| = |B|$.

Just note that the map $f : A \to B$ defined as $f(x) = x$ is an injection. Now apply the Cantor-Bernstein-Schröeder theorem.

3. Let $\mathscr{F}$ be the set of all functions $\mathbb{N} \to \{0,1\}$. Show that $|\mathbb{R}| = |\mathscr{F}|$.

Because $|\mathbb{R}| = |\mathscr{P}(\mathbb{N})|$, it suffices to show that $|\mathscr{F}| = |\mathscr{P}(\mathbb{N})|$. To do this, we will exhibit a bijection $f : \mathscr{F} \to \mathscr{P}(\mathbb{N})$. Define f as follows. Given a function $\varphi \in \mathscr{F}$, let $f(\varphi) = \{n \in \mathbb{N} : \varphi(n) = 1\}$. To see that f is injective, suppose $f(\varphi) = f(\theta)$. Then $\{n \in \mathbb{N} : \varphi(n) = 1\} = \{n \in \mathbb{N} : \theta(n) = 1\}$. Put $X = \{n \in \mathbb{N} : \varphi(n) = 1\}$. Now we see that if $n \in X$, then $\varphi(n) = 1 = \theta(n)$. And if $n \in \mathbb{N} - X$, then $\varphi(n) = 0 = \theta(n)$. Consequently $\varphi(n) = \theta(n)$ for any $n \in \mathbb{N}$, so $\varphi = \theta$. Thus f is injective. To see that f is surjective, take any $X \in \mathscr{P}(\mathbb{N})$. Consider the function $\varphi \in \mathscr{F}$ for which $\varphi(n) = 1$ if $n \in X$ and $\varphi(n) = 0$ if $n \notin X$. Then $f(\varphi) = X$, so f is surjective.

5. Consider the subset $B = \{(x,y) : x^2 + y^2 \leq 1\} \subseteq \mathbb{R}^2$. Show that $|B| = |\mathbb{R}^2|$.

This will follow from the Cantor-Bernstein-Schröeder theorem provided that we can find injections $f : B \to \mathbb{R}^2$ and $g : \mathbb{R}^2 \to B$. The function $f : B \to \mathbb{R}^2$ defined as $f(x,y) = (x,y)$ is clearly injective. For $g : \mathbb{R}^2 \to B$, consider the function

$$g(x,y) = \left(\frac{x^2 + y^2}{x^2 + y^2 + 1} x, \; \frac{x^2 + y^2}{x^2 + y^2 + 1} y \right).$$

Verify that this is an injective function $g : \mathbb{R}^2 \to B$.

7. Prove or disprove: If there is a injection $f : A \to B$ and a surjection $g : A \to B$, then there is a bijection $h : A \to B$.

This is true. Here is an outline of a proof. Define a function $g' : B \to A$ as follows. For each $b \in B$, choose an element $x_b \in g^{-1}(\{x\})$. (That is, choose an element $x_b \in A$ for which $g(x_b) = b$.) Now let $g' : B \to A$ be the function defined as $g'(b) = x_b$. Check that g' is injective and apply the the Cantor-Bernstein-Schröeder theorem.

Index